Statistical Plasma Physics

Volume II: Condensed Plasmas

Setsuo Ichimaru

Advanced Book Program

CRC Press
Taylor & Francis Group
Boca Raton London New York

CRC Press is an imprint of the
Taylor & Francis Group, an **informa** business

First published 2004 by Westview Press

Published 2018 by CRC Press
Taylor & Francis Group
6000 Broken Sound Parkway NW, Suite 300
Boca Raton, FL 33487-2742

CRC Press is an imprint of the Taylor & Francis Group, an informa business

Copyright © 2004 by Taylor & Francis Group LLC

No claim to original U.S. Government works

Visit the Taylor & Francis Web site at
http://www.taylorandfrancis.com

and the CRC Press Web site at
http://www.crcpress.com

A Cataloging-in-Publication data record for this book is available from the Library of Congress.

ISBN 13: 978-0-8133-4179-8 (pbk)

Frontiers in Physics
David Pines, Editor

Volumes of the Series published from 1961 to 1973 are not officially numbered. The parenthetical numbers shown are designed to aid librarians and bibliographers to check the completeness of their holdings.

Titles published in this series prior to 1987 appear under either the W. A. Benjamin or the Benjamin/Cummings imprint; titles published since 1986 appear under the Westview Press imprint.

Frontiers in Physics

Volumes published from 1974 onward are being numbered as an integral part of the bibliography.

Frontiers in Physics

Frontiers in Physics

Editor's Foreword

The problem of communicating in a coherent fashion recent developments in the most exciting and active fields of physics continues to be with us. The enormous growth in the number of physicists has tended to make the familiar channels of communication considerably less effective. It has become increasingly difficult for experts in a given field to keep up with the current literature; the novice can only be confused. What is needed is both a consistent account of a field and the presentation of a definite "point of view" concerning it. Formal monographs cannot meet such a need in a rapidly developing field, while the review article seems to have fallen into disfavor. Indeed, it would seem that the people who are most actively engaged in developing a given field are the people least likely to write at length about it.

Frontiers in Physics was conceived in 1961 in an effort to improve the situation in several ways. Leading physicists frequently give a series of lectures, a graduate seminar, or a graduate course in their special fields of interest. Such lectures serve to summarize the present status of a rapidly developing field and may well constitute the only coherent account available at the time. One of the principal purposes of the *Frontiers in Physics* series is to make notes on such lectures available to the wider physics community.

As *Frontiers in Physics* has evolved, a second category of book, the informal text/monograph, an intermediate step between lecture notes and formal text or monographs, has played an increasingly important role in the series. In an informal text or monograph an author has reworked his or her lecture notes to the point at which the manuscript represents a coherent summation of a newly developed field, complete with references and problems, suitable for either classroom teaching or individual study.

In the first volume, which is self-contained, the basic principles are introduced for equilibrium situations and then applied to nonequilibrium conditions in which transient processes, instabilities, and turbulence may play a role. This second volume is devoted to topics in the fields of condensed plasma physics, with applications to subfields which range from condensed matter physics to astrophysics. It gives me great pleasure to welcome Professor Ichimaru once more to the ranks of authors for *Frontiers in Physics*.

David Pines

Urbana, Illinois
January 1994

Preface

Plasma physics is concerned with the equilibrium and nonequilibrium properties of a statistical system containing many charged particles. In Volume I of *Statistical Plasma Physics*, basic principles in statistical mechanics and electrodynamics were applied to elucidation of the physical processes mostly in tenuous, high-temperature plasmas. The role that the long-range Coulombic forces play in establishing the collective phenomena has been particularly emphasized.

As the density is increased, the plasma begins exhibiting features characteristic of a condensed matter, where short-range as well as long-range forces conspire to endow the plasma with a character of a *strongly coupled many-particle system*. As the temperature is lowered, quantum statistic and dynamic effects start to play a dominant part in the plasma, so that interplay with atomic, molecular, and nuclear physics becomes a major issue. Associated with these is the emergence of features such as insulator-to-metal transition, order-disorder transition, and chemical separation. Microscopic properties of the plasma depend delicately on these phase transitions, which in turn affect the macroscopic properties of the plasma through the rates of elementary and transport processes.

The present volume, *Statistical Plasma Physics: II. Condensed Plasmas*, aims at elucidating a number of basic topics in the physics of dense plasmas interfacing with condensed matter physics, atomic physics, nuclear physics, and astrophysics. Key phrases on the contents include equations of states, phase transitions, thermodynamic properties, transport processes, atomic, optical, and nuclear processes, all influenced by the strong exchange and Coulomb correlations in condensed plasmas. Astrophysical dense plasmas are those which we find in the interiors, surfaces, and outer envelopes of such astronomical objects as neutron stars, white dwarfs, the Sun, brown dwarfs, and giant planets. Condensed plasmas in laboratory settings include metals and alloys (solid, amorphous, liquid, and compressed), semiconductors (electrons, holes, and their droplets), and various realizations of dense plasmas (shock-compressed, diamond-anvil cell, metal vaporization, pinch discharges, and so on).

In Chapter 1, fundamentals of condensed plasma physics are exposed by way of an introduction. Basic parameters characterizing physical properties of those various condensed plasmas in astronomical and terrestrial settings are introduced. Basics of theoretical approaches to the treatment of strong exchange and Coulomb coupling effects are elucidated.

Chapter 2 considers the interparticle correlations and thermodynamic properties for dense classical one-component plasmas as well as binary-ionic-mixture plasmas; those are investigated by computer-simulation methods and through integral-equation approaches. Path-integral Monte Carlo simulation study on the quantum-mechanical Coulomb solids is also described. Issues on dynamic correlations in classical plasmas are taken up. Finally a number of examples for inhomogeneous ordered structures and order-disorder transitions are considered; these include the analyses on the phase diagrams for dense binary-ionic-mixture plasmas.

Chapter 3 treats the issues related to *strongly coupled quantum plasmas*, which are quantum-theoretic counterparts to the subjects considered in Chapter 2. Theoretical methods of taking into account the effects of strong exchange and Coulomb correlations for the descriptions of thermodynamic and dynamic properties in such a quantum plasma are elucidated. Spin-inhomogeneous ground states for two-dimensional systems of layered electrons are studied; magnetic order structures arise from antiferromagnetic spin-spin correlations between nearest neighbor electrons.

In Chapter 4, the issues of dense electron-ion plasma materials are considered. In such a two-component plasma, the attractive interaction between electrons and ions, an essential ingredient for formation of atoms, brings about a novel feature in the physics of dense plasmas in that the strong correlations between electrons and ions be taken into account on an equal footing to the atomic and molecular processes in such a condensed environment. Atomic levels and their existence should be influenced strongly by the statistical properties of the dense plasmas. The *strong electron-ion coupling* thus opens up new dimensions in the condensed plasma physics, where an outstanding issue will be the interplay between the atomic and molecular physics, on the one hand, and the plasma and condensed-matter physics, on the other. By taking those strong-coupling effects into account, we investigate the degrees of various stages of ionization, the resultant equations of state, and transport processes in dense plasma materials.

In Chapter 5, we study current stages of understanding with regard to those effects of enhancement in the rates of nuclear reactions that are expected from the correlation and thermodynamic effects, for various realizations of the condensed plasmas both in the astrophysical and laboratory settings. Consequently we shall be concerned with the aspects of statistical condensed-matter physics rather than with those of nuclear reaction physics per se. The subject of *nuclear fusion in dense plasmas* is viewed here as a forum in which the interplay between the nuclear physics and the statistical physics may be studied usefully through the concept of correlation functions.

Every effort has been expended to make the presentation in this volume as self-contained as possible. Minimum but significant overlaps, therefore, exist between the materials in this volume and in Vol. I, with inclusion of the Appendices. A Bibliography is provided mainly for the purpose of assisting the reader in finding relevant expositions extending and supplementing the materials presented here; no assessment on priorities is implied.

This volume is intended as a graduate-level textbook on the subjects of condensed plasma physics, material sciences, and/or condensed-matter astrophysics, with 39 to 45 lecture hours. It will be useful also to researchers in the fields of plasma physics, condensed-matter physics, atomic physics, nuclear physics, and astrophysics.

The encouragement and pertinent suggestions extended by D. Pines have been a constant motive for the realization of this volume. For the completion of this volume, the research activities that the author has enjoyed jointly with H. Iyetomi, S. Ogata, and H.M. Van Horn during these years through the Japan-US Cooperative Science Programs on astrophysical dense plasmas have been extremely useful. In addition, the author has benefited from illuminating discussions with many fellow scientists, including R. Abe, N. Ashcroft, M. Baus, D. Baldwin, H. Böhm, C. Deutsch, H. DeWitt, D. Dubin, D. DuBois, D. Ceperley, Ph. Choquard, R. Davidson, W. Ebeling, V. Fortov, Y. Furutani, M. Goldman, J.-P. Hansen, W. Horton, W. Hubbard, K. Husimi, A. Isihara, N. Itoh, B. Jancovici, R. Kalia, T. Kihara, W. Kohn, W. Kraeft, R. Kubo, J. Malmberg (late), J. Meyer-ter-Vehn, A. Nakano, K. Nomoto, T. O'Neil, F. Perrot, C. Pethick, F. Rogers, M. Rosenbluth, N. Rostoker, K. Singwi (late), A. Sjölander, T. Tajima, S. Tanaka, D. ter Haar, F.-T. Thielemann, W. Thompson, H. Totsuji, K. Utsumi, P. Vashishta, Y. Wada, M. Watabe, M. Yamada, and F. Yonezawa. Most of the drawings in this volume were prepared by H. Iyetomi, S. Ogata, and K. Tsuruta; the figures on the cover were originally produced by K. Tsuruta. Finally, the author's wife, Tomoko Ichimaru, rendered continual help to him in every way throughout the project. All of these are gratefully acknowledged.

Setsuo Ichimaru

Aspen, Colorado
January 1994

Contents

CHAPTER 1

Fundamentals

1.1 CONDENSED PLASMAS IN NATURE

Plasmas are any statistical systems containing mobile charged particles. When such a system is condensed, interaction between the particles becomes very effective so that the system may undergo various changes in the internal states or the phase transitions. The thermodynamic properties and the rates of elementary processes, for instance, are influenced significantly by the state of such a plasma. The purpose of the present volume is to elucidate the physical properties of various condensed plasmas.

Astrophysical dense plasmas are those we find in the interiors, surfaces, and outer envelopes of such astronomical objects as neutron stars, white dwarfs, the Sun, brown dwarfs, and giant planets. Condensed plasmas in laboratory settings include metals and alloys (solid, amorphous, liquid, and compressed), semiconductors (electrons, holes, and their droplets), and various realizations of dense plasmas (shock-compressed, diamond-anvil cell, metal vaporization, pinch discharges).

Elementary processes in condensed plasmas include photon transfer and opacities, electron transports, and nuclear reaction processes. The rates of these processes in condensed plasmas may depend sensitively on the changes in microscopic, macroscopic, thermodynamic, dielectric, and/or magnetic states of the matter. These changes of states may be associated with freezing transitions, chemical separations between the compositions, supercritical fluids, ionization or insulator-to-metal transitions, magnetic transitions, and transitions between normal to superconductive states.

A. Astrophysical Condensed Plasmas

Interiors of the *main sequence stars* such as the *Sun* are dense plasmas constituted mostly of hydrogen. The Sun has radius, $R_S \cong 6.96 \times 10^{10}$ cm, and mass, $M_S \cong 1.99 \times 10^{33}$ g; its mass density is 1.41 g/cm^3 on average. The total luminosity is $L_S \cong 3.85 \times 10^{26}$ W, and the average luminosity per mass is $L_S/M_S \cong 1.93 \times 10^{-7}$ W/g. The central part of the Sun has a mass density of approximately 1.56×10^2 g/cm^3, a temperature of approximately 1.5×10^7 K, and a pressure of approximately 3.4×10^5 Mbar. The mass fraction of hydrogen takes on a value of 0.36 near the center and 0.73 near the surface. The rates of nuclear reactions, photon transport and opacities, conductivities, atomic states, and their miscibilities are all essential elements in setting a model for the Sun [Bahcall et al. 1982; Bahcall & Pinsonneault 1992]. The solar luminosities are to be accounted for, in particular, by the rates of proton-proton reactions.

The atmospheric region of the Sun, approximately 2000 km in depth, is called the *chromosphere*. It has a plasma of 10^{10} to 10^{11} cm^{-3} average electron density and a temperature near 6000 K. The dense plasma effects are crucial to the analyses of the opacities and the atomic states for those "impurities" starting with helium.

Very low mass stars $(0.08 M_S < M < 0.3 M_S)$ dominate the solar neighborhood $(< 10$ pc)* and constitute the most numerous stellar component of the galaxy [Kumar 1963; Liebert & Probst 1987; Burrows, Hubbard, & Lunine 1989]. *Brown dwarfs* [D'Antona & Mazzitelli 1985; Kafatos, Harrington, & Maran 1986; Burrows, Hubbard, & Lunine 1989] may be defined as those astrophysical objects having masses insufficient for achieving thermal equilibrium through hydrogen burning $(M < 0.08 M_S)$ but with masses sufficiently large to be supported primarily by thermal or electron-degeneracy pressure $[M > 0.01 M_S;$ Nelson, Rappaport, & Joss 1986]. Because of the absence of hydrogen burning, expected surface temperatures of brown dwarfs are low $(T < 2000$ K), so that they could possibly be observed in infrared. Although the observational evidence for their existence still remains to be confirmed, the issues of brown dwarfs carry astrophysical consequences of considerable interest [Stevenson 1991]: A possibility has been suggested that these objects may comprise a significant fraction of the local missing mass in the galactic disk [Kafatos, Harrington, & Maran 1986]. Observation of a brown dwarf would provide useful clues as to the formation of planetary systems. The mass density and temperature near the center of a brown dwarf may range from 10^2 to 10^3 g/cm^3 and $(2-3) \times 10^6$ K, respectively. Equation of state, opacities, and rates of nuclear reactions are essential elements in theoretical prediction for the critical masses, structures, and evolution of brown dwarfs [Stevenson 1991; Burrows & Liebert 1993].

---

*1 pc (Parsec) = 3.26 light years = 3.08×10^{18} cm.

The materials inside *giant planets* (Jupiter, Saturn, Uranus, Neptune) offer important objects of study in the dense plasma physics [e.g., Hubbard 1980, 1984; Stevenson 1982]. Typically, Jupiter has a radius $R_J \cong 0.103R_S \cong 7.14 \times 10^9$ cm, and a mass $M_J \cong 0.95 \times 10^{-3}M_S \cong 1.90 \times 10^{30}$ g. The mass density, temperature, and pressure of its interior (outside the central "rock"), consisting of hydrogen plasmas with a few percent (in molar fraction) admixture of helium, are estimated to range from 2 to 5 g/cm^3, 5×10^3 to 2×10^4 K, and 3 to 30 Mbar, respectively.

Jupiter has been known to emit radiation energy in the infrared range at an effective temperature of approximately 130 K, approximately 2.7 times as intense as the total amount of radiation that it receives from the Sun. By observa-. tion through terrestrial atmospheric transmission windows at 8–14 mm [Menzel, Coblentz, & Lampland 1926] and 17.5–25 mm [Low 1966], Jupiter was known to be an unexpectedly bright infrared radiator. This feature has been accurately reconfirmed quantitatively by a telescope airborne at an altitude of 15 km [Armstrong, Harper, & Low 1972] and through flyby measurements with *Pioneers 10* and *11* [Ingersoll et al. 1976]. For Jupiter, the effective surface temperature determined from integrated infrared power over 8 to 300 mm was 129 ± 4 K, while the surface temperature calculated from equilibration with the absorbed solar radiation was 109.4 K [Hubbard 1980]; the balance needs to be accounted for by internal power generation.

To account for the source of such an excess infrared luminosity, theoretical models such as "adiabatic cooling" [Hubbard 1968; Graboske et al. 1975; Stevenson & Salpeter 1976], "gravitational unmixing" [Smoluchowski 1967; Stevenson & Salpeter 1976], and "latent heat due to metallization" [Saumon et al. 1992] have been considered. Theoretical predictions on the evolution, internal structures, and gravitational harmonics [Hubbard & Marley 1989] for such giant planets depend to a great extent on the thermodynamic, transport, and optical properties of the dense plasma materials. It is essential in these connections to explore possible nuclear reactions of deuterons, which may remain in a giant planet at an atomic abundance of approximately 0.003% [Anders & Grevesse 1989].

The *white dwarf* [e.g., Schatzman 1958; Luyten 1971; Liebert 1980; Shapiro & Teukolsky 1983] represents a final stage of stellar evolution, corresponding to a star of about one solar mass compressed to a characteristic radius of 5000 km and an average density of 10^6 g/cm^3. Its interior consists of a multi-ionic condensed matter composed of C and O as the main elements and Ne, Mg,. . ., Fe as trace elements. Condensed matter problems in white dwarfs include the assessment of the possibilities of chemical separation, or the phase diagrams, associated with the freezing transitions of the multi-ionic plasmas [Stevenson 1980; Van Horn 1991; Ogata et al. 1993]. These problems are related to the internal structures and cooling rates of the white dwarf [D'Antona & Mazzitelli

1990] as well as the rates of nuclear reactions [Ichimaru 1993a], evolution and nucleosynthesis [Clayton 1968], detailed mechanisms of supernova explosion, and possible formation of a neutron star [Canal, Isern, & Labay 1990; Nomoto & Kondo 1991].

As a progenitor of type I supernova, a white dwarf with an interior consisting of a carbon-oxygen mixture can be considered a kind of *binary-ionic mixture* (BIM), with a central mass density of 10^7 to 10^{10} g/cm^3 and a temperature of 10^7 to 10^9 K [Starrfield et al. 1972; Whelen & Iben 1973; Canal & Schatzman 1976]. Thermonuclear runaway leading to supernova explosion is expected to take place when the thermal output due to nuclear reactions exceeds the rate of energy losses. Assuming that neutrino losses are the major effects in the latter, one estimates [e.g., Arnett & Truran 1969; Nomoto 1982] that a nuclear runaway should take place when the nuclear power generated exceeds 10^{-9}–10^{-8} W/g. These values give approximate measures against which the rates of nuclear reactions may be compared.

Helium burning is one of the major reaction processes in stellar evolution and in accreting white dwarfs and neutron stars in close binary systems [Nomoto, Thielemann, & Miyaji 1985]. In the latter cases, helium burning is so explosive as to give rise to remarkable astronomical phenomena, such as x-ray bursts in neutron stars [Lewin & Joss 1983] and type I supernovae in white dwarfs [Nomoto 1982].

The *neutron star* [e.g., Baym & Pethick 1975; Shapiro & Teukolsky 1983], another of the final stages in the stellar evolution, is a highly degenerate star corresponding approximately to a compression of a solar mass into a radius of approximately 10 km. According to theoretical model calculations, it has an *outer crust*, consisting mostly of iron, with a thickness of several hundred meters and a mass density in the 10^4–10^7 g/cm^3 range. At these densities, iron atoms are completely ionized, so each contributes 26 conduction electrons to the system. At temperatures near 10^7 K, the thermal de Broglie wavelengths of the resultant Fe nuclei are substantially shorter than the average internuclear separations; the iron nuclei may be regarded as forming a classical plasma of ions.

When the mass density exceeds a critical value near 10^7 g/cm^3 for the electron captures, neutron-rich "inflated" nuclei begin to emerge. At approximately 4×10^7 g/cm^3, the neutron drip density, the estimated atomic and mass numbers for such nuclei are $Z = 36$ and $A = 118$; at approximately 2×10^{14} g/cm^3, which defines the inner edge of an inner crust, one calculates $Z = 201$ and $A = 2500$ [Baym, Pethick, & Sutherland 1971].

Over the bulk of the crustal parts, the nuclei are considered to form *Coulomb solids*. A neutron star may be appropriately looked upon as a "three-component star," consisting of an ultradense interior of neutron fluids with fractional constituents of protons and electrons, a crust of Coulomb solids, and a thin layer of "ocean" fluids. Nuclear reactions are expected in these surface layers on accreting neutron stars in close binary systems [Lewin & Joss 1983]. Electron transports and photon opacities in the outer crust and in the surface layer play the crucial parts [Gudmundsson, Pethick, & Epstein 1982] in the cooling rates of neutron stars [Nomoto & Tsuruta 1981]. Nonvanishing shear moduli associated with the crustal solids [Fuchs 1936; Ogata & Ichimaru 1990] lead to a prediction of rich spectra in the oscillations of a neutron star [McDermott, Van Horn, & Hansen 1988; Strohmayer et al. 1991]. In the interior of the neutron star, the physics of matter at nuclear and subnuclear densities is the subject of foremost importance [e.g., Hansen, C.J. 1974].

B. Dense Plasmas in the Laboratory

The states of plasmas for the *inertial confinement fusion* (ICF) research [e.g., Motz 1979; Hora 1991] are similar to those of the solar interior mentioned in the preceding subsection; the projected temperatures in the ICF plasmas are on the order of 10^7 to 10^8 K. Those materials that drive implosion of the fuel consist of high-Z elements, such as C, Al, Fe, Au, Pb, which after ionization form plasmas with charge numbers substantially greater than unity. The atomic physics of such a high Z element is influenced strongly by the correlated behaviors of charged particles in a dense plasma [e.g., Goldstein et al. 1991].

Laboratory realization of a condensed plasma includes those produced by shock compression [e.g., Fortov 1982], in pinch discharges [e.g., Pereira, Davis, & Rostoker 1989], and through metal vaporization [Mostovych et al. 1991]. Ultrahigh-pressure metal physics studied in the shock compression experiments [Nellis et al. 1988] aims at detecting changes in the equation of state through transitions between the electronic states in compressed metals such as Al, Cu, and Pb. Another scheme of the ultrahigh-pressure experiments utilizing the diamond-anvil cells [Mao, Hemley, & Hanfland 1990] strives for the ultimate realization of *metallic hydrogen* through insulator-to-metal transitions [Wigner & Huntington 1935] by compression to multimegabar pressures [Hemley & Mao 1991]. The dielectric and optical properties depend sensitively on the states of the resultant dense plasmas.

Metals and *alloys* (solid, amorphous, liquid, and compressed) are the most typical examples of condensed laboratory plasmas [e.g., Mott & Jones 1936; Ashcroft & Stroud 1978; March & Tosi 1984; Endo 1990]. The conduction electrons in metals form *quantum plasmas*, where the wave nature of the electrons as fermions plays an essential part [e.g., Pines & Nozières 1966]. At room temperatures, their number densities range over 10^{22} to 10^{23} cm^{-3} for most of the simple metals such as Al, Li, and Na. In these density regimes, the effects arising from exchange and Coulomb coupling between electrons become significant;

thus conduction electrons in metals are referred to as *strongly coupled* quantum plasmas [e.g., Ichimaru 1982, 1992]. Owing to the presence of the core electrons, the ion-ion and electron-ion interactions are described by the pseudopotentials, deviating from the pure Coulombic form. The strong coupling effects between the conduction electrons have a strong influence on the determination of those pseudopotentials [e.g., Singwi & Tosi 1981; Hafner 1987].

Certain metals such as palladium, titanium, and vanadium possess a remarkable ability for absorbing sizable amounts of hydrogen [Alefeld & Völkl 1978]. Nuclear reactions between hydrogen isotopes trapped in a metal hydride, such as palladium deuteride (PdD) and titanium deuteride (TiD_2), offer a unique opportunity for studying reaction processes in microscopically inhomogeneous metallic environments of regular or irregular (e.g., due to the atomic defects) lattice fields produced by the metal atoms. The experiments are usually carried out in nonequilibrium situations such as electrolysis and absorption-desorption processes; no fusion yields have been confirmed as yet. [For examples of earlier experiments, see Jones et al. 1989; Ziegler et al. 1989; Gai et al. 1989.] It is essential here to estimate the fusion rates in metal hydrides in order to help provide an objective assessment of these possibilities [Leggett & Baym 1989; Ichimaru, Ogata, & Nakano 1990].

Pressurized liquid metals offer another interesting environment in which to study nuclear reactions [Ichimaru 1991, 1993a]. It will be shown that $d(p, \gamma)$ ^{3}He and ^{7}Li (p, α) ^{4}He reactions can take place at a power-producing level on the order of a few kW/cm^3 if such a material is brought to a liquid metallic state under an ultrahigh pressure on the order of 10^2 to 10^3 Mbar at a mass density of 10 to 10^2 g/cm^3 and a temperature of $(1-2) \times 10^3$ K, in the vicinity of the estimated melting conditions for hydrogen. Such a range of physical conditions may be accessible through extension of those ultrahigh-pressure metal technologies [e.g., Nellis et al. 1988; Mao, Hemley, & Hanfland 1990; Ruoff et al. 1990; Hemley & Mao 1991].

Electrons and holes in *semiconductors* are good examples of plasmas in solids [e.g., Glicksman 1971]. Particularly of interest in conjunction with the condensed plasma physics are the electron-hole droplets and the exciton molecular liquids, generated by photoexcitation in Ge and Si [e.g., Rice 1977; Hensel, Phillip, & Thomas 1977; Jeffries & Keldish 1983; Vashishta, Kalia, & Singwi 1983]. Related physics issues include the critical conditions and the phase diagrams associated with the *metal-insulator transitions* [Sah, Comberscot, & Dayem 1977; Smith & Wolfe 1986] and observation of the Bose–Einstein condensation in exciton gases [Snoke, Wolfe, & Mysyrowicz 1990].

Some of the *nonneutral plasmas* [Davidson 1990] cooled to sub-Kelvin temperatures may likewise qualify as condensed plasmas. Penning-trapped, pure electron [Driscoll & Malmberg 1983] or pure ion [Bollinger et al. 1990] plasmas rotate around the magnetic axis due in part to space-charge fields in the radial directions. In the frame co-rotating with bulk of the plasma, such a system of charged particles may be regarded effectively as a one-component plasma. The Penning-trapped plasmas have been stably maintained at cryogenic temperatures, 10^{-2} to 10^0 K, for many hours, exhibiting ordered structures in the configurations of ions reminiscent of a freezing transition.

Systems of charged particles in the laboratories with effective spatial dimensions lower than three may constitute a significant class of condensed plasmas [Isihara 1989]. Those electrons (or holes) trapped in the surface states of liquid helium [Grimes 1978] or in the interfaces of the metal-oxide-semiconductor systems form a pseudo–two-dimensional system [Ando, Fowler, & Stern 1982]. The electrons on the liquid-helium surfaces are characterized by the areal densities and temperatures in the ranges of $n = 10^7$–2×10^9 cm^{-2} and $T = 0.1$–1 K; these are the two-dimensional, classical one-component plasmas. Grimes and Adams [1979] found crystallization of such a system at $(\pi n)^{1/2} e^2 / k_B T \sim 137$, where k_B is the Boltzmann constant.

In a two-dimensional model of the cuprate-oxide high-temperature superconductors [Bednorz & Mueller 1986], the magnetic-order structures in the electrons may play a fundamental role in elucidating microscopic mechanisms of the superconductivity. Analyses of the spin-dependent correlations in such a layered system predict that a *spin-density wave* (SDW) instability with wave number $k \cong 2k_F = 2\sqrt{2\pi n}$ may take place in the paramagnetic electrons; associated with the onset of such an SDW instability, an attractive interaction may emerge effectively between the nearest-neighbor electrons with antiparallel spins. The SDW instability may thus be followed by the structural transitions into an inhomogeneous, antiferromagnetically spin-ordered, conductive ground state, and may lead eventually to an antiferromagnetic Mott-insulator [Mott 1990] state.

C. Basic Parameters of Dense Plasmas

We begin by modeling a plasma at a temperature T as consisting of the atomic nuclei (which will be called the *ions*) with an electric charge Ze and a rest mass M $(= Am_N)$, and of electrons with electric charge $-e$ and rest mass m. Physical quantities associated with these separate constituents are distinguished by the suffixes "i" and "e." The atomic mass number is denoted by A, and m_N represents the average mass per nucleon.

In some cases, salient features of a plasma can be understood through investigation of the *one-component plasma* (OCP) properties. This model consists of a single species of charged particles embedded in a uniform background of neutralizing charges. The conduction electrons in the jellium model of metals [e.g., Pines & Nozières 1966] offer an example of such an electron OCP, where all effects of the lattice periodicity in the ion distributions are ignored. Another example is the outer crustal matter of a neutron star, where one neglects polarizability of the dense, relativistic electrons and treats the Fe^{26+} nuclei approximately as forming an ion OCP.

Modeling for a dense plasma under different circumstances may call for consideration of cases containing multiple species of ions, or *multi-ionic plasmas*, the constituents of which will be distinguished by such subscripts as i and j; the BIM, mentioned in Section 1.1A, is an example of such a multi-ionic plasma. Thus, for macroscopic neutrality of the electric charges, the number densities n_i and n_e are assumed to satisfy

$$\sum_i Z_i n_i = n_e . \tag{1.1}$$

A dimensionless parameter characterizing the system of electrons is

$$r_s = \left(\frac{3}{4\pi n_e}\right)^{1/3} \frac{me^2}{\hbar^2} \cong \left(\frac{n_e}{1.611 \times 10^{24} \text{ cm}^{-3}}\right)^{-1/3} . \tag{1.2}$$

It is the *Wigner–Seitz radius* of the electrons,

$$a_e = (3/4\pi n_e)^{1/3} , \tag{1.3}$$

in units of the *Bohr radius*,

$$a_B = \hbar^2/me^2 = 0.5292 \times 10^{-8} \text{ cm} , \tag{1.4}$$

and depends only on the electron density.

The *Fermi wave number* of the electrons in the *paramagnetic state*—that is, a state with an equal number of electrons in the two spin states—is given by

$$k_F = (3\pi^2 n_e)^{1/3} = \frac{3.627 \times 10^8 \text{ cm}^{-1}}{r_s} . \tag{1.5}$$

A parameter characterizing the relativistic effects is defined by the ratio

$$x_F = \frac{\hbar k_F}{mc} = \frac{1.400 \times 10^{-2}}{r_s} . \tag{1.6}$$

The *Fermi energy* of electrons is then given by

$$E_F = mc^2[(1 + x_F^2)^{1/2} - 1]$$
$$= mc^2 \left\{[1 + 1.96 \times 10^{-4} r_s^{-2}]^{1/2} - 1\right\} \tag{1.7}$$

with inclusion of the relativistic effects; these effects are significant in the high-density regime such that $r_s < 0.1$.

The degree of Fermi degeneracy for the electrons at a temperature T is measured by the ratio

$$\Theta = \frac{k_B T}{E_F} , \tag{1.8}$$

where k_B denotes the Boltzmann constant. When $\Theta < 0.1$, the electrons are in the state of complete Fermi degeneracy; thermal effects are small. The condition, $0.1 \leq \Theta \leq 10$, corresponds to a state of intermediate degeneracy; quantum and thermal effects coexist. When $\Theta > 10$, we may regard the system of electrons as being in a nondegenerate, classical state; quantum mechanical interference effects are negligible except in short-range collisions.

Relativistic effects are negligible in most of the electron gases at finite temperatures, that is, those with $\Theta \geq 0.1$. The thermodynamic functions for such a nonrelativistic electron gas are expressible in terms of the Coulomb coupling parameter for the electrons,

$$\Gamma_e = \frac{e^2}{a_e k_B T} , \tag{1.9}$$

and the degeneracy parameter Θ. A useful relation is

$$\Gamma_e = 2\left(\frac{4}{9\pi}\right)^{2/3} \frac{r_s}{\Theta} . \tag{1.10}$$

Let us consider several parameters characterizing the dense semiclassical multi-ionic systems. The Wigner–Seitz radius for the ions of the i-species,

$$a_i = \left(\frac{3Z_i}{4\pi n_e}\right)^{1/3} , \tag{1.11a}$$

is called the *ion-sphere radius*. (For an OCP, the subscripts such as i and j will sometimes be omitted.) The ion-sphere radius between i and j is then given by

$$a_{ij} = \frac{a_i + a_j}{2} . \tag{1.11b}$$

In dense plasmas, where the density of electrons is extremely high and may therefore be regarded as incompressible, Eq. (1.11b) offers an appropriate scaling for the interionic spacings. Such will be referred to as the *constant electron density, ion-sphere scaling*.

The ratio between the thermal de Broglie wavelength and the ionic spacing,

$$\Lambda_{ij} = \frac{\hbar\sqrt{2\pi}}{a_{ij}\sqrt{2\mu_{ij}k_B T}} , \qquad (1.12)$$

measures the degree to which wave-mechanical effects enter a description of the ion fluids, where

$$\mu_{ij} = \frac{M_i M_j}{M_i + M_j} \qquad (1.13)$$

is the reduced mass. In the case of an OCP, Eq. (1.12) may be computed as

$$\Lambda \cong 0.094 \left(\frac{A}{12}\right)^{-5/6} \left(\frac{\rho_m}{10^6 \text{ g/cm}^3}\right)^{1/3} \left(\frac{T}{10^7 \text{ K}}\right)^{-1/2} , \qquad (1.14)$$

where ρ_m designates the mass density.

In an ion fluid with $\Lambda_{ij} < 1$, the wave mechanical effects are negligible; the Coulomb coupling parameter for such a classical plasma is given by

$$\Gamma_{ij} = \frac{Z_i Z_j e^2}{a_{ij} k_B T} . \qquad (1.15)$$

For an OCP, we compute Eq. (1.15) as

$$\Gamma \cong 36 \left(\frac{Z}{6}\right)^2 \left(\frac{A}{12}\right)^{-1/3} \left(\frac{\rho_m}{10^6 \text{ g/cm}^3}\right)^{1/3} \left(\frac{T}{10^7 \text{ K}}\right)^{-1} . \qquad (1.16)$$

A *weakly coupled plasma* corresponds to cases with $\Gamma_{ij} << 1$, where the Coulomb interaction can be treated perturbation theoretically. A *strongly coupled plasma* refers to cases with $\Gamma_{ij} \geq 1$, where a perturbation theory is no longer valid and the system begins to exhibit features of microscopic correlations characteristic of a liquid.

At still higher densities and lower temperatures, an OCP undergoes a freezing transition into a *body-centered cubic* (bcc) crystalline state [Brush, Sahlin, & Teller 1966]. It has been shown by a Monte Carlo simulation method [Slattery, Doolen, & DeWitt 1982; Ogata 1992] that a dense classical OCP freezes into a crystalline solid as the Coulomb coupling parameter exceeds a critical value, $\Gamma_m \sim 180$. Here it is necessary in general to deal with a quantum mechanical Coulomb solid [Wigner 1934, 1938; Carr 1961; Iyetomi, Ogata, & Ichimaru 1993]; the quantum effects are measured through the Einstein frequency ω_E of the Wigner–Seitz solid [e.g., Pines 1963] in its dimensionless form,

$$Y = \frac{\hbar \omega_{\mathrm{E}}}{k_{\mathrm{B}} T} = \frac{\hbar}{k_{\mathrm{B}} T} \sqrt{\frac{4\pi \left(en_{\mathrm{e}}\right)^2}{3\rho_{\mathrm{m}}}} . \tag{1.17}$$

Thus far we have considered the basic parameters characterizing the individual constituents—electrons and ions—of a plasma. In an ultradense plasma with $r_{\mathrm{s}} \leq 0.01$, the Fermi energy of electrons is relativistically high [cf. Eq. (1.7)], so that their coupling with the ions is indeed weak. The kinematic effects of relativistic degenerate electrons soften the electrons against compression, however, and thus act to enhance their polarizations, to an extent qualitatively different from those in the nonrelativistic electrons [e.g., Landau & Lifshitz 1969; Ichimaru & Utsumi 1983].

In nonrelativistic plasmas near the boundaries of the *metal insulator transitions* [Mott 1990], effects of the strong coupling between electrons and ions become pronounced, giving rise to interesting facets in the plasma physics problems interlinking with atomic and molecular physics. Noteworthy among these is the emergence of an "*incipient Rydberg state* (IRS)" for the short-range, electron-ion correlations in the *metallic* (plasma) phase [Tanaka, Yan, & Ichimaru 1990; Ichimaru 1993b]; the IRS thereby accounts for the mutual scattering between electrons and ions beyond the Born approximation. The IRS can likewise modify the short-range interionic potentials and may influence the nuclear reaction rates in dense plasmas through such a modification.

The strong electron-ion coupling in the liquid metallic hydrogen affects it in a number of ways to influence the degree of ionization for the "impurity" atoms immersed in it. The effects include a shift (shallowing) and disappearance of the impurity-atomic levels, a change in the strength of coupling between the impurity-atoms and the hydrogen plasma, and a modification in the equation of state for the hydrogen plasma. Solutions to these problems bear important consequences to the opacities, internal structures, and evolution in various stellar objects, including giant planets, brown dwarfs, and the Sun.

1.2   THEORETICAL BACKGROUND

The physical properties of condensed plasmas are analyzed by a number of theoretical methods. In this section, some of the fundamental theories are considered for introduction. As a specific example of multicomponent systems, we shall treat the case of an electron gas where dependence on spin orientations is particularly singled out. Extension of these theories to other cases of multicomponent plasmas is straightforward.

A. Density-Fluctuation Excitations

Consider the creation and annihilation operators, $c_{p\sigma}^\dagger$ and $c_{p\sigma}$, for the electrons with momentum $\mathbf{p}$ and spin σ. The operators satisfy the anticommutator relations for the fermions [e.g., Fetter & Walecka 1971]:

$$\{c_{p\sigma}^\dagger, c_{p'\sigma'}^\dagger\} = \{c_{p\sigma}, c_{p'\sigma'}\} = 0 \,, \tag{1.18a}$$

$$\{c_{p\sigma}^\dagger, c_{p'\sigma'}\} = \delta_{pp'}\delta_{\sigma\sigma'} \,. \tag{1.18b}$$

In these equations, $\{A, B\} = AB + BA$, meaning an anticommutator; $\delta_{pp'}$ and $\delta_{\sigma\sigma'}$ are Kronecker's deltas.

For a treatment of the density-fluctuation excitations, it is convenient to work with operators representing electron-hole pairs [cf. Wigner 1932; Brittin & Chappell 1962],

$$\rho_{p\mathbf{k}\sigma} = c_{p\sigma}^\dagger c_{\mathbf{p}+\hbar\mathbf{k}\sigma} \,. \tag{1.19}$$

The Fourier component of spin-dependent, density-fluctuation excitations with wave vector $\mathbf{k}$ is then given by

$$\rho_{\mathbf{k}\sigma} = \sum_{\mathbf{p}} \rho_{p\mathbf{k}\sigma} \,, \tag{1.20a}$$

and the total density fluctuation is calculated as

$$\rho_{\mathbf{k}} = \sum_{p\sigma} \rho_{p\mathbf{k}\sigma} \,. \tag{1.20b}$$

The Fourier components of the charge- and spin-density-fluctuation excitations are

$$\rho_{\mathbf{k}}^{(c)} = -e \sum_{p\sigma} \rho_{p\mathbf{k}\sigma} \,, \tag{1.21a}$$

$$\rho_{\mathbf{k}}^{(s)} = \sum_{\mathbf{p}} (\rho_{p\mathbf{k}\uparrow} - \rho_{p\mathbf{k}\downarrow}) \,. \tag{1.21b}$$

The occupation number operator for the state $(\mathbf{p}, \sigma)$ is expressed as

$$n_{p\sigma} = \rho_{\mathbf{p},\mathbf{k}=0,\sigma} \,. \tag{1.22}$$

Response functions of such an electron gas may be formulated by the application of an external potential field $\Phi_\sigma^{\text{ext}}(\mathbf{k}, \omega)$ that couples with the density fluctuations (1.19) [e.g., Ichimaru 1992]. The total Hamiltonian of the system is given as a sum of the unperturbed and external Hamiltonians,

$$H_{tot} = H + H_{ext} , \qquad (1.23)$$

so that

$$H = \sum_{p\sigma} \frac{p^2}{2m} c_{p\sigma}^\dagger c_{p\sigma} + \frac{1}{2} \sum_{\substack{pp'k \\ \sigma,\sigma'}}' v(k) c_{p+\hbar k,\sigma}^\dagger c_{p'-\hbar k,\sigma'}^\dagger c_{p',\sigma'} c_{p,\sigma} , \qquad (1.24)$$

$$H_{ext} = -e \sum_{p\sigma} \Phi_\sigma^{ext}(k,\omega) c_{p,\sigma}^\dagger c_{p-\hbar k,\sigma} \exp(-i\omega t + 0t) + hc . \qquad (1.25)$$

Here

$$v(k) = \frac{4\pi e^2}{k^2} \qquad (1.26)$$

is the Fourier transform of the Coulomb interaction e^2/r, the prime at the summation implies omission of the terms with $k = 0$, hc stands for the Hermitian conjugate, and the 0 in Eq. (1.25) denotes a positive infinitesimal.

The Heisenberg equation of motion for $\rho_{pk}\sigma$ is

$$i\hbar \frac{\partial}{\partial t} \rho_{pk\sigma} = \left[\rho_{pk\sigma}, H_{tot} \right] , \qquad (1.27)$$

where $[A, B] = AB - BA$ denotes a commutator. Explicit calculation with the aid of Eqs. (1.24) and (1.25) yields

$$i\hbar \frac{\partial}{\partial t} \rho_{pk\sigma} = \hbar\omega_{pk}\rho_{pk\sigma} \qquad (1.28a)$$

$$+ \frac{1}{2} v(k) \left\{ \rho_k, n_{p\sigma} - n_{p+\hbar k,\sigma} \right\} \qquad (1.28b)$$

$$+ \frac{1}{2} \sum_{q(\neq k)}' v(q) \left\{ \rho_q, \rho_{p,k-q,\sigma} - \rho_{p+\hbar q,k-q,\sigma} \right\} \qquad (1.28c)$$

$$- e\Phi_\sigma^{ext}(k,\omega)(n_{p\sigma} - n_{p+\hbar k,\sigma}) \exp(-i\omega t + 0t) , \qquad (1.28d)$$

where

$$\omega_{pk} = \frac{k \cdot p}{m} + \frac{\hbar k^2}{2m} \qquad (1.29)$$

is the excitation frequency of an electron-hole pair.

The four terms on the right-hand side of Eq. (1.28) govern the evolution of density-fluctuation excitations in the electron gas. Equation (1.28a) describes free motions of the electron-hole pairs. The terms (1.28b) and (1.28c), on the other hand, stem from the Coulomb interaction. The former represents a mean field contribution linear in the density-fluctuation excitations, while the latter describes the effects of nonlinear coupling between the density fluctuations. Finally, the term (1.28d) accounts for the coupling between the density fluctuations and the external potential.

B. Dielectric Formulation

The dielectric response functions are a class of linear response functions that stem from the exchange and Coulomb interactions between particles; these functions describe the dielectric properties associated with various density-fluctuation excitations in the plasma [e.g., Ichimaru 1992].

An external electrostatic potential field of infinitesimal strength, applied to a uniform and stationary plasma, introduces an external Hamiltonian expressed as

$$H_{\text{ext}} = \sum_{\mathbf{p},\sigma} Z_\sigma e \Phi_\sigma^{\text{ext}}(\mathbf{k}, \omega) c_{\mathbf{p},\sigma}^\dagger c_{\mathbf{p}-\hbar\mathbf{k},\sigma} \exp(-i\omega t + 0t) + \text{hc} . \qquad (1.30)$$

Here we consider a multicomponent system of fermions with electric charges $Z_\sigma e$ for generality. (In the case of electron gas, one sets $Z_\sigma = -1$.) The potential disturbs the plasma and thereby creates various density fluctuations. We are here concerned only with a linear response, so the induced density fluctuations given by

$$\delta n_{\mathbf{k}\sigma}(t) = \left\langle \sum_{\mathbf{p}} \rho_{\mathbf{p}\mathbf{k}\sigma} \right\rangle , \qquad (1.31a)$$

$$\delta n_\sigma(\mathbf{k}, \omega) = \int_{-\infty}^{\infty} dt \, \delta n_{\mathbf{k}\sigma}(t) \exp(i\omega t) \qquad (1.31b)$$

are expressed as

$$\delta n_\sigma(\mathbf{k}, \omega) = \sum_\tau \chi_{\sigma\tau}(\mathbf{k}, \omega) Z_\tau e \Phi_\tau^{\text{ext}}(\mathbf{k}, \omega) , \qquad (1.32)$$

which define the *density-density response functions*, $\chi_{\sigma\tau}(\mathbf{k}, \omega)$, between the particles of σ and τ species. In Eq. (1.31a), $\langle \cdots \rangle$ means a statistical average over the states of the plasma.

The *dielectric function*, $\varepsilon(\mathbf{k}, \omega)$, representing the ratio between the externally applied potential and the total potential in the plasma (see Problem 1.6), is then given by

$$\frac{1}{\varepsilon(\mathbf{k}, \omega)} = 1 + v(k) \sum_{\sigma, \tau} Z_\sigma Z_\tau \chi_{\sigma\tau}(\mathbf{k}, \omega) . \tag{1.33}$$

With the aid of the fluctuation-dissipation theorem summarized in Appendix A, the *dynamic structure factors* $S_{\sigma\tau}(\mathbf{k}, \omega)$, the *static structure factors* $S_{\sigma\tau}(\mathbf{k})$, and the *radial distribution functions* $g_{\sigma\tau}(\mathbf{r})$ are calculated as

$$S_{\sigma\tau}(\mathbf{k}, \omega) = -\frac{\hbar}{2\pi} \coth\left(\frac{\hbar\omega}{2k_{\mathrm{B}}T}\right) \operatorname{Im} \chi_{\sigma\tau}(\mathbf{k}, \omega) , \tag{1.34}$$

$$S_{\sigma\tau}(\mathbf{k}) = \frac{1}{\sqrt{n_\sigma n_\tau}} \int_{-\infty}^{\infty} d\omega \, S_{\sigma\tau}(\mathbf{k}, \omega) , \tag{1.35}$$

$$g_{\sigma\tau}(\mathbf{r}) = 1 + \frac{1}{\sqrt{n_\sigma n_\tau}} \int_{-\infty}^{\infty} \frac{d\mathbf{k}}{(2\pi)^3} [S_{\sigma\tau}(\mathbf{k}) - \delta_{\sigma\tau}] \exp(i\mathbf{k} \cdot \mathbf{r}) . \tag{1.36}$$

These functions describe the two-particle joint distributions in various versions and play the essential parts in formulating the equation of state, thermodynamic functions, and transport properties [e.g., Ichimaru 1992].

The density-density response formalism (1.32) may be transformed to a description of the spin-density responses for an electron gas as follows: Let an external magnetic field, $H^{\mathrm{ext}}(\mathbf{k}, \omega)\exp[i(\mathbf{k} \cdot \mathbf{r} - \omega t)]$, be applied to the electron gas in an arbitrarily chosen direction, which will then induce spin-density fluctuations $\delta n_\sigma(\mathbf{k}, \omega)$. The *spin susceptibility,* $\chi^s(\mathbf{k}, \omega)$, is defined in accordance with

$$-\frac{g\mu_{\mathrm{B}}}{2} \left[\delta n_\uparrow(\mathbf{k}, \omega) - \delta n_\downarrow(\mathbf{k}, \omega)\right] = \chi^s(\mathbf{k}, \omega) H^{\mathrm{ext}}(\mathbf{k}, \omega) , \tag{1.37}$$

where $g = 2.0023$ is the g-factor and $\mu_{\mathrm{B}} = e\hbar/2mc$ is the Bohr magneton. Since Eq. (1.32) may be re-expressed in these circumstances as

$$\delta n_\sigma(\mathbf{k}, \omega) = \frac{g\mu_{\mathrm{B}}}{2} \sum_\tau \tau \chi_{\sigma\tau}(\mathbf{k}, \omega) H^{\mathrm{ext}}(\mathbf{k}, \omega) , \tag{1.38}$$

we find

$$\chi^s(\mathbf{k}, \omega) = -\left(\frac{g\mu_{\mathrm{B}}}{2}\right)^2 \left[\chi_{\uparrow\uparrow}(\mathbf{k}, \omega) + \chi_{\downarrow\downarrow}(\mathbf{k}, \omega) - \chi_{\uparrow\downarrow}(\mathbf{k}, \omega) - \chi_{\downarrow\uparrow}(\mathbf{k}, \omega)\right] . \tag{1.39}$$

Spin-dependent, two-particle distribution functions can be calculated in terms of those spin-dependent response functions with the aid of the fluctuation-dissipation theorem (Appendix A) in a way analogous to Eqs. (1.34)–(1.36).

C. The Random-Phase Approximation

The density-density response functions of the previous section may be calculated through a solution to the equation of motion (1.28), which governs evolution of the density-fluctuation excitations in the electron gas in the presence of the external Hamiltonian (1.25). In the *Hartree–Fock approximation*, Coulomb-interaction terms (1.28b) and (1.28c) are ignored in the calculation of the response functions. In this approximation, the density responses are decoupled between different species and take a form,

$$\chi_{\sigma\tau}^{HF}(\mathbf{k}, \omega) = \sum_{\rho} \delta_{\sigma\rho} \delta_{\tau\rho} \chi_{\rho}^{(0)}(\mathbf{k}, \omega) \, , \tag{1.40}$$

where the free-electron polarizability of the σ species is given by

$$\chi_{\sigma}^{(0)}(\mathbf{k}, \omega) = \frac{1}{\hbar} \sum_{\mathbf{p}} \frac{1}{\omega - \omega_{\mathbf{pk}} + i0} [f_{\sigma}(\mathbf{p}) - f_{\sigma}(\mathbf{p} + \hbar\mathbf{k})] \tag{1.41}$$

with the Fermi distribution

$$f_{\sigma}(\mathbf{p}) = \frac{1}{\exp\left[(\varepsilon_{\mathbf{p}} - \mu_{\sigma})/k_B T\right] + 1} \, . \tag{1.42}$$

Here

$$\varepsilon_{\mathbf{p}} = \frac{p^2}{2m} \tag{1.43}$$

is the kinetic energy of an electron and μ_{σ} is the chemical potential for the free-electron gas of the σ species, which is determined by the normalization,

$$\int \frac{d\mathbf{p}}{(2\pi\hbar)^3} f_{\sigma}(\mathbf{p}) = n_{\sigma} \, , \tag{1.44}$$

with n_{σ} denoting the number density of the spin-σ electrons.

In the treatment of a free-electron gas at finite temperatures, it is useful to define the Fermi integrals,

$$I_{\nu}(\alpha) = \int_{0}^{\infty} dx \frac{x^{\nu}}{\exp(x - \alpha) + 1} \, . \tag{1.45}$$

Mathematical properties of the Fermi integrals are summarized in Appendix B. When the electron gas is in the ground state ($T = 0$), the Fermi distribution takes a step function; Eq. (1.41) evaluated in these circumstances is called the

Lindhard polarizability [Lindhard 1954; Pines & Nozières 1966]. In the classical limit of high temperature and low density, the Fermi distribution approaches the Maxwell–Boltzmann form and then Eq. (1.41) becomes the *Vlasov polarizability* [Vlasov 1967; Ichimaru 1992].

The *random-phase approximation* (RPA) [Bohm & Pines 1953] is an approach that goes beyond the Hartree–Fock approximation. In the RPA, one takes account of the Coulomb interaction, neglected in the Hartree–Fock approximation, through the mean field term (1.28b); the nonlinear coupling term (1.28c) between the density-fluctuation excitations remains neglected. Since the mean field term is linear in the fluctuations, the RPA density-density response functions are calculated as

$$\chi_{\sigma\tau}^{\text{RPA}}(\mathbf{k}, \omega) = \sum_{\rho} \delta_{\sigma\rho}\delta_{\tau\rho}\chi_{\rho}^{(0)}(\mathbf{k}, \omega) + \frac{Z_\sigma Z_\tau v(k)\chi_\sigma^{(0)}(\mathbf{k}, \omega)\chi_\tau^{(0)}(\mathbf{k}, \omega)}{\varepsilon_0(\mathbf{k}, \omega)} , \quad (1.46)$$

where

$$\varepsilon_0(\mathbf{k}, \omega) = 1 - \sum_{\sigma} Z_\sigma^2 v(k)\chi_\sigma^{(0)}(\mathbf{k}, \omega) \qquad (1.47)$$

is the *RPA dielectric function*.

For an electron gas in the ground state, the static (i.e., $\omega = 0$) values of the Lindhard polarizability are evaluated as

$$\chi_\sigma^{(0)}(k, 0) = -\frac{3n_\sigma m}{(\hbar k_{\text{F}\sigma})^2} \left\{ \frac{1}{2} + \frac{k_{\text{F}\sigma}}{2k}\left(1 - \frac{k^2}{4k_{\text{F}\sigma}^2}\right) \ln\left|\frac{k + 2k_{\text{F}\sigma}}{k - 2k_{\text{F}\sigma}}\right| \right\} , \qquad (1.48)$$

where

$$k_{\text{F}\sigma} = (6\pi^2 n_\sigma)^{1/3} \qquad (1.49)$$

is the *Fermi wave number* appropriate to the fermions of the σ species. In the limit of long wavelengths, the static RPA screening function, given by $\varepsilon_0(k, 0)$, is then expressed as

$$\varepsilon_0(k, 0) \underset{(k\to 0)}{\to} 1 + \frac{k_{\text{TF}}^2}{k^2} \equiv 1 + \frac{1}{k^2}\sum_{\sigma} \frac{6\pi n_\sigma (Z_\sigma e)^2}{(\hbar k_{\text{F}\sigma})^2/2m} , \qquad (1.50)$$

which defines the *Thomas–Fermi screening parameter* k_{TF}.

For a high-temperature and low-density classical plasma, the static (i.e., $\omega = 0$) values of the Vlasov polarizability are evaluated as

$$\chi_\sigma^{(0)}(k, 0) = -\frac{n_\sigma}{k_{\text{B}}T} . \qquad (1.51)$$

The static RPA screening function is therefore expressed as

$$\varepsilon_0(k, 0) = 1 + \frac{k_{\mathrm{D}}^2}{k^2} \equiv 1 + \frac{1}{k^2} \sum_{\sigma} \frac{4\pi n_\sigma (Z_\sigma e)^2}{k_{\mathrm{B}} T} , \tag{1.52}$$

which defines the *Debye–Hückel screening parameter* k_D.

D. Strong Coupling Effects

In the RPA of the previous subsection, the strong exchange and Coulomb coupling effects represented by the nonlinear term (1.28c) of the fluctuations have not been taken into consideration. A theoretical method of going beyond such an RPA description and thereby accounting for the strong interparticle-correlation effects rigorously in the framework of the dielectric formulation has been provided by a *polarization potential approach* [Pines 1966; Ichimaru 1992], which we shall summarize in this section.

In this approach, the linear density response relations stemming from the solution to Eqs. (1.28) may be written as

$$\delta n_\sigma(\mathbf{k}, \omega) =$$
$$\chi_\sigma^{(0)}(\mathbf{k}, \omega) \left\{ Z_\sigma e \Phi_\sigma^{\mathrm{ext}}(\mathbf{k}, \omega) + \sum_\tau Z_\sigma Z_\tau v(k) \left[1 - G_{\sigma\tau}(\mathbf{k}, \omega)\right] \delta n_\tau(\mathbf{k}, \omega) \right\} . \tag{1.53}$$

This relation thus implies that the induced density fluctuations, $\delta n_\tau(\mathbf{k}, \omega)$, of the particles of the τ species may produce an effective potential field of the strength

$$Z_\sigma e \Phi_{\sigma\tau}^{\mathrm{pol}}(\mathbf{k}, \omega) = Z_\sigma Z_\tau v(k) \left[1 - G_{\sigma\tau}(\mathbf{k}, \omega)\right] \delta n_\tau(\mathbf{k}, \omega) , \tag{1.54}$$

which acts on the particles of the σ species; $\Phi_{\sigma\tau}^{\mathrm{pol}}(\mathbf{k}, \omega)$ in Eq. (1.54) may be looked upon as a polarization potential in these circumstances. Such a polarization potential generally differs from the RPA mean field, $Z_\sigma Z_\tau v(k) \delta n_\tau(\mathbf{k}, \omega)$, since the exchange and Coulomb correlations between the particles in effect modify the potentials due to the induced space charges. The differences are measured by the *dynamic local field corrections* $G_{\sigma\tau}(\mathbf{k}, \omega)$ in Eq. (1.54), which originates from the nonlinear term (1.28c) in the equation of motion for the density-fluctuation excitations. The density-density response functions may then be calculated explicitly from the solution to Eq. (1.53) in accordance with Eq. (1.32).

In certain cases of the condensed plasma problems, one adopts an approximation whereby the dynamic local field corrections are assumed to be independent of the frequency variables and are replaced by their static evaluations,

$$G_{\sigma\tau}(\mathbf{k}, \omega) \to G_{\sigma\tau}(\mathbf{k}) \equiv G_{\sigma\tau}(\mathbf{k}, \omega = 0) . \tag{1.55}$$

The functions $G_{\sigma\tau}(\mathbf{k})$ are called the *static local field corrections*, and a theoretical scheme of treating the strong coupling effects via $G_{\sigma\tau}(\mathbf{k})$ will be referred to as the *static local field correction approximation*.

Let us now revisit the spin-density responses in such a static local field correction approximation for an electron gas in the paramagnetic state, in which

$$n_\uparrow = n_\downarrow = \frac{n}{2} . \tag{1.56}$$

Since the system is isotropic, it is appropriate to define the local field corrections for parallel and antiparallel spins, $G_p(k)$ and $G_a(k)$, and the total free-electron polarizability $\chi_0(k, \omega)$ via

$$G_p(k) = G_{\uparrow\uparrow}(k) = G_{\downarrow\downarrow}(k) , \tag{1.57}$$

$$G_a(k) = G_{\uparrow\downarrow}(k) = G_{\downarrow\uparrow}(k) , \tag{1.58}$$

$$\chi_0(k, \omega) = 2\chi_\uparrow^{(0)}(k, \omega) = 2\chi_\downarrow^{(0)}(k, \omega) \tag{1.59}$$

[e.g., Ichimaru, Iyetomi, & Tanaka 1987].

The dielectric function and the spin susceptibility are then given by

$$\varepsilon(k, \omega) = 1 - \frac{v(k)\chi_0(k, \omega)}{1 + v(k)G(k)\chi_0(k, \omega)} , \tag{1.60}$$

$$\chi^s(k, \omega) = -\left(\frac{g\mu_B}{2}\right)^2 \frac{\chi_0(k, \omega)}{1 + J(k)\chi_0(k, \omega)} . \tag{1.61}$$

Here

$$G(k) = \frac{1}{2}\left[G_p(k) + G_a(k)\right] , \tag{1.62}$$

$$J(k) = \frac{v(k)}{2}\left[G_p(k) - G_a(k)\right] . \tag{1.63}$$

In the limit of long wavelengths, these functions are related to the *isothermal compressibility* κ_T and the *static spin susceptibility* χ_P via the thermodynamic sum rules treated in Appendix A. Since one defines

$$\kappa_T = \frac{1}{n}\left(\frac{\partial n}{\partial P}\right)_T , \tag{1.64}$$

$$\chi_P = \left(\frac{\partial M}{\partial B}\right)_T , \tag{1.65}$$

with P, M, and B denoting the pressure, the density of spin magnetization, and the strength of the magnetic field, respectively, the *compressibility sum rule* and the *spin-susceptibility sum rule* read

$$\lim_{k \to 0} v(k) G(k) = -\frac{1}{\chi_0(0,0)} - \frac{1}{n^2 \kappa_T} , \tag{1.66}$$

$$\lim_{k \to 0} J(k) = -\frac{1}{\chi_0(0,0)} - \left(\frac{g\mu_B}{2}\right)^2 \frac{1}{\chi_P} . \tag{1.67}$$

These sum rule relations provide a crucial linkage between the dielectric formulation and the *Landau theory of Fermi liquids* [Landau 1956, 1957; Pines & Nozières 1966]. Their extension to the regime of finite wavelengths offers an essential ingredient of the density-functional theory, which we shall consider in the next subsection.

E. The Density-Functional Theory

Consider a system of N_σ electrons contained in a box of volume V under the influence of external potentials $\phi_\sigma(\mathbf{r})$, where σ denotes the spin coordinate. The field operators, which are constructed by coherent superposition of the creation and annihilation operators of Section 1.2A, may be expressed as

$$\psi_\sigma^\dagger(\mathbf{r}) = \frac{1}{\sqrt{V}} \sum_{\mathbf{p}} c_{\mathbf{p}\sigma}^\dagger \exp\left(-\frac{i}{\hbar}\mathbf{p} \cdot \mathbf{r}\right) , \tag{1.68a}$$

$$\psi_\sigma(\mathbf{r}) = \frac{1}{\sqrt{V}} \sum_{\mathbf{p}} c_{\mathbf{p}\sigma} \exp\left(\frac{i}{\hbar}\mathbf{p} \cdot \mathbf{r}\right) . \tag{1.68b}$$

The density operators are given by

$$\hat{n}_\sigma(\mathbf{r}) = \psi_\sigma^\dagger(\mathbf{r})\psi_\sigma(\mathbf{r}) . \tag{1.69a}$$

We assume that the system of electrons under consideration has a unique, nondegenerate ground state Ψ. Clearly, then, Ψ is a unique functional of $\phi_\sigma(\mathbf{r})$ and therefore so are the expectation values of the electron densities,

$$n_\sigma(\mathbf{r}) = (\Psi, \hat{n}_\sigma(\mathbf{r})\Psi) . \tag{1.69b}$$

The important conclusion derived by Hohenberg and Kohn [1964] is that $\phi_\sigma(\mathbf{r})$ and Ψ, in turn, are uniquely determined by the knowledge of the density distributions, $n_\sigma(\mathbf{r})$. This conclusion lays the theoretical foundation to the *density-functional theory* [e.g., Kohn & Vashishta 1983; Callaway & March 1984], which we shall summarize in this subsection.

The total Hamiltonian operator of the system may be written as a sum of the kinetic, internal energy, and external contributions,

$$H_{\text{tot}} = H_K + H_{\text{int}} + H_{\text{ext}} , \tag{1.70}$$

where

$$H_K = \frac{\hbar^2}{2m} \sum_\sigma \int d\mathbf{r} \, \nabla \psi_\sigma^\dagger(\mathbf{r}) \cdot \nabla \psi_\sigma(\mathbf{r}) , \tag{1.71a}$$

$$H_{\text{int}} = \frac{1}{2} \sum_{\sigma,\tau} \int d\mathbf{r} \, d\mathbf{r}' \, \frac{Z_\sigma Z_\tau e^2}{|\mathbf{r} - \mathbf{r}'|} \psi_\sigma^\dagger(\mathbf{r}) \psi_\tau^\dagger(\mathbf{r}') \psi_\tau(\mathbf{r}') \psi_\sigma(\mathbf{r}) , \tag{1.71b}$$

$$H_{\text{ext}} = \sum_\sigma \int d\mathbf{r} \, Z_\sigma e \phi_\sigma(\mathbf{r}) \psi_\sigma^\dagger(\mathbf{r}) \psi_\sigma(\mathbf{r}) . \tag{1.71c}$$

Expectation values, designated by angular brackets, of these contributions and therefore the total Hamiltonian are the functionals of the densities, $n_\sigma(\mathbf{r})$. We write, in particular,

$$\langle H_K + H_{\text{int}} \rangle = E_K [n_\sigma(\mathbf{r})]$$
$$+ \frac{1}{2} \sum_{\sigma,\tau} \int d\mathbf{r} \, d\mathbf{r}' \, \frac{Z_\sigma Z_\tau e^2}{|\mathbf{r} - \mathbf{r}'|} n_\sigma(\mathbf{r}) n_\tau(\mathbf{r}') + E_{\text{xc}} [n_\sigma(\mathbf{r})] , \tag{1.72}$$

where $E_K[n_\sigma(\mathbf{r})]$ refers to the kinetic energy of a *noninteracting* electron gas of the densities $n_\sigma(\mathbf{r})$ in its ground state, the second is the Hartree interaction term, and the remainder, $E_{\text{xc}}[n_\sigma(\mathbf{r})]$, is called the *exchange and correlation energy*. The expectation value of the total Hamiltonian is thus expressed as

$$E_{\phi_\sigma} [n_\sigma(\mathbf{r})] = E_K [n_\sigma(\mathbf{r})] + \frac{1}{2} \sum_{\sigma,\tau} \int d\mathbf{r} \, d\mathbf{r}' \, \frac{Z_\sigma Z_\tau e^2}{|\mathbf{r} - \mathbf{r}'|} n_\sigma(\mathbf{r}) n_\tau(\mathbf{r}') + E_{\text{xc}} [n_\sigma(\mathbf{r})]$$
$$+ \sum_\sigma \int d\mathbf{r} \, Z_\sigma e \phi_\sigma(\mathbf{r}) n_\sigma(\mathbf{r}) . \tag{1.73}$$

Its dependence on the external potentials $\phi_\sigma(\mathbf{r})$ has been denoted explicitly in Eq. (1.73).

In the ground state, the expectation values (1.73) take on the minimal values with respect to variations of the densities around themselves, subject to the conditions

$$\int d\mathbf{r} \, n_\sigma(\mathbf{r}) = N_\sigma . \tag{1.74}$$

The resultant Euler equations ensuing from the density-functional derivatives (see Appendix C) of Eq. (1.73) are

$$\frac{\delta E_K [n_\sigma]}{\delta n_\sigma (\mathbf{r})} + Z_\sigma e \phi_\sigma^{tot}(\mathbf{r}) + v_\sigma^{xc}(\mathbf{r}) - \mu_\sigma = 0 \, . \tag{1.75}$$

Here

$$\phi_\sigma^{tot}(\mathbf{r}) = \phi_\sigma (\mathbf{r}) + \sum_\tau \int d\mathbf{r}' \, \frac{Z_\tau e}{|\mathbf{r} - \mathbf{r}'|} n_\tau (\mathbf{r}') \tag{1.76}$$

refer to the total classical potentials,

$$v_\sigma^{xc}(\mathbf{r}) \equiv \frac{\delta E_{xc} [n_\sigma]}{\delta n_\sigma (\mathbf{r})} \tag{1.77}$$

are the *exchange-correlation potentials* defined via the functional derivatives, and μ_σ are the Lagrange parameters associated with the subsidiary conditions (1.74).

The *Kohn–Sham* [1965] *self-consistent equations* for the single-particle wave functions $\psi_i(\mathbf{r})$ are derived in terms of those potentials as

$$\left[-\frac{\hbar^2}{2m} \nabla^2 + v_\sigma^{eff}(\mathbf{r}) \right] \psi_i(\mathbf{r}) = \varepsilon_i \psi_i(\mathbf{r}) \, , \tag{1.78}$$

where

$$v_\sigma^{eff}(\mathbf{r}) = Z_\sigma e \phi_\sigma^{tot}(\mathbf{r}) + v_\sigma^{xc}(\mathbf{r}) \, . \tag{1.79}$$

The self-consistency is enforced via

$$n_\sigma (\mathbf{r}) = \sum_{i=1}^{N_\sigma} |\psi_i(\mathbf{r})|^2 \, , \tag{1.80}$$

where the sum is to be carried out over the N_σ lowest occupied eigenstates.

Analytic theories accounting for the thermodynamic properties of a strongly coupled plasma are concerned with the derivation of relevant expressions for the local field corrections, $G_{\sigma\tau}(k)$, as explained in Section 1.2D; these functions measure the extent to which the particle interactions affect the static correlational properties in such a plasma. The local field corrections can be formulated in the density-functional theory extended to the cases of finite temperatures by Mermin [1965].

The thermodynamic potentials of an interacting inhomogeneous plasma in external potentials $\phi_\sigma(\mathbf{r})$ may be expressed as

$$\Omega_{\phi_\sigma}[n_\sigma(\mathbf{r})] = F_0[n_\sigma(\mathbf{r})] + \frac{1}{2}\sum_{\sigma,\tau}\int d\mathbf{r}\,d\mathbf{r}'\,\frac{Z_\sigma Z_\tau e^2}{|\mathbf{r}-\mathbf{r}'|}n_\sigma(\mathbf{r})n_\tau(\mathbf{r}') + F_{xc}[n_\sigma(\mathbf{r})]$$

$$+ \sum_\sigma\int d\mathbf{r}\,(Z_\sigma e\phi_\sigma(\mathbf{r}) - \mu_\sigma)n_\sigma(\mathbf{r})\;. \qquad (1.81)$$

Here $F_0[n_\sigma(\mathbf{r})]$ denotes the free-energy functional of a free-particle system with density distributions $n_\sigma(\mathbf{r})$ and $F_{xc}[n_\sigma(\mathbf{r})]$ refers to the remaining exchange-correlation free-energy functional for the interacting system. Assuming that the densities have the form

$$n_\sigma(\mathbf{r}) = n_\sigma + \delta n_\sigma(\mathbf{r})\;, \qquad (1.82a)$$

with $n_\sigma = N_\sigma/V$, $|\delta n_\sigma(\mathbf{r})|/n_\sigma << 1$, and

$$\int d\mathbf{r}\,\delta n_\sigma(\mathbf{r}) = 0\;, \qquad (1.82b)$$

one finds the Euler equations through minimization of the thermodynamic potential (1.81) with respect to the density variations $\delta n_\sigma(\mathbf{r})$. That is,

$$\frac{\delta F_0[n_\sigma]}{\delta n_\sigma(\mathbf{r})} + v_\sigma^{xc}(\mathbf{r}) + Z_\sigma e\phi_\sigma^{tot}(\mathbf{r}) = 0\;, \qquad (1.83)$$

where

$$v_\sigma^{xc}(\mathbf{r}) \equiv \frac{\delta F_{xc}[n_\sigma]}{\delta n_\sigma(\mathbf{r})} \qquad (1.84)$$

are the exchange-correlation potentials at finite temperatures.

To the lowest order in $\delta n_\sigma(\mathbf{r})$, the first two terms in Eq. (1.83) should be linear in $\delta n_\sigma(\mathbf{r})$ on account of Eq. (1.82b), so that one writes

$$\sum_\tau\int d\mathbf{r}'\left[K_{\sigma\tau}^{(0)}(\mathbf{r}-\mathbf{r}';n_\sigma) + K_{\sigma\tau}^{xc}(\mathbf{r}-\mathbf{r}';n_\sigma) + \frac{Z_\sigma Z_\tau e^2}{|\mathbf{r}-\mathbf{r}'|}\right]\delta n_\tau(\mathbf{r}') + Z_\sigma e\phi_\sigma^{ext}(\mathbf{r})$$

$$= 0\;. \qquad (1.85)$$

The kernels, $K_{\sigma\tau}^{(0)}(\mathbf{r})$ and $K_{\sigma\tau}^{xc}(\mathbf{r})$, introduced in Eq. (1.85), are the second density-functional derivatives of $F_0[n_\sigma(\mathbf{r})]$ and $F_{xc}[n_\sigma(\mathbf{r})]$ around $n_\sigma(\mathbf{r}) = n_\sigma$ and depend only on $\mathbf{r} - \mathbf{r}'$ and n_σ, the average number densities. The spatial Fourier transformation of Eq. (1.85) yields

$$\sum_\tau \left[\tilde{K}_{\sigma\tau}^{(0)}(\mathbf{k}; n_\sigma) + \tilde{K}_{\sigma\tau}^{\text{xc}}(\mathbf{k}; n_\sigma) + Z_\sigma Z_\tau v(k) \right] \delta\tilde{n}_\tau(\mathbf{k}) + Z_\sigma e \Phi_\sigma^{\text{ext}}(\mathbf{k}, 0) = 0 \, .$$

(1.86)

Here, for example,

$$\tilde{K}_{\sigma\tau}^{(0)}(\mathbf{k}; n_\sigma) = \int d\mathbf{r} \, K_{\sigma\tau}^{(0)}(\mathbf{r}; n_\sigma) \exp(-i\mathbf{k} \cdot \mathbf{r}) \, .$$

(1.87)

Equation (1.86) is a relation for the static linear response in the plasma. Direct density-functional calculations and comparison with Eqs. (1.40) and (1.53) yield

$$\sum_\rho \tilde{K}_{\sigma\rho}^{(0)}(\mathbf{k}; n_\sigma) \chi_{\rho\tau}^{\text{HF}}(\mathbf{k}, 0) = -\delta_{\sigma\tau} \, ,$$

(1.88)

$$\tilde{K}_{\sigma\tau}^{\text{xc}}(\mathbf{k}; n_\sigma) = -Z_\sigma Z_\tau v(k) G_{\sigma\tau}(\mathbf{k}) \, ,$$

(1.89)

where $G_{\sigma\tau}(\mathbf{k})$ are the static local field corrections defined in Eq. (1.55).

F. Green's Function Formalism

Green's functions describe the propagation of one or more specified particles in a many-particle system [e.g., Martin & Schwinger 1959; Kadanoff & Baym 1962; Nozières 1964; Rickayzen 1980]. The static and dynamic properties of the system may be revealed in the features of the multiparticle responses and correlation functions through the ways that those specific particles interact with one another as well as with the rest of the particles in the process of the propagation.

The single-particle Green's functions $G(1, 2)$ may be defined as

$$\begin{aligned} G(1, 2) &= -i \left(\Psi, \mathbf{T} \left\{ \psi_{\sigma_1}(\mathbf{r}_1, t_1) \psi_{\sigma_2}^\dagger(\mathbf{r}_2, t_2) \right\} \Psi \right) \\ &= -i \left(\Psi, \mathbf{T} \left\{ \psi(1) \psi^\dagger(2) \right\} \Psi \right) \, . \end{aligned}$$

(1.90)

Here $\psi(1)$ and $\psi^\dagger(2)$ are the fermion field operators introduced in Eqs. (1.68), which evolve through the Heisenberg equation of motion (see Eq. 1.27)—that is, the *Heisenberg operators*—with the unperturbed grand-canonical Hamiltonian,

$$\begin{aligned} H = &-\sum_\sigma \int d\mathbf{r} \left[\frac{\hbar^2}{2m} \nabla^2 + \mu_\sigma \right] \psi_\sigma^\dagger(\mathbf{r}) \psi_\sigma(\mathbf{r}) \\ &+ \frac{1}{2} \sum_{\sigma,\tau} \int d\mathbf{r} d\mathbf{r}' \frac{Z_\sigma Z_\tau e^2}{|\mathbf{r} - \mathbf{r}'|} \psi_\sigma^\dagger(\mathbf{r}) \psi_\tau^\dagger(\mathbf{r}') \psi_\tau(\mathbf{r}') \psi_\sigma(\mathbf{r}) \, ; \end{aligned}$$

(1.91)

μ_σ is the chemical potential; a shorthand such as

$$1 \equiv (\mathbf{r}_1, \sigma_1, t_1) \tag{1.92}$$

has been used for conciseness; and **T** is a "chronological" operator, which has the effect of ordering the factors within the brackets so that the time variable increases from right to left. Hence

$$\mathbf{T}\{\psi(1)\psi^\dagger(2)\} = \begin{cases} \psi(1)\psi^\dagger(2), & \text{if } t_2 < t_1 \\ -\psi^\dagger(2)\psi(1), & \text{if } t_2 > t_1. \end{cases} \tag{1.93}$$

In Eq. (1.90), Ψ refers to the state vector of the N-particle system, which we assume to be uniform in space and stationary in time; then $G(1, 2)$ are functions of $\mathbf{r}_1 - \mathbf{r}_2$ and $t_1 - t_2$ only, as far as the space-time variables are concerned.

The physical meaning of Green's functions is clear: For $t_1 > t_2$, $G(1, 2)$ represent the inner products between the state vectors, $\psi^\dagger(1)\Psi$ and $\psi^\dagger(2)\Psi$, and as such describe the propagation of *one extra particle* introduced in the original N-particle system, that is, the evolution of the $(N + 1)$-particle system. For $t_2 > t_1$, on the other hand, they are the inner products between the state vectors, $\psi(1)\Psi$ and $\psi(2)\Psi$, and thus describe the propagation of *one extra hole* introduced in the original N-particle system, that is, the evolution of the $(N - 1)$-particle system. In light of the equal-time commutator relations (1.18), one finds

$$\left(\Psi, \mathbf{T}\{\psi_\sigma(\mathbf{r}_1, t + 0)\psi_\sigma^\dagger(\mathbf{r}_2, t)\}\Psi\right) - \left(\Psi, \mathbf{T}\{\psi_\sigma(\mathbf{r}_1, t - 0)\psi_\sigma^\dagger(\mathbf{r}_2, t)\}\Psi\right)$$
$$= \delta(\mathbf{r}_1 - \mathbf{r}_2),$$

where $\delta(\mathbf{r})$ is the three-dimensional delta function; the discontinuity of Green's functions at $t_2 = t_1$ thus measures the number density.

One likewise defines *two-particle Green's functions* as

$$K(1, 2, 3, 4) = \left(\Psi, \mathbf{T}\{\psi(1)\psi(2)\psi^\dagger(3)\psi^\dagger(4)\}\Psi\right). \tag{1.94}$$

An important class of Green's functions, the *density-density correlation functions* (see Eq. A.7), are obtained as

$$C_{\sigma_1\sigma_2}(\mathbf{r}_1, t_1; \mathbf{r}_2, t_2) = K(1, 2, 2^+, 1^+)$$
$$= \left(\Psi, \mathbf{P}\{\hat{n}_{\sigma_1}(\mathbf{r}_1, t_1)\hat{n}_{\sigma_2}(\mathbf{r}_2, t_2)\}\Psi\right), \tag{1.95}$$

where $\hat{n}_\sigma(\mathbf{r}, t)$ is the Heisenberg operator for Eq. (1.69a), **P** is a chronological operator involving no change of sign,

$$1^+ \equiv (\mathbf{r}_1, \sigma_1, t_1 + 0) \,, \tag{1.96}$$

and 0 is a positive infinitesimal.

Multiparticle responses and correlations can be analyzed through the functional derivatives of Green's functions [Nakano & Ichimaru 1989a, 1990]. For this purpose, we reformulate the single-particle Green's functions in the presence of an external potential $\phi(1, 1')$, nonlocal in space and time, as

$$G(1, 1') = \frac{\left(\Psi, \mathbf{T} \left\{ \psi(1)\psi^\dagger(1')S \right\} \Psi \right)}{i (\Psi, S\Psi)} . \tag{1.97}$$

Here the scattering matrix is given by

$$S = \mathbf{T} \exp \left[-\frac{i}{\hbar} \int d1 \int d1' \, \psi^\dagger(1)\phi(1, 1')\psi(1') \right] , \tag{1.98}$$

$$1' \equiv (\mathbf{r}'_1, \sigma_1, t'_1) \,, \tag{1.99}$$

and integrations in Eq. (1.98) involve a summation over the common spin index σ_1.

The ν-body response functions $\chi^{(\nu)}(1, 1'; \ldots; \nu, \nu')$ are then defined in terms of the functional derivatives of $G(1, 1')$ with respect to $\phi(j, j')$ as

$$\chi^{(\nu)}(1, 1'; \ldots; \nu, \nu') = -i \left. \frac{\delta^{\nu-1}}{\delta\phi(\nu, \nu') \ldots \delta\phi(2, 2')} G(1, 1') \right|_{\phi \to 0} . \tag{1.100}$$

In particular, the density-density response between spin components, σ_1 and σ_2, may be calculated from Eq. (1.100) as

$$\chi^{(2)}(1, 1^+; 2^+, 2) = \frac{1}{\hbar} \left[\left(\Psi, \mathbf{P} \left\{ \hat{n}_{\sigma_1}(\mathbf{r}_1, t_1) \hat{n}_{\sigma_2}(\mathbf{r}_2, t_2) \right\} \Psi \right) \right. $$
$$\left. - (\Psi, \hat{n}_{\sigma_1}(\mathbf{r}_1, t_1)\Psi)(\Psi, \hat{n}_{\sigma_2}(\mathbf{r}_2, t_2)\Psi) \right] . \tag{1.101}$$

With the aid of the equation of motion for the field operators, one can derive the hierarchical relations between the space-time Fourier transforms of multiparticle Green's functions and thereby analyze the quasiparticle properties and correlation effect s in the many-body system. In Chapter 3, we shall consider these problems explicitly for electron gas.

G. Computer Simulation Methods

A *Monte Carlo* (MC) *method* is any method making use of random numbers to solve a problem [James 1980]. The power of the MC method lies basically in its ability to carry through multidimensional integrations with improved accuracy through techniques of importance sampling and with increased capacity of the modern computers [e.g., Binder 1979, 1992].

In the Metropolis algorithm [Metropolis et al. 1953], one works with a statistical ensemble at constant temperature, volume, and number of particles, that is, the *canonical ensemble*. Monte Carlo steps (configurations) are generated by random displacements of particles in the many-particle system under study. Configurations so created will be accepted or rejected with the probability of acceptance

$$
P = \begin{cases} \exp\left(-\dfrac{\Delta E}{k_{\mathrm{B}}T}\right), & \text{if } \Delta E > 0, \\ 1, & \text{if } \Delta E \le 0, \end{cases} \tag{1.102}
$$

where ΔE is the increment of interaction energy in the system at the creation of a configuration. A Marcoff chain representing the canonical ensemble is thereby generated, with its thermalization usually monitored through evaluation of the interaction energy. In their pioneering work, Brush, Sahlin, and Teller [1966] performed numerical experiments on the strongly coupled OCPs by such an MC method.

In the method of *molecular dynamics* (MD), pioneered by Alder and Wainwright [1959], the classical equations of motion for a system of interacting particles are solved by integration in discrete time steps and equilibrium properties are determined from time averages taken over a sufficiently long time interval [e.g., Ciccontti, Frenkel, & McDonald 1987; Yonezawa 1992]. In the MD simulation, one works with the *microcanonical ensemble*, in which the total energy, volume, and number of particles are kept constant; the MD methods are therefore deterministic in principle. There exist provisions [Andersen 1980; Hoover et al. 1980; Nosé & Klein 1983; Nosé 1984] in the framework of the MD method that can simulate *isothermal* and/or *isobaric* ensembles. In the approach proposed by Car and Parrinello [1985], dynamic aspects of an MD method are accommodated in the treatment of simulated annealing for a quantum many-body system through a density-functional theory explained in Section 1.2E.

Quantum many-body problems are approached by various computer simulation techniques [e.g., Ceperley & Kalos 1979; Binder 1979, 1992; Yonezawa

1992]. In the *variational MC method* [McMillan 1965], one calculates an expectation value,

$$E_{\mathrm{T}} = \frac{\int d\mathbf{R}\, \psi_{\mathrm{T}}(\mathbf{R}) H(\mathbf{R}) \psi_{\mathrm{T}}(\mathbf{R})}{\int d\mathbf{R}\, |\psi_{\mathrm{T}}(\mathbf{R})|^2}\,, \tag{1.103}$$

of the N-particle Hamiltonian, $H(\mathbf{R})$, with a trial wave function, $\psi_{\mathrm{T}}(\mathbf{R})$, by multidimensional MC integrations, where

$$\mathbf{R} = (\mathbf{r}_1, \ldots, \mathbf{r}_N) \tag{1.104}$$

refers to the coordinates of the N particles. Variational trial functions for a system of fermions are sometimes chosen in the form of a Jastrow [1955] type,

$$\psi_{\mathrm{J}}(\mathbf{R}) = \psi_0(\mathbf{R}) \exp\left[-\frac{1}{2}\sum_{i<j} u(r_{ij})\right], \tag{1.105}$$

where $\psi_0(\mathbf{R})$ is the ideal Fermi gas wave function and $u(r_{ij})$ corresponds to a variational "pseudopotential" accounting for interparticle correlation with $r_{ij} = |\mathbf{r}_i - \mathbf{r}_j|$. Solution to the Schrödinger equation,

$$H(\mathbf{R})\psi(\mathbf{R}) = E\psi(\mathbf{R})\,, \tag{1.106}$$

is obtained for the ground state by minimizing the energy (Eq. 1.103) through variations of the trial functions.

In *Green's function Monte Carlo* (GFMC) method [e.g., Ceperley & Kalos 1979], one seeks an integral formulation of the Schrödinger equation and considers Green's function for the left-hand side of Eq. (1.106), that is,

$$\left[-\frac{\hbar^2}{2m}\nabla_{\mathbf{R}}^2 + V(\mathbf{R}) + V_0\right] G(\mathbf{R}, \mathbf{R}_0) = \delta(\mathbf{R} - \mathbf{R}_0)\,. \tag{1.107}$$

Here the Hamiltonian has been split into the kinetic and potential energy contributions and $-V_0$ refers to a minimum of $V(\mathbf{R})$. It is then possible to devise an MC method, in the general sense of a random sampling algorithm, that produces populations drawn from the successive $\psi^{(n)}(\mathbf{R})$ generated by

$$\psi^{(n+1)}(\mathbf{R}) = (E + V_0) \int d\mathbf{R}'\, G(\mathbf{R}, \mathbf{R}')\psi^{(n)}(\mathbf{R}')\,. \tag{1.108}$$

As the iterations converge, solution to the Schrödinger equation may be obtained. Ceperley and Alder [1980] calculated the ground-state energy of electron gas by the GFMC method.

Path-integral formulation of the quantum statistical mechanics [Feynman & Hibbs 1965; Feynman 1972] enables one to express the partition function Ξ_N of the N-particle system as

$$\Xi_N = \int d\overline{\mathbf{R}} \int_{\mathbf{R}(0)=\mathbf{R}(\hbar\beta)} D\mathbf{R}(\xi) \exp\left\{-\frac{1}{\hbar}\int_0^{\hbar\beta} d\xi \, H\left[\mathbf{R}(\xi)\right]\right\} , \qquad (1.109)$$

where

$$\overline{\mathbf{R}} = \frac{1}{\hbar\beta}\int_0^{\hbar\beta} d\xi \, \mathbf{R}(\xi) . \qquad (1.110)$$

The paths $\mathbf{R}(\xi)$ in Eq. (1.109) are periodic functions of an "imaginary time" ξ with periodicity $\hbar\beta = \hbar/k_B T$ and the path integral along $D\mathbf{R}(\xi)$ is carried out over the whole periodic paths with a given "time average" $\overline{\mathbf{R}}$ defined by Eq. (1.110).

In the *path-integral Monte Carlo method* [e.g., Creutz & Freedman 1981], one calculates the partition function by introducing a reference system with known properties, which will accurately mimic the original system. Multidimensional MC integrations are involved in the calculations of various averages in this reference system. In executing the path-integral computations in practice, the continuous paths, $\mathbf{R}(\xi)$, are replaced by discrete paths with L segments (Trotter decomposition),

$$\mathbf{R}(\xi) \rightarrow \{\mathbf{R}(0), \mathbf{R}(1), \ldots, \mathbf{R}(L) = \mathbf{R}(0)\} ,$$

and one thereby establishes an isomorphism between quantum and classical systems [Chandler & Wolynes 1981]. One looks upon each quantum particle with coordinate $\mathbf{r}_i$ as a classical polymer chain consisting of L atoms with coordinates $\{\mathbf{r}(1), \mathbf{r}(2), \ldots, \mathbf{r}(L)\}$. Trotter's theorem then assures that the statistical properties of the classical polymer chain asymptotically approach those of a quantum particle as L increases [Trotter 1959].

An equation of state for quantum Coulomb solids at finite temperatures has been obtained recently by such a path-integral Monte Carlo method [Iyetomi, Ogata, & Ichimaru 1993].

PROBLEMS

1.1 Consider completely ionized hydrogen plasmas, representing interiors of the Sun (S), a brown dwarf (BD), and a giant planet (GP), with mass density and temperature as given in the table:

	ρ_m (g/cm^3)	$T(K)$
S	100	1.5×10^7
BD	200	2.0×10^6
GP	3	1.0×10^4

Compute the corresponding values of r_s, x_F, Θ, Λ, and Γ. In terms of these values, discuss the physical features of those plasmas.

1.2   Consider completely ionized carbon ($Z = 6$, $A = 12$) materials, representing the interior of a white dwarf, in the range of mass density, $10^6 \leq \rho_m \leq 10^9$ (g/cm^3), and temperature, $10^6 \leq T \leq 10^9$ (K). Plot on the $\log_{10} \rho_m$–$\log_{10} T$ plane the lines corresponding to $r_s = 0.01$, $\Theta = 0.1$, $\Lambda = 1$, $\Gamma = 172$, and $Y = 1$.

1.3   Consider completely ionized iron ($Z = 26$, $A = 56$) materials, representing outer-crustal matter of a neutron star, in the range of mass density, $10^4 \leq \rho_m \leq 10^7$ (g/cm^3), and temperature, $10^6 \leq T \leq 10^9$ (K). Plot on the $\log_{10} \rho_m$–$\log_{10} T$ plane the lines corresponding to $r_s = 0.01$, $\Theta = 0.1$, $\Lambda = 0.1$, $\Gamma = 172$, and $Y = 1$.

1.4   Complete the following table for the number density

Elements	ρ_m	$\langle A \rangle$	Z_v	n_e	r_s
Al	2.69	26.98	3		
Mg	1.74	24.31	2		
Li	0.534	6.94	1		
Na	0.97	22.99	1		
K	0.86	39.10	1		
Cs	1.87	132.9	1		

$$n_e \cong Z_v \frac{\rho_m}{m_N \langle A \rangle} (\text{cm}^{-3})$$

of the conduction electrons and the r_s parameter, with the values of the mass density ρ_m (g/cm^3), the valence number Z_v, and the average mass number $\langle A \rangle$, for various metal elements; take $m_N = 1.67 \times 10^{-24}$ g for the average mass per nucleon.

1.5   Consider liquid-metallic sodium at $\rho_m = 1.0$ g/cm^3 and $T = 5.0 \times 10^2$ K where sodium atoms may be regarded as singly ionized. Calculate the values of r_s, Θ, Λ, Γ, and the appropriate RPA screening parameter for the conduction electrons. In terms of these values, discuss the physical features of such a plasma.

1.6   The dielectric function, $\varepsilon(\mathbf{k}, \omega)$, is defined in terms of the linear response relation between the external and induced electrostatic potentials, $\Phi^{ext}(\mathbf{k}, \omega)$ and $\Phi^{ind}(\mathbf{k}, \omega)$, expressed as

$$\Phi^{ind}(\mathbf{k}, \omega) = \left[\frac{1}{\varepsilon(\mathbf{k}, \omega)} - 1 \right] \Phi^{ext}(\mathbf{k}, \omega)$$

[see Ichimaru 1992]. Use this definition for derivation of Eq. (1.33).

1.7   Derive Eq. (1.39).

1.8   Substitute Eq. (1.46) in Eq. (1.33), and verify Eq. (1.47).

1.9   Derive Eqs. (1.50) and (1.52).

1.10   Derive Eqs. (1.60) and (1.61).

1.11   Verify Eqs. (1.66) and (1.67).

1.12   Derive Eqs. (1.88) and (1.89).

1.13   Derive Eq. (1.101).

CHAPTER 2

Dense Classical Plasmas

2.1 COMPUTER SIMULATION STUDIES

The interparticle correlations and thermodynamic properties for dense classical plasmas have been investigated successfully by computer simulation methods. The radial distribution functions for various realizations of dense plasmas have been sampled by the MC method with the Metropolis algorithm. The short-range correlations, which are not accessible in such a method, have been approached by the analyses combining the short-range expansion of the screening potential and direct sampling of the potential-field distributions at properly constructed test charges. We consider in this section the essential information on the radial distribution functions and the associated screening potentials that has been obtained through such a simulation study. These results provide vital data for the equations of state and the correlation characters on those various plasmas and set rigorous criteria for a subsequent formulation of integral equation theories.

A. Radial Distribution Functions

In classical plasmas, the joint distribution function between two particles separated at a distance r_{12}—that is, the *radial distribution function $g_{12}(r_{12})$*—is defined and calculated as

$$g_{12}(r_{12}) = \left(1 - \frac{1}{N}\right) \frac{V^2}{Q_N} \int d\mathbf{r}_3 \cdots d\mathbf{r}_N \, \exp(-\beta H_{\text{int}}). \qquad (2.1)$$

Here the potential energy for N particles of charges $Z_i e$ at positions $\mathbf{r}_i$, plus the background (with the charge density $-\rho_e$), in a volume V is given by

$$
\begin{aligned}
H_{\text{int}} &= \sum_{1 \le i < j \le N} \frac{Z_i Z_j e^2}{r_{ij}} - \rho_e \sum_{1 \le i \le N} \int d\mathbf{r} \, \frac{Z_i e}{|\mathbf{r}_i - \mathbf{r}|} + \frac{1}{2} \int d\mathbf{r} d\mathbf{r}' \, \frac{\rho_e^2}{|\mathbf{r} - \mathbf{r}'|} \\
&= \frac{Z_1 Z_2 e^2}{r_{12}} + W(\mathbf{r}_1, \mathbf{r}_2, \ldots, \mathbf{r}_N),
\end{aligned} \tag{2.2}
$$

$\beta = 1/k_B T$ is the inverse temperature in energy units, $r_{ij} = |\mathbf{r}_i - \mathbf{r}_j|$,

$$
Q_N = \int d\mathbf{r}_1 d\mathbf{r}_2 \cdots d\mathbf{r}_N \, \exp(-\beta H_{\text{int}}) \tag{2.3}
$$

is the *configuration integral* for the N-particle system, and the function W in Eq. (2.2) represents the sum of all interactions except the one between particles 1 and 2 [e.g., Jancovici 1977; Hansen & McDonald 1986]. The radial distribution functions are the probability densities of finding another particle 2 at a distance r_{12} away from a given particle 1 and are normalized so that they approach unity as the interparticle separations tend to infinity; particles separated at extremely large distances are not correlated with each other.

Interparticle correlations in dense classical plasmas have been approached by the MC simulation methods [Brush, Sahlin, & Teller 1966; Hansen 1973] with the Metropolis algorithm (cf. Section 1.2G). Figure 2.1 plots the MC data of the OCP radial distribution functions at various values of Γ [Iyetomi, Ogata, & Ichimaru 1992]; for an OCP, we delete the subscripts, such as 1, 2, i, and j, which distinguish between the charge species. The radial distribution functions in Fig. 2.1 clearly exhibit the effects of strong Coulombic repulsion at short distances in creating the excluded *Coulomb holes* around given particles. Physically these Coulomb holes and the ion sphere model [Salpeter 1954], introduced in Section 1.3C of Volume I, are the same in content; these and the oscillatory behaviors are the most salient features in the interionic correlations for dense plasmas with uniform neutralizing background of electrons.

Figures 2.2 and 2.3 show examples of the radial distribution functions $g_{ij}(r)$ sampled in dense BIMs by the MC simulations [Ogata et al. 1993; Ichimaru 1993a]. The Coulomb holes and the oscillations are the noteworthy features in these BIMs as well. In addition, we observe that the peak heights and positions are delicate functions of the charge ratio,

$$
R_Z = \frac{Z_2}{Z_1} \qquad (> 1), \tag{2.4}
$$

and the Coulomb coupling parameters, Eq. (1.15). The positions r_{ij}^{PK}, of the first

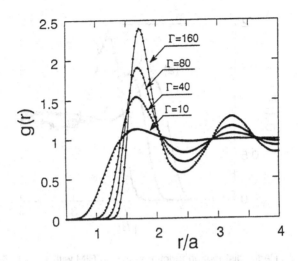

Figure 2.1 Radial distribution functions of OCP fluids obtained by MC simulations (dots) with $N = 1024$ at various values of Γ. The number of the MC configurations generated for each run was 7×10^6; $g(r)$ was calculated with 200 bins in the range $0 < r < L/2$, a half of the cubic MC cell with size $L = 16.2a$. The solid curves represent the results calculated with the extracted bridge functions (2.56). [Iyetomi, Ogata, & Ichimaru 1992]

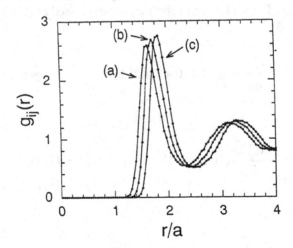

Figure 2.2 Radial distribution functions $g_{ij}(r)$ in BIM with $Z_1 = 6$, $Z_2 = 8$, $x = 0.5$, and $\Gamma_{11} = 163.5$; $a = \langle Z \rangle a_e$ with $\langle Z \rangle = (1 - x)Z_1 + xZ_2$. The number of MC particles is 1024, and the number of MC configurations generated, 7×10^6. (a) $g_{11}(r)$, (b) $g_{12}(r)$, and (c) $g_{22}(r)$. [Ichimaru 1993a]

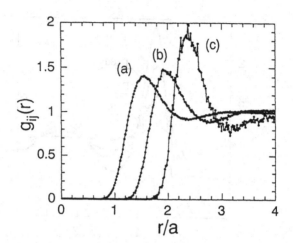

Figure 2.3 Radial distribution functions $g_{ij}(r)$ in BIM with $Z_1 = 1$, $Z_2 = 3$, $x = 0.1$, and $\Gamma_{11} = 20$; $a = \langle Z \rangle a_e$ with $\langle Z \rangle = (1-x)Z_1 + xZ_2$. The number of MC particles is 1000, and the number of MC configurations generated, 7×10^6. (a) $g_{11}(r)$, (b) $g_{12}(r)$, and (c) $g_{22}(r)$. [Ichimaru 1993a]

peaks, in particular, have been found to obey the constant electron density, ion sphere scaling of Eqs. (1.11) closely; they are expressed as [Ogata, Ichimaru, & Van Horn 1993]

$$\frac{r_{ij}^{\mathrm{PK}}}{a_{ij}} = \xi_{ij}(1.610 + 0.025 \ln \Gamma_{ij}) \qquad \Gamma_{ij} \geq 5, \tag{2.5}$$

with

$$\xi_{ij} = \begin{cases} 1 - \dfrac{0.135}{\Gamma_{11}} \sqrt{x(R_Z - 1)}, & \text{for } (i, j) = (1, 1), \\ 1, & \text{for } (i, j) = (1, 2) \text{ or } (2, 2), \end{cases} \tag{2.6}$$

and

$$x = \frac{N_2}{N} \tag{2.7}$$

denoting the *molar fraction* of the larger Z species. A weak but nonnegligible departure from the ion sphere scaling is exhibited in Eq. (2.6) only for $(i, j) = (1, 1)$ as x and R_Z increase; this feature will lead to a certain modification in the calculation of thermonuclear reaction rates for dense BIMs.

B. Screening Potentials

The *screening potential* for a multi-ionic plasma is defined as

$$H_{ij}(r) = \frac{Z_i Z_j e^2}{r} + \frac{1}{\beta} \ln\left[g_{ij}(r)\right]. \tag{2.8}$$

It is the difference between the bare potential, the first term on the right-hand side, and the *potential of mean force* represented by the second term. The latter potential stems from a sum of a certain class of Mayer diagrams in the theory of liquids [e.g., Hansen & McDonald 1986] and does not mean a real potential in the sense of particle dynamics. The screening potential is related closely to the bridge function, to be considered in Section 2.2D, and plays an essential part in the theoretical treatment of an enhancement factor for the nuclear reaction rate due to many-body correlation.

It is instructive to investigate the short-range behavior of the screening potentials by expanding Eq. (2.8) in a power series of r^2 as

$$H_{ij}(r) = H_{ij}(0) + H_{ij}^{(1)}\left(\frac{r}{a_{ij}}\right)^2 + H_{ij}^{(2)}\left(\frac{r}{a_{ij}}\right)^4 + \cdots \tag{2.9}$$

[Widom 1963; Jancovici 1977]. When calculating the coefficients in this expression, one carries out an expansion of the following function around $\mathbf{r}_1 = 0$ and $\mathbf{r}_2 = 0$ as

$$\exp\left(\frac{\beta Z_1 Z_2 e^2}{r_{12}}\right) g_{12}(r_{12})$$

$$= \left(1 - \frac{1}{N}\right) \frac{V^2}{Q_N} \int d\mathbf{r}_3 \cdots d\mathbf{r}_N \, \exp\left[-\beta W(0, 0, \mathbf{r}_3, \ldots, \mathbf{r}_N)\right]$$

$$\times \left\{1 - \frac{\beta}{6}\Delta_1 W\left(r_1^2 + \frac{Z_2}{Z_1}r_2^2\right) + \frac{\beta^2}{6}\left(\frac{\partial W}{\partial \mathbf{r}_1}\right)^2\left(\mathbf{r}_1 + \frac{Z_2}{Z_1}\mathbf{r}_2\right)^2 + \cdots\right\}. \tag{2.10}$$

Here the differentials with respect to $\mathbf{r}_1$, such as the Laplacian Δ_1 and $\partial/\partial\mathbf{r}_1$, are performed on the function W and then the limits of $\mathbf{r}_1 \to 0$ and $\mathbf{r}_2 \to 0$ are approached.

The value of the screening potential is obtained by noting that the excess part F_{ex} of the Helmholtz free energy—that is, the interaction free energy—for the N-particle system is given in terms of the configuration integral Eq. (2.3) as

$$F_{ex} = -\frac{1}{\beta} \ln\left(\frac{Q_N}{V^N}\right). \tag{2.11}$$

For an OCP, one thus derives a relation,

$$H(0) = F_{ex}^{BIM}(N, 0) - F_{ex}^{BIM}(N - 2, 1) , \qquad (2.12)$$

where $F_{ex}^{BIM}(N_1, N_2)$ denotes the excess free energy of a BIM consisting of N_1 ions with charge number Z and N_2 ions with charge number $2Z$. Extension of Eq. (2.12) to a multi-ionic plasma is straightforward [Ogata et al. 1993].

The coefficients $H_{ij}^{(1)}$ can be calculated for the Coulombic system explicitly by observing

$$\Delta_1 W(0, 0, \mathbf{r}_3, \ldots, \mathbf{r}_N) = -4\pi Z_1 e \left[\sum_j Z_j e \delta(\mathbf{r}_j) - \rho_e \right] \qquad (2.13a)$$

and the configuration average,

$$\left\langle \left(\frac{\partial W}{\partial \mathbf{r}_1} \right)^2 \right\rangle = \frac{4\pi Z_1^2 e \rho_e}{(Z_1 + Z_2)\beta} , \qquad (2.13b)$$

which may be derived by partial integrations of the average (Problem 2.2). Thus for an OCP, one finds

$$\frac{\beta H^{(1)}}{\Gamma} = -\frac{1}{4} . \qquad (2.14a)$$

For a BIM, one has

$$\frac{\beta H_{ij}^{(1)}}{\Gamma_{ij}} = -\frac{(Z_i^{1/3} + Z_j^{1/3})^3}{16(Z_i + Z_j)} . \qquad (2.14b)$$

The coefficient $H_{ij}^{(2)}$ is related to a mean square value of the microscopic forces acting on a composite test particle with charge $(Z_i + Z_j)e$, which can be sampled by an MC method. For simplicity, we here treat the case of an OCP and define $\Phi(\mathbf{r})$ as the potential energy (in units of $k_B T$) for that test particle with charge $2Ze$ at $\mathbf{r}$ interacting with all other N particles (charge Ze) which form the OCP. The coefficient $H^{(2)}$ has then been calculated in the ensemble of 2.5×10^5 MC generated configurations [Ogata, Iyetomi, & Ichimaru 1991] as

$$\frac{\beta H^{(2)}}{\Gamma} = \frac{a^4}{384\Gamma} \left\langle \left[\left(\frac{d\Phi}{dr_1} \right)^2 - 2 \frac{d^2\Phi}{dr_1^2} \right]^2 \right\rangle - \frac{\Gamma}{32}, \qquad (2.15)$$

where r_1 refers to one of the Cartesian components of $\mathbf{r}$. Within the accuracy of

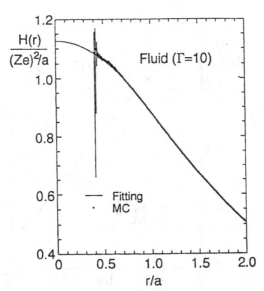

Figure 2.4 Screening potential of OCP fluid at $\Gamma = 10$, with the number of MC particles, $N = 432$, and the number of MC configurations generated, 2.5×10^8. The maximum extent of uncertainties in the MC sampled points is 10^{-4} unless explicitly shown by vertical bars. [Ogata, Iyetomi, & Ichimaru 1991]

these MC sampling, it has been concluded that *

$$\frac{\beta H^{(2)}}{\Gamma} = 0.00 \pm 0.01 \tag{2.16}$$

[Ichimaru 1993a]. The computed value of $H^{(2)}$, smaller in magnitude than the extent of errors, is far smaller than $H^{(1)}$.

The screening potentials may be sampled directly through Eq. (2.8) with the Metropolis algorithm. Examples of such samplings for the OCP are exhibited in Figs. 2.4 and 2.5. As these figures illustrate, the screening potentials cannot be obtained directly by the MC method at short distances, since the strong Coulomb repulsion makes it impossible to sample $g(r)$ at $r \approx 0$. The analysis in terms of Eq. (2.9) is useful for the evaluation of screening potentials in such a short range.

The screening potential for the OCP fluid is thus expressed for $5 < \Gamma \leq 180$ as

$$\frac{\beta H(r)}{\Gamma} = \begin{cases} A_1 - B_1^2 - \dfrac{1}{4}x^2, & \text{for } x \leq 2B_1, \\[2ex] A_1 - B_1 x + \dfrac{1}{x}\exp(C_1\sqrt{x} - D_1), & \text{for } 2B_1 < x < 2. \end{cases} \tag{2.17}$$

---

* The MC sampling calculations of the coefficients, h_1 and h_2, have been reexamined subsequently [Ichimaru, Ogata, & Tsuruta 1994] in a larger system with extended sequence of the configurations generated; the statistical errors have now been reduced drastically. The assessment (2.16) has remained consistent with these improved data as well as with the old data [Ogata, Iyetomi, & Ichimaru 1991].

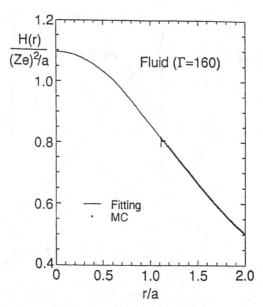

Figure 2.5 Screening potential of OCP fluid at $\Gamma = 160$, with the number of MC particles, $N = 432$, and the number of MC configurations generated, 1.0×10^8. The maximum extent of uncertainties in the MC sampled points is 10^{-5} unless explicitly shown by vertical bars. [Ogata, Iyetomi, & Ichimaru 1991]

Here $x = r/a$, and the fitting parameters are given by

$$\begin{cases} A_1 = 1.356 - 0.0213 \ln \Gamma, & B_1 = 0.456 - 0.013 \ln \Gamma, \\ C_1 = 9.29 + 0.79 \ln \Gamma, & D_1 = 14.83 + 1.31 \ln \Gamma. \end{cases} \tag{2.18}$$

Figures 2.4 and 2.5, compare the MC values and the fitting formula (2.17); agreement is excellent. Combining Eqs. (2.17) and (2.18), one obtains MC values of the screening potential at zero separation as

$$\frac{\beta H(0)}{\Gamma} = 1.148 - 0.00944 \ln \Gamma - 0.000168 (\ln \Gamma)^2. \tag{2.19}$$

This quantity is essential for calculating the enhancement factor for the nuclear reaction rates in dense plasmas, as we shall consider in Chapter 5. Estimated error in the evaluation in (2.19) is a few tenths of a percent.

The functional form of the screening potentials for dense BIM fluids can be determined accurately, as in the cases of dense OCP fluids, through combined analysis of the MC sampling at intermediate distances ($0.4 < r/a_{ij} < 2$) and the short-range expansion (Ogata, Ichimaru, & Van Horn 1993):

$$\frac{\beta H_{ij}(r)}{\Gamma_{ij}} =$$

$$\begin{cases} A_{ij} - \dfrac{\Gamma_{ij} B_{ij}^2}{4\beta H_{ij}^{(1)}} - \dfrac{\beta H_{ij}^{(1)}}{\Gamma_{ij}}\left(\dfrac{r}{a_{ij}}\right)^2, & \text{for } \dfrac{r}{a_{ij}} \le \dfrac{\Gamma_{ij} B_{ij}}{2\beta H_{ij}^{(1)}}, \\[4mm] A_{ij} - B_{ij}\dfrac{r}{a_{ij}} + \dfrac{a_{ij}}{r}\exp\left(C_{ij}\sqrt{\dfrac{r}{a_{ij}}} - D_{ij}\right), & \text{for } \dfrac{\Gamma_{ij} B_{ij}}{2\beta H_{ij}^{(1)}} < \dfrac{r}{a_{ij}} < 2, \end{cases}$$

$$(2.20)$$

and

$$A_{ij} = 1.356 - 0.0213 \ln \Gamma_{ij}, \qquad B_{ij} = 0.456 - 0.013 \ln \Gamma_{ij},$$
$$C_{ij} = 9.29 + 0.79 \ln \Gamma_{ij}, \qquad D_{ij} = 14.83 + 1.31 \ln \Gamma_{ij}. \qquad (2.21)$$

Figures 2.6 and 2.7, compare the MC values and the fitting formula (2.20); agreement is again excellent. Combining Eqs. (2.17) and (2.18), one obtains MC values of the screening potential at zero separation as

$$\frac{\beta H_{ij}(0)}{\Gamma_{ij}} = 1.148 - 0.00944 \ln \Gamma_{ij} - 0.000168(\ln \Gamma_{ij})^2. \qquad (2.22)$$

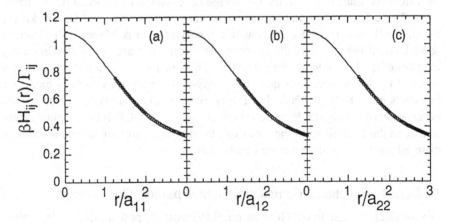

Figure 2.6 Screening potentials of BIM fluid under the same conditions as those in Fig. 2.2. The open circles are MC sampling values; the solid curve, fitting formula (2.20). [Ichimaru 1993a]

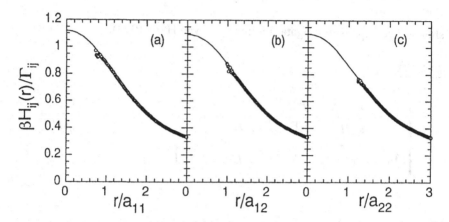

Figure 2.7 Screening potentials of BIM fluid under the same conditions as those in Fig. 2.3. The open circles are MC sampling values; the solid curve, fitting formula (2.20). [Ichimaru 1993a]

2.2   INTEGRAL EQUATION APPROACHES

Hierarchical relations between multiparticle distributions take the form of integral equations in the phase space. For analysis of multiparticle correlations in dense classical plasmas, we begin with a theory based on the density-functional formalism of Section 1.2E. This theory leads to a description appropriate to a plasma in a state of strong coupling, in that a renormalization of the interparticle potential arising from strong multiparticle correlations is realized in a natural way. The central roles in such a renormalization are played by such quantities as the direct correlation functions, the bridge functions, and the local field corrections. These features are in contrast to the case of a conventional Ursell–Mayer expansion theory or the nodal expansion theory [Mayer & Mayer 1940; Hansen & McDonald 1986], where the interparticle correlations are formulated basically in terms of the bare interaction potentials. The density-functional theory was developed for inhomogeneous quantum many-particle systems such as electrons in metals [e.g., Kohn & Vashishta 1983]. Here we shall apply the theory to the description of multiparticle correlations in a classical OCP. It is straightforward to extend the formalism to the cases of plasmas consisting of several species of charged particles, or the multi-ionic plasmas.

A. Density-Functional Formulation of Multiparticle Correlations

We consider the case in which an external electrostatic potential, $\phi(\mathbf{r})$, is applied to an OCP that would be uniform at the number density, $n = N/V$, and the temperature T if the external potential were absent; the system is inhomogeneous due to the presence of the external potential. If, in particular,

$$\phi(\mathbf{r}) = \frac{(Ze)^2}{r} \tag{2.23}$$

is assumed so that the potential produced by a plasma particle at the origin ($\mathbf{r} = 0$) is singled out, then the resultant density variation, $n(\mathbf{r})$, amounts to a calculation of the radial distribution function or the pair distribution function, $g(\mathbf{r}) = n(\mathbf{r})/n$, in the uniform OCP.

The thermodynamic potential of the inhomogeneous plasma, Eq. (1.81), is expressed as a functional of the density variation, $\delta n(\mathbf{r}) = n(\mathbf{r}) - n$,

$$\Omega_\phi[n(\mathbf{r})] = F_0[n(\mathbf{r})] + \frac{1}{2} \int d\mathbf{r} d\mathbf{r}' \frac{(Ze)^2}{|\mathbf{r} - \mathbf{r}'|} \delta n(\mathbf{r}) \delta n(\mathbf{r}') + F_c[n(\mathbf{r})]$$
$$+ \int d\mathbf{r} \, [\phi(\mathbf{r}) - \mu] n(\mathbf{r}) \tag{2.24}$$

[Evans 1979]. The term $F_0[n(\mathbf{r})]$ denotes the free-energy functional of the corresponding noninteracting system,

$$F_0[n(\mathbf{r})] = k_B T \int d\mathbf{r} \, n(\mathbf{r}) \left\{ \ln \left[(2\pi\hbar^2/Mk_B T)^{3/2} n(\mathbf{r}) \right] - 1 \right\}, \tag{2.25}$$

where M is the mass of a particle. The remaining contribution $F_c[n(\mathbf{r})]$ on the right-hand side of Eq. (2.24) stems from the interparticle correlations. The stationary condition of $\Omega_\phi[n(\mathbf{r})]$ with respect to variations of $n(\mathbf{r})$, that is,

$$\frac{\delta\Omega_\phi[n(\mathbf{r})]}{\delta n(\mathbf{r})} = \frac{\delta F_0[n(\mathbf{r})]}{\delta n(\mathbf{r})} + \int d\mathbf{r}' \frac{(Ze)^2}{|\mathbf{r} - \mathbf{r}'|} \delta n(\mathbf{r}') + \frac{\delta F_c[n(\mathbf{r})]}{\delta n(\mathbf{r})} + \phi(\mathbf{r}) - \mu$$
$$= 0, \tag{2.26}$$

determines the equilibrium ion density distribution $n(\mathbf{r})$ in the presence of $\phi(\mathbf{r})$. This is a basic equation for the density-functional analyses of multiparticle correlation functions. The complexities arising from the interparticle correlations are all contained in the third term on right-hand side of Eq. (2.26) that depends on $F_c[n(\mathbf{r})]$.

To proceed further, we expand $F_c[n(\mathbf{r})]$ around the uniform state as

$$F_c[n(\mathbf{r})] =$$
$$F_c[n] + \frac{1}{2!} \int d\mathbf{r} d\mathbf{r}' \, K_c^{(2)}(|\mathbf{r} - \mathbf{r}'|) \, \delta n(\mathbf{r}) \delta n(\mathbf{r}')$$
$$+ \frac{1}{3!} \int d\mathbf{r} d\mathbf{r}' d\mathbf{r}'' \, K_c^{(3)}(\mathbf{r} - \mathbf{r}', \mathbf{r} - \mathbf{r}'') \delta n(\mathbf{r}) \delta n(\mathbf{r}') \delta n(\mathbf{r}'') + \cdots . \tag{2.27}$$

The kernels $K_c^{(v)}$ of the integrations, which we shall call the *v-body correlation potentials*, are expressed as the functional derivatives (see Appendix C) of $F_c[n(\mathbf{r})]$ with respect to $n(\mathbf{r})$, that is,

$$K_c^{(v)}(\mathbf{r}_1, \ldots, \mathbf{r}_v) = \left.\frac{\delta^v F_c[n(\mathbf{r})]}{\delta n(\mathbf{r}_1)\cdots\delta n(\mathbf{r}_v)}\right|_{n(\mathbf{r})\to n}. \qquad (2.28)$$

These derivatives are therefore evaluated in the homogeneous state, where $\phi(\mathbf{r}) = 0$. The correlation potentials with $v \geq 2$ physically describe the v-body interactions among the induced density fluctuations stemming from the interparticle correlations.

The one-body correlation potential is calculated by taking the limit of Eq. (2.26) for $\phi(\mathbf{r}) \to 0$ and $\delta n(\mathbf{r}) \to 0$ as

$$K_c^{(1)}(\mathbf{r}) = \mu - k_B T \ln\left[n(2\pi\hbar^2/Mk_B T)^{3/2}\right]. \qquad (2.29)$$

Successive functional differentials of Eq. (2.27) with respect to $n(\mathbf{r})$ then yield the amenable formulas for $K_c^{(v)}(\mathbf{r}_1, \ldots, \mathbf{r}_v)$ for $v \geq 2$,

$$K_c^{(v)}(\mathbf{r}_1, \ldots, \mathbf{r}_v) = -K_0^{(v)}(\mathbf{r}_1, \ldots, \mathbf{r}_v) + K^{(v)}(\mathbf{r}_1, \ldots, \mathbf{r}_v) - \frac{(Ze)^2}{|\mathbf{r}_1 - \mathbf{r}_2|}\delta_{v,2}, \qquad (2.30)$$

where

$$K_0^{(v)}(\mathbf{r}_1, \ldots, \mathbf{r}_v) = \left.\frac{\delta^v F_0[n(\mathbf{r})]}{\delta n(\mathbf{r}_1)\cdots\delta n(\mathbf{r}_v)}\right|_{n(\mathbf{r})\to n}$$

$$= (-1)^v (v - 2)! n^{1-v} k_B T \delta(\mathbf{r}_1 - \mathbf{r}_2)\cdots\delta(\mathbf{r}_1 - \mathbf{r}_v), \qquad (2.31)$$

$$K^{(v)}(\mathbf{r}_1, \ldots, \mathbf{r}_v) = -\left.\frac{\delta^{v-1}\phi[n(\mathbf{r}_1)]}{\delta n(\mathbf{r}_2)\cdots\delta n(\mathbf{r}_v)}\right|_{n(\mathbf{r})\to n}. \qquad (2.32)$$

In Eqs. (2.29) and (2.30), δ_{ij} is Kronecker's delta and $\delta(\mathbf{r})$ represents the three-dimensional delta function.

The *grand partition function* for the plasma under consideration is given by [Mayer & Mayer 1940; Hansen & McDonald 1986]

$$\Xi = \sum_{N=0}^{\infty}\frac{1}{N!}\int d\mathbf{r}_1\cdots d\mathbf{r}_N \prod_{i=1}^{N} z^*(\mathbf{r}_i) \prod_{i<j}^{N}\exp\left[-\frac{\beta(Ze)^2}{|\mathbf{r}_i - \mathbf{r}_j|}\right], \qquad (2.33)$$

where the *activity* with inclusion of the external potential is

$$z^*(\mathbf{r}) = \left(\frac{2\pi\hbar^2}{Mk_BT}\right)^{-3/2} \exp\left\{-\beta\left[\phi(\mathbf{r}) - \mu\right]\right\} . \tag{2.34}$$

The single-particle distribution is then calculated as (Problem 2.3)

$$n(\mathbf{r}) = \frac{\delta \ln \Xi}{\delta \ln z^*(\mathbf{r})} . \tag{2.35}$$

For construction of hierarchical relations between those multiparticle functions, it is useful to introduce the *Ursell functions*, defined for $\nu \geq 2$ as

$$U^{(\nu)}(\mathbf{r}_1, \ldots, \mathbf{r}_\nu) = \left(-\frac{1}{\beta}\right)^{\nu-1} \frac{\delta^{\nu-1} n(\mathbf{r}_1)}{\delta\phi(\mathbf{r}_2)\cdots\delta\phi(\mathbf{r}_\nu)}\bigg|_{n(\mathbf{r})\to n} . \tag{2.36}$$

Clearly, the Ursell functions are the inverse of the K functions in Eq. (2.32).

The functional derivative of Eq. (2.35) with respect to the external potential yields

$$U^{(2)}(|\mathbf{r}_1 - \mathbf{r}_2|) = n^2 h(|\mathbf{r}_1 - \mathbf{r}_2|) + n\delta(\mathbf{r}_1 - \mathbf{r}_2) , \tag{2.37}$$

where

$$h(r) = g(r) - 1 \tag{2.38}$$

is the *pair correlation function* in the homogeneous system. Analogously $U^{(\nu)}(\mathbf{r}_1, \ldots, \mathbf{r}_\nu)$ are expressible in the forms involving the correlation functions up to the νth order.

B. Direct Correlation Functions

For investigation of Eq. (2.30) at $\nu = 2$ in more detail, we introduce the *direct correlation function*:

$$c(r) = -\beta\left[\frac{(Ze)^2}{r} + K_c^{(2)}(r)\right] . \tag{2.39}$$

It is clear from this definition that $-c(r)$ has a physical meaning of an effective potential in units of the thermal energy, k_BT, produced by the sum of the bare potential, $(Ze)^2/r$, and the binary correlation potential, $K_c^{(2)}(r)$. We then note the chain rule Eq. (C.5) of the functional derivatives,

$$\int d\mathbf{r}_2\, K^{(2)}(\mathbf{r}_1, \mathbf{r}_2) U^{(2)}(\mathbf{r}_2, \mathbf{r}_3) = \beta^{-1}\delta(\mathbf{r}_1 - \mathbf{r}_3) , \tag{2.40}$$

and find that at $\nu = 2$ Eq. (2.30) is re-expressed as

$$h(r) = c(r) + n \int d\mathbf{r}' \, h\left(|\mathbf{r} - \mathbf{r}'|\right) c(r') . \tag{2.41}$$

This is the *Ornstein–Zernike relation* between the pair and direct correlation functions.

Using the expansion (2.27) as well as Eqs. (2.29), (2.32), and (2.41), we may rewrite Eq. (2.26) with Eq. (2.23) as

$$g(r) = \exp\left[-\beta \frac{(Ze)^2}{r} + h(r) - c(r) + B(r)\right] , \tag{2.42}$$

where

$$B(r) =$$
$$-\sum_{\nu=3}^{\infty} \frac{n^{\nu-1}}{(\nu-1)!} \int d\mathbf{r}_1 \cdots d\mathbf{r}_{\nu-1} \, \beta K_c^{(\nu)}(\mathbf{r}, \mathbf{r}_1, \ldots, \mathbf{r}_{\nu-1}) h(r_1) \cdots h(r_{\nu-1}) , \tag{2.43}$$

and

$$h(r) = \delta n(\mathbf{r})/n . \tag{2.44}$$

Equation (2.42) is an exact expression for $g(r)$ and can be derived also by means of the Mayer–Ursell cluster expansion theory [e.g., Hansen & McDonald 1986]. The term $B(r)$ in the logarithm of $g(r)$ corresponds to the *bridge function*, representing the sum of all the elementary diagram contributions. In the density-functional theory, bridge functions are expressed in terms of the correlation functions and the correlation potentials, as in Eq. (2.43). This equation thus provides an essential link between the diagrammatic expansion theory and the density-functional formalism.

To close the set of integral equations (2.41) to (2.43), one must find a way to evaluate the bridge functions explicitly. This generally calls for the knowledge of the correlation potentials with $\nu \geq 3$. Those higher-order correlation potentials can be calculated through successive functional derivatives of Eq. (2.40) with the aid of Eqs. (2.28), (2.30)–(2.32), and (2.36) [e.g., Percus 1964]. Figure 2.8 depicts the first few of the functions $K^{(\nu)}(\mathbf{r}_1, \ldots, \mathbf{r}_\nu)$ obtained by such a method [Stell 1976; Iyetomi 1984]. Here the white circles refer to the particle coordinates under consideration, the filled circles are those to be integrated, the wavy line joining two particles represents the two-body function $K^{(2)}$, and the ν-polygon implies minus the ν-body Ursell function, $U^{(\nu)}$. In terms of the diagrammatic

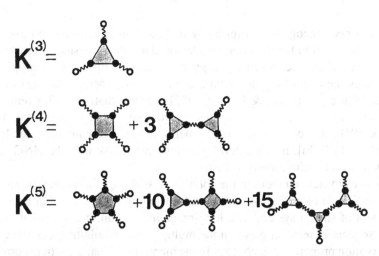

Figure 2.8 Diagrammatic representation of $K^{(\nu)}$ for $\nu = 3$, 4, and 5. See the text for the meaning of the white circles, the wavy lines, and the polygons. [Ichimaru, Iyetomi, & Tanaka 1987]

language [Wu & Chien 1970; Wu 1971], it can be said that $K^{(\nu)}(\mathbf{r}_1, \ldots, \mathbf{r}_\nu)$ corresponds to a collection of all topologically distinct normal ν-rooted Cayley tree graphs, provided that the degree of each vertex is three or more. A *Cayley tree* is a connected graph containing no circles; that is, one cannot return to a point on a Cayley tree by following a sequence of lines. Here the vertex functions are minus the Ursell functions and the bonds are the two-body functions, $K^{(2)}$.

The bridge functions are closely related to the screening potentials, as Eq. (2.8) illustrates. Explicit evaluation of the bridge functions through combined analyses of MC simulated data and solution to the integral equations will be considered in Section 2.2D.

C. The Hypernetted-Chain Scheme

The simplest way to close the set of Eqs. (2.41)–(2.43) is to assume $B(r) = 0$ in Eq. (2.42) and thereby to neglect the contributions of ternary and higher-order correlation potentials in the logarithm of $g(r)$. The resulting approximate equation,

$$g^{\text{HNC}}(r) = \exp\left[-\beta\frac{(Ze)^2}{r} + h(r) - c(r)\right], \qquad (2.45)$$

coupled with Eq. (2.41) makes a closed set of integral equations, called the *hypernetted-chain* (HNC) scheme [van Leeuwen, Groeneveld, & De Boer 1959; Morita 1960].

It has been recognized empirically that such an HNC scheme provides an accurate description for the interparticle correlations in classical plasmas [Baus & Hansen 1980]; the internal energy calculated in the HNC approximation reproduces the exact MC data within errors on the order of 1% over the fluid regime [Slattery, Doolen, & DeWitt 1982]. These situations differ remarkably from those in the cases of short-ranged hard-core systems, where the Percus–Yevick (PY) equation is known to be superior to the HNC equation [e.g., Hansen & McDonald 1986]. It is a natural question then to ask why the HNC scheme works so well for Coulombic systems.

A key physical element in the solution to such a problem may be sought in the significance of charge conservation for Coulombic systems. For an accurate treatment of such a system with long-range interaction, it is essential to maintain a sequential relation between the multiparticle distributions, or the charge-conservation property, at each stage in the hierarchy of many-particle correlation functions. Charge conservation is physically equivalent to the condition for perfect screening in the structure of Coulombic correlations; inaccuracy that would have resulted from spurious space-charge effects could thereby be avoided. The issue of charge conservation has been considered through diagrammatic analyses of the bridge functions [Iyetomi 1984], which are the neglected terms in the HNC approximation. Charge neutrality conditions ensure that the bridge functions be short-ranged to all orders. The multiparticle correlation functions constructed in the *convolution approximation* (CA) exactly satisfy such sequential relations and make $K_c^{(v)} = 0$ for $v \geq 3$, leading to the HNC approximation.

Let $g^{(v)}(\mathbf{r}_1, \ldots, \mathbf{r}_v)$ denote the v-particle joint distribution functions (see Section 2.1C in Volume I), normalized so that $g^{(v)}(\mathbf{r}_1, \ldots, \mathbf{r}_v) = 1$ when the radial coordinates of the v particles are separated far away from each other. The v-particle correlation functions $h^{(v)}(\mathbf{r}_1, \ldots, \mathbf{r}_v)$ are then defined successively in accordance with

$$h(\mathbf{r}_1, \mathbf{r}_2) = g(\mathbf{r}_1, \mathbf{r}_2) - 1, \qquad (2.46a)$$

$$h^{(3)}(\mathbf{r}_1, \mathbf{r}_2, \mathbf{r}_3) = g^{(3)}(\mathbf{r}_1, \mathbf{r}_2, \mathbf{r}_3) - h(\mathbf{r}_1, \mathbf{r}_2) - h(\mathbf{r}_2, \mathbf{r}_3) - h(\mathbf{r}_3, \mathbf{r}_1) - 1, \qquad (2.46b)$$

and so on; we omit the superscript (2) for the two-particle functions in these equations, as in Eq. (2.38).

The correlation functions satisfy the sequential relations,

$$\int d\mathbf{r}_v h^{(v)}(\mathbf{r}_1, \ldots, \mathbf{r}_v) n(\mathbf{r}_v) = -(v - 1) h^{(v-1)}(\mathbf{r}_1, \ldots, \mathbf{r}_{v-1}) , \qquad (2.47)$$

with the initial condition, $h^{(1)} = 1$. Equivalently the overall charge-neutrality

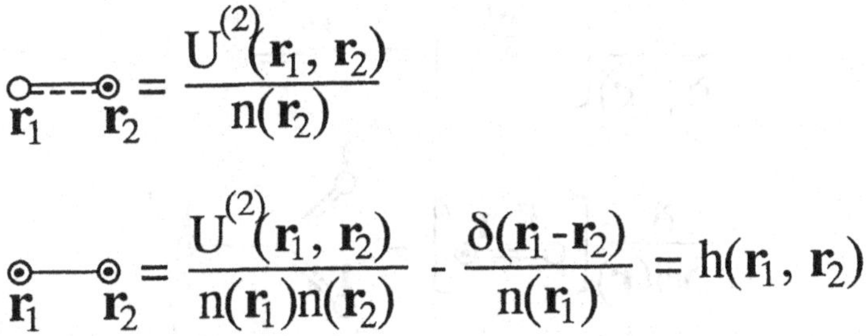

$$\underset{\mathbf{r}_1 \quad \mathbf{r}_2}{\circ\text{-}\text{-}\text{-}\bullet} = \frac{U^{(2)}(\mathbf{r}_1, \mathbf{r}_2)}{n(\mathbf{r}_2)}$$

$$\underset{\mathbf{r}_1 \quad \mathbf{r}_2}{\circ\text{---}\circ} = \frac{U^{(2)}(\mathbf{r}_1, \mathbf{r}_2)}{n(\mathbf{r}_1)n(\mathbf{r}_2)} - \frac{\delta(\mathbf{r}_1 - \mathbf{r}_2)}{n(\mathbf{r}_1)} = h(\mathbf{r}_1, \mathbf{r}_2)$$

Figure 2.9 Definition of the single and double lines.

conditions are expressed in terms of the Ursell functions as

$$\int d\mathbf{r}_i \, U^{(\nu)}(\mathbf{r}_1, \ldots, \mathbf{r}_\nu) = 0, \qquad i = 1, \ldots, \nu . \tag{2.48}$$

At $\nu = 2$, Eq. (2.47) reads

$$\int d\mathbf{r}_2 \, n(\mathbf{r}_2) \left[g(\mathbf{r}_1, \mathbf{r}_2) - 1 \right] = -1 . \tag{2.49}$$

Combining Eq. (2.49) with generalization of Eq. (2.37) to inhomogeneous cases yields Eq. (2.48) at $\nu = 2$.

We now introduce the CA in terms of diagrammatic representation for the multiparticle functions: Let the pair correlation function and the two-body Ursell function be represented by the single and double lines, as in Fig. 2.9. Higher-order Ursell functions are generated successively through the functional derivatives, in accordance with Eq. (2.36). In the CA, the functional derivatives operate in the ways depicted in Fig. 2.10; these operations ensure a one-to-one correspondence between the graphs of $\delta U_{CA}^{(\nu)}(\mathbf{r}_1, \ldots, \mathbf{r}_\nu)/\delta\phi(\mathbf{r}_{\nu+1})$ and those of $U_{CA}^{(\nu+1)}(\mathbf{r}_1, \ldots, \mathbf{r}_{\nu+1})$. The first few Ursell functions so derived in the CA are illustrated in Fig. 2.11. In diagrammatic language, it can be said that $U_{CA}^{(\nu)}(\mathbf{r}_1, \ldots, \mathbf{r}_\nu)$ corresponds to the collection of all topologically distinct normal ν-rooted Cayley tree graphs, provided that each double black point is a node. All terminal lines are double (solid and dotted), and all internal lines, in contrast, are single (solid). The double black point refers to the particle coordinate to be integrated with a weight of density $n(\mathbf{r})$. Since all the terminal lines of $U_{CA}^{(\nu)}(\mathbf{r}_1, \ldots, \mathbf{r}_\nu)$ are double, the convolution form of $U_{CA}^{(\nu)}(\mathbf{r}_1, \ldots, \mathbf{r}_\nu)$ satisfies the overall charge-neutrality condition (2.48) exactly, as long as the two-body Ursell function $U^{(2)}(\mathbf{r}_1, \mathbf{r}_2)$ satisfies the corresponding condition at $\nu = 2$.

One can readily confirm that $K_c^{(3)}(\mathbf{r}_1, \mathbf{r}_2, \mathbf{r}_3)$ vanishes identically in the CA by substituting $U_{CA}^{(3)}(\mathbf{r}_1, \mathbf{r}_2, \mathbf{r}_3)$ of Fig. 2.11 in place of $U^{(3)}(\mathbf{r}_1, \mathbf{r}_2, \mathbf{r}_3)$ in Fig. 2.8

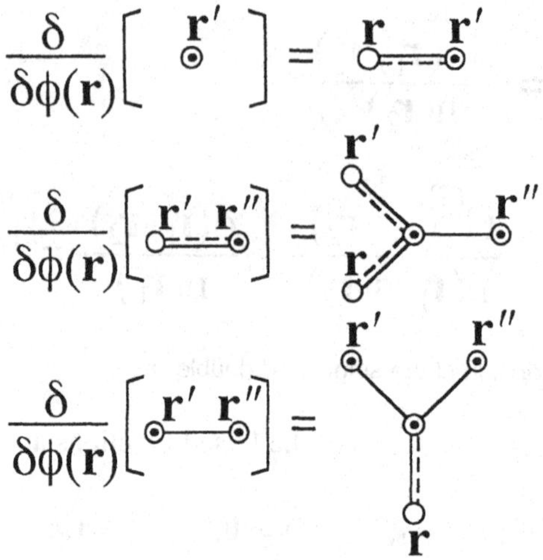

Figure 2.10 Fundamental operations for the calculations of $\delta U_{CA}^{(\nu)}(r_1, \ldots, r_\nu)/\delta\phi(r_{\nu+1})$ in the convolution approximation.

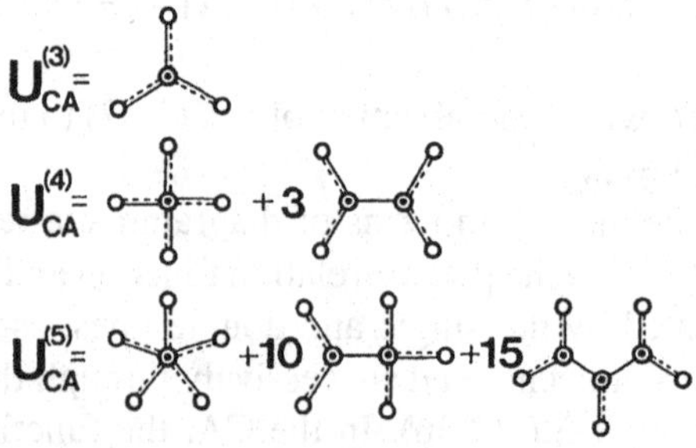

Figure 2.11 Diagrammatic representation of $U_{CA}^{(\nu)}(r_1, \ldots, r_\nu)$ for $\nu = 3, 4,$ and 5.

and thereby calculating Eq. (2.30) at $\nu = 3$. It is then possible to prove by mathematical induction that all the higher-order correlation potentials vanish identically in the CA because the recursive relation (2.28) remains valid in the CA; the CA thus leads to the HNC approximation. Conversely, one proves that the Ursell functions can be constructed uniquely in the CA scheme, which results in $K_c^{(\nu)}(\mathbf{r}_1, \ldots, \mathbf{r}_\nu) = 0$ for $\nu \geq 3$; hence $B(r) = 0$ in light of Eq. (2.43). The CA on which the HNC scheme is based satisfies exactly the overall charge-neutrality condition (2.48), crucial in the Coulombic systems.

The HNC scheme has a unique feature in that the chemical potential and hence the Helmholtz free energy can be calculated directly, without resorting to the usual coupling-constant integrations. A general prescription for the calculation of a free energy in an interacting many-particle system with a pair potential

$\phi(r)$ is to perform a coupling-constant integration of the interaction energy (see Section 3.5E in Volume I). The excess chemical potential, which is the excess contribution arising from interaction between the particles, is written in the form of a coupling-constant integration as

$$\mu_{ex} = n \int_0^1 d\lambda \int d\mathbf{r}\, \phi(r) g(r; \lambda) . \tag{2.50}$$

Here $g(r; \lambda)$ is the pair distribution function between a single "test particle" and the system particles interacting via the potential $\lambda\phi(r)$ with $0 \le \lambda \le 1$. For an OCP, $g(r; \lambda)$ in Eq. (2.50) is replaced by the corresponding pair correlation function $h(r; \lambda)$ to take into account the presence of the neutralizing background.

The excess part of the chemical potential μ_{ex}^{HNC} for an OCP in the HNC approximation, however, is calculated in a form *without* such a coupling-constant integration as

$$\beta\mu_{ex}^{HNC} = \frac{n}{2} \int d\mathbf{r}\, h(r) \left[h(r) - c(r) \right] - n \lim_{k \to 0} \left[\tilde{c}(k) + \frac{4\pi\beta(Ze)^2}{k^2} \right] , \tag{2.51}$$

where $\tilde{c}(k)$ is the Fourier transform of $c(r)$; derivation of Eq. (2.51) is offered as Problem 2.5. This particular feature in the HNC scheme can be utilized advantageously in some cases, since a lesser dimension in the numerical integrations usually means an improved accuracy in the evaluations.

D. Bridge Functions

Bridge functions represent contributions from a set of closely connected Mayer diagrams, called the bridge diagrams [e.g., Hansen & McDonald 1986]; the simplest of such a diagram, the Wheatstone bridge diagram, is depicted in Fig. 2.12. In the density-functional formalism of Section 2.2A, a bond (a solid line) corresponds to a pair correlation function, $h(r) = \delta n(r)/n$, so that the Wheatstone bridge contribution may be evaluated as

$$B_{CK}(r) = \frac{n^2}{2} \int d\mathbf{r}_1 d\mathbf{r}_2\, h\left(|\mathbf{r} - \mathbf{r}_1|\right) h\left(|\mathbf{r}_1 - \mathbf{r}_2|\right) h\left(|\mathbf{r}_2 - \mathbf{r}|\right) h\left(|\mathbf{r}_1|\right) h\left(|\mathbf{r}_1|\right) . \tag{2.52}$$

In the cluster-expansion theory of liquids [e.g., Hansen & McDonald 1986], one alternatively formulates the diagrammatic expansion in terms of f-bonds,

$$f(r) = exp[-\beta\phi(r)] - 1 , \tag{2.53}$$

where $\phi(r)$ refers to the potential of binary interaction. Since $h(r)$ or $f(r)$ quickly vanishes as $r \to \infty$, bridge functions are usually short-ranged.

Figure 2.12 Wheatstone bridge diagram.

Rosenfeld and Ashcroft [1979] assumed that the bridge functions would not depend on details of the potentials and thus should have a nearly universal functional form (*universality ansatz*). The bridge functions of the OCP were thereby replaced by those of hard-sphere systems, which were short-ranged and stayed negative (repulsive) over the whole range of interparticle separations. The bridge functions for the OCP, a system with a softest interparticle potential, thus provide a crucial test for such a universality ansatz. Breakdown of the universality ansatz in the vicinity of the first peak of the radial distribution function $g(r)$ was noted earlier through a calculation of the lowest-order bridge diagrams for the OCP [Iyetomi & Ichimaru 1983].

In a strongly coupled plasma, one can use the set of equations, (2.41) and (2.42), for a rigorous determination of $B(r)$, once $g(r)$ is known by some means. The bridge functions are related to the screening potentials via

$$B(r) = \beta H(r) - h(r) + c(r). \tag{2.54}$$

The screening potentials can be sampled through the Metropolis algorithm (see Section 2.1B). The MC simulation data alone are not sufficient to determine $B(r)$ over the entire regime of interparticle separations, however, because the size of the MC cell is finite and because the strong Coulomb repulsion virtually makes $g(r) = 0$ at short distances. Such an issue of "extracting" the bridge functions out of the MC samplings has been solved by Iyetomi, Ogata, and Ichimaru [1992]: Extrapolation to the short ranges has been achieved by the Widom expansion (2.9). In the long ranges, the compressibility sum rule for the Fourier transform $\tilde{c}(k)$ of the direct correlation function, that is,

$$n \lim_{k \to 0} \left[\tilde{c}(k) + \frac{4\pi\beta(Ze)^2}{k^2} \right] = 1 - \frac{\kappa_0}{\kappa_T}, \tag{2.55}$$

has been taken into accouunt. Here κ_T and κ_0 are the isothermal compressibility (see Eq. 1.64) and its ideal-gas value β/n for the OCP.

The bridge function so extracted has been expressed in an analytic formula as

$$\frac{B(r)}{\Gamma} = (-b_0 + c_1 x^4 + c_2 x^6 + c_3 x^8) \exp\left(-\frac{b_0}{b_1} x^2\right), \qquad (2.56)$$

with the parameters given by

$$b_0 = 0.258 - 0.0612 \ln \Gamma + 0.0123 (\ln \Gamma)^2 - 1/\Gamma, \qquad (2.57a)$$
$$b_1 = 0.0269 + 0.0318 \ln \Gamma + 0.00814 (\ln \Gamma)^2, \qquad (2.57b)$$
$$c_1 = 0.498 - 0.280 \ln \Gamma + 0.0294 (\ln \Gamma)^2, \qquad (2.58a)$$
$$c_2 = -0.412 + 0.219 \ln \Gamma - 0.00251 (\ln \Gamma)^2, \qquad (2.58b)$$
$$c_3 = 0.0988 - 0.0534 \ln \Gamma + 0.00682 (\ln \Gamma)^2, \qquad (2.58c)$$

and $x = r/a$. This form guarantees that Eq. (2.56) recovers the short-range expansion (2.17) to the quartic order in x with the parameters b_0 and b_1 given by Eqs. (2.57). The remaining parameters c_i ($i = 1, 2, 3$) in Eq. (2.56) account for the attractive part near the first peak of $g(r)$ for large Γ values. Accuracy of the parametrized bridge function has been confirmed to stay within the limit of computational errors in the MC simulations over the entire regime of the interparticle separation for $5 < \Gamma \le 180$.

The HNC scheme is good at portraying long-range correlations in a Coulombic system, while the bridge functions account for strong correlations at short distances. Explicit evaluation of the bridge functions plays the key role in an attempt to improve the HNC scheme, which we shall consider in the next subsection.

E. Improved Hypernetted-Chain Schemes

The HNC equation (2.45), though satisfactory in many of the Coulombic problems, contains "errors" due to its neglect of the bridge functions (2.43); the HNC results exhibit systematic departures from the exact MC results in the strong coupling regime. Developing an analytic scheme to improve the HNC approximation is a worthy theoretical project. Such a theory will be useful in the study of phase properties and miscibility problems in multi-ionic plasmas, for example. Evaluation of thermodynamic quantities with high precision on the order of 0.1%, over a wide range of parameters for the mixtures, is required when solving such a problem; they are sometimes inaccessible in practice by computer simulations alone.

An improved HNC scheme begins with rewriting the exact formula (2.42) in the form

$$g(r) = \exp\left[-\beta \Phi^{\text{tot}}(r) + h(r) - c(r)\right] \qquad (2.59)$$

with

$$\Phi^{\text{tot}}(r) = \frac{(Ze)^2}{r} - \frac{B(r)}{\beta} . \tag{2.60}$$

Once $B(r)$ is formulated explicitly, an effective HNC scheme is obtained with Eqs. (2.41) and (2.59), in which Eq. (2.60) plays the role of the effective binary potential. Numerical complexity in a solution to the improved HNC scheme does not exceed that in the ordinary HNC scheme. Various schemes for improvement of the HNC approximation have been proposed [Ichimaru, Iyetomi, & Tanaka 1987].

Rosenfeld and Ashcroft [1979] developed a semiempirical scheme on the basis of the universality ansatz for the bridge functions: It is assumed that the OCP bridge functions are given by those of an equivalent hard-sphere (HS) reference system with the packing fraction η:

$$B(r) = B^{\text{HS}}(r; \eta) . \tag{2.61}$$

HS systems are characterized by the effective diameter d of a particle with the number density n, so that the packing fraction is given by

$$\eta = \pi n d^3 / 6 . \tag{2.62}$$

In the Percus–Yevick (PY) approximation [e.g., Hansen & McDonald 1986], the bridge function is evaluated as

$$B^{\text{PY}}(r) = c(r) - h(r) + \ln[1 + h(r) - c(r)] , \tag{2.63}$$

which has the analytic solution $B^{\text{HS}}(r; \eta)$ for the hard-sphere system [Thiele 1963; Wertheim 1963, 1964]. The Thiele–Wertheim solution is expressed most compactly with the HS direct correlation function in the PY approximation; that is,

$$c^{\text{HS}}(r) = \begin{cases} -\lambda_1 - 6\eta\lambda_2 \dfrac{r}{d} - \dfrac{1}{2}\eta\lambda_2 \left(\dfrac{r}{d}\right)^3, & r < d \\ 0, & r > d , \end{cases} \tag{2.64}$$

where

$$\lambda_1 = \frac{(1 + 2\eta)^2}{(1 - \eta)^4} , \tag{2.65a}$$

$$\lambda_2 = \frac{-(1 + \frac{1}{2}\eta)^2}{(1 - \eta)^4} . \tag{2.65b}$$

The HS pair correlation function is then obtained by substitution of Eq. (2.64) in the Ornstein–Zernike relation (2.41); evaluation of $B^{HS}(r; \eta)$ follows from Eqs. (2.38) and (2.42).

The Rosenfeld–Ashcroft scheme thus utilizes such a parametrized HS bridge function for a closure of the hierarchy. The packing fraction η is regarded as a free parameter, which is to be determined from a self-consistency condition such as the compressibility sum rule. This scheme, called the *modified HNC*, has been applied successfully for solving various condensed-matter problems.

Rogers and Young [1984] proposed an integral equation that hybridizes the PY and HNC equations, so that the short-range bridge effects are included in the PY approximation. The bridge function is expressed as

$$B^{RY}(r) = -\gamma(r) + \ln\left\{\frac{\exp[f(r)\gamma(r)] - 1}{f(r)} + 1\right\}, \qquad (2.66)$$

where $\gamma(r) = h(r) - c(r)$. The mixing function $f(r)$ has been chosen in a simple form as

$$f(r) = 1 - \exp(-\alpha r), \qquad (2.67)$$

where α is the adjustable parameter for securing thermodynamic consistency. In the vicinity of $r = 0$, Eq. (2.66) reduces to the PY bridge function (2.63); on the other hand, Eq. (2.66) approaches the HNC approximation as r increases. This

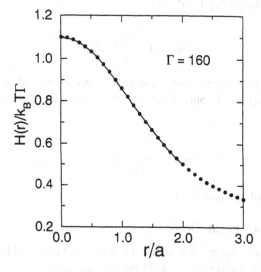

Figure 2.13   Screening potential of the OCP at $\Gamma = 160$. The solid curve represents Eq. (2.17); the dots are the improved HNC values with Eq. (2.56).

improved HNC scheme has been actually applied to various fluids with inverse power potentials, including the HS and OCP systems. With the exception of the OCP, it has been found that the parameter α remains almost constant for all densities; in the OCP, α depends on the coupling parameter Γ.

Substitution of the explicit OCP bridge function (2.56) in Eq. (2.60) finally enables one to close the hierarchy and thereby establishes an improved HNC scheme. The validity and accuracy of such a scheme have been confirmed through various points of examination: the correlation functions, the thermodynamic functions, and the compressibility sum rule [Iyetomi, Ogata, & Ichimaru 1992]. Solid curves in Fig. 2.1, obtained through a solution to this improved scheme, reproduce the MC data almost identically. Figure 2.13 shows that the screening potential in the improved HNC scheme excellently agrees with the MC and Widom expansion values.

2.3 THERMODYNAMIC PROPERTIES

In a multi-ionic plasma, the interaction energy per unit volume is given by

$$E_{\text{int}} = \sum_{\sigma,\tau} \int \frac{d\mathbf{k}}{(2\pi)^3} \frac{2\pi Z_\sigma Z_\tau e^2 \sqrt{n_\sigma n_\tau}}{k^2} [S_{\sigma\tau}(k) - \delta_{\sigma\tau}] , \qquad (2.68)$$

where $S_{\sigma\tau}(k)$ are the static structure factors formulated in Section 1.2B. The excess Helmholtz free energy per unit volume is then calculated with the coupling-constant integration of Eq. (2.68) as

$$F_{\text{ex}} = \int_0^1 \frac{d\eta}{\eta} E_{\text{int}}(\eta) . \qquad (2.69)$$

Here $E_{\text{int}}(\eta)$ refers to the interaction energy (2.68) evaluated in a system where the strength of Coulomb coupling e^2 is replaced by ηe^2. The excess pressure is given by

$$P_{\text{ex}} = -V \left(\frac{\partial F_{\text{ex}}}{\partial V} \right)_{T,N_\sigma} , \qquad (2.70)$$

where V is the volume and $N_\sigma = n_\sigma V$.

A. Dense Semiclassical One-Component Plasmas

The internal energy (per particle in units of $k_B T$) of a classical OCP is expressed as a sum of the ideal-gas part ($= 3/2$) and the excess:

$$u = 3/2 + u_{\text{ex}} . \qquad (2.71)$$

The excess part is calculated by the use of Eqs. (1.34) and (1.35) in Eq. (2.68). In the RPA, one uses Eqs. (1.46) and (1.51) to find

$$u_{\text{ex}}^{\text{DH}} = -\frac{\sqrt{3}}{2}\Gamma^{3/2} . \tag{2.72}$$

This is the Debye–Hückel value for the excess internal energy.

The next terms in the coupling-constant (Γ) expansion of u_{ex} were calculated by Abe [1959] exactly in the giant cluster-expansion theory, with the result:

$$u_{\text{ex}}^{\text{ABE}} = -\frac{\sqrt{3}}{2}\Gamma^{3/2} - 3\Gamma^3 \left[\frac{3}{8}\ln(3\Gamma) + \frac{\gamma}{2} - \frac{1}{3} \right] , \tag{2.73}$$

where $\gamma = 0.57721\ldots$ is Euler's constant. This formula has been rederived through various methods: O'Neil and Rostoker [1965] analyzed the triple correlation in a plasma by expanding the BBGKY hierarchy (see, e.g., Section 2.1D, Volume I) in powers of the plasma coupling parameter and thereby derived Eq. (2.73). Totsuji and Ichimaru [1973] calculated the structure factors in the CA of Section 2.2C and thereby evaluated Eq. (2.68) through a split of the k integration into long- and short-wavelength domains (Problem 2.7). Abe's formula (2.73) accurately represents the values of excess internal energy for $\Gamma < 0.1$.

In the strong coupling regime, $1 \leq \Gamma < 180$, the excess internal energy has been accurately evaluated by computer simulations [Slattery, Doolen, & DeWitt 1982; Ogata & Ichimaru 1987], with the result

$$u_{\text{ex}}^{\text{OCP}}(\Gamma) = -0.898004\Gamma + 0.96786\Gamma^{1/4} + 0.220703\Gamma^{-1/4} - 0.86097 . \tag{2.74}$$

In the intermediate-coupling regime, $0.1 \leq \Gamma < 1$, the excess internal energy has been calculated [Slattery, Doolen, & DeWitt 1980] through a solution to the HNC integral equations (Section 2.2C). Using these HNC values, one finds a formula connecting Eqs. (2.73) and (2.74) as

$$u_{\text{ex}}(\Gamma) = \frac{u_{\text{ex}}^{\text{ABE}}(\Gamma) + (3 \times 10^3)\Gamma^{5.7}u_{\text{ex}}^{\text{OCP}}(\Gamma)}{1 + (3 \times 10^3)\Gamma^{5.7}} . \tag{2.75}$$

This formula is applicable for a classical OCP fluid in the range $\Gamma < 180$ with the accuracy better than 0.1%.

The Helmholtz free energy per ion in units of $k_B T$ is again expressed as the sum of the ideal part (see Eq. 2.25) and the excess

$$f(\Gamma) = f_0 + f_{\text{ex}}(\Gamma) \tag{2.76a}$$

with

$$f_0 = 3\ln(\Lambda) - 1 + \ln\left(\frac{3}{4\pi}\right) \tag{2.76b}$$

where $e = 2.71828\ldots$, and Λ is the thermal de Broglie wavelength parameter defined generally in Eq. (1.12).

The excess free energy may be obtained by a coupling-constant integration of Eq. (2.69). In the weak coupling regime, one thus finds

$$f_{ex}^{ABE}(\Gamma) = -\frac{\Gamma^{3/2}}{\sqrt{3}} - \frac{\Gamma^3}{2}\left[\frac{3}{4}\ln(3\Gamma) + \gamma - \frac{11}{12}\right]. \tag{2.77a}$$

In the strong coupling regime $1 \le \Gamma < 180$, one likewise obtains from Eq. (2.74) as

$$\begin{aligned}
f_{ex}^{OCP}(\Gamma) = &-0.898004\Gamma + 3.87144\Gamma^{1/4} - 0.882812\Gamma^{-1/4} \\
&- 0.86097\ln\Gamma - 2.52692
\end{aligned} \tag{2.77b}$$

by taking $f_{ex}^{OCP}(\Gamma = 1) = -0.4363$ [Slattery, Doolen, & DeWitt 1982].

As Λ increases in an OCP plasma with higher densities and lower temperatures, the free energy is expressed by the Wigner–Kirkwood expansion [e.g., Landau & Lifshitz 1969; Hansen & McDonald 1986] in powers of $\hbar^2$ as

$$f_{FL} = f^{(0)} + f^{(1)} + f^{(2)} + \cdots. \tag{2.78}$$

Here

$$f^{(0)} = 3\ln(\Lambda) - 1 + \ln\left(\frac{3}{4\pi}\right) + f_{ex}(\Gamma) \tag{2.79}$$

is the classical term as given by Eqs. (2.76) and (2.77), and the quantum correction terms are

$$f^{(1)} = H(\Gamma)\frac{\Gamma\Lambda^2}{4\pi}, \tag{2.80a}$$

$$f^{(2)} = -\left[\frac{J(\Gamma)}{20} + \frac{3K(\Gamma)}{80} + \frac{1}{240}\right]\left(\frac{\Gamma\Lambda^2}{4\pi}\right)^2, \tag{2.80b}$$

with

$$H(\Gamma) = \frac{\Gamma a^4}{6N}\left\langle\sum_{i\ne j}^{N}\frac{1}{r_{ij}^4}\right\rangle, \tag{2.81a}$$

$$J(\Gamma) = \frac{a^6}{3N} \left\langle \sum_{i \neq j}^{N} \frac{1}{r_{ij}^6} \right\rangle , \tag{2.81b}$$

$$K(\Gamma) = \frac{a^6}{9N} \left\langle \sum_{i \neq j \neq k}^{N} \left[3 \frac{(\mathbf{r}_{ij} \cdot \mathbf{r}_{ik})^2}{r_{ij}^5 r_{ik}^5} - \frac{1}{r_{ij}^3 r_{ik}^3} \right] \right\rangle . \tag{2.81c}$$

The statistical average in Eq. (2.81a) is calculated as described in Problem 2.2 yielding

$$H(\Gamma) = 1/4 . \tag{2.82a}$$

The calculations of (2.81b) and (2.81c) call for knowledge on two- and three-particle distributions, which are available through the MC simulations. The current evaluations are

$$\begin{aligned} J(\Gamma) = &\ 0.13573 + 0.17362/\Gamma^{1/2} + 0.92707/\Gamma \\ &- 0.09740/\Gamma^{3/2} + 1.7824/\Gamma^2 + 1.9878/\Gamma^3 , \end{aligned} \tag{2.82b}$$

and

$$K(\Gamma) = -0.091964 . \tag{2.82c}$$

[Hansen & Vieillefosse 1975; Iyetomi, Ogata, & Ichimaru 1993].

B. Binary-Ionic-Mixture Fluids

The thermodynamic functions for dense BIM fluids have been investigated extensively for the construction of phase-separation diagrams associated with possible freezing transitions in the interiors of white dwarfs [Stevenson 1980; Barrat, Hansen, & Mochkovitch 1988; Ichimaru, Iyetomi, & Ogata 1988]. Substantial progress has been achieved due primarily to progress in MC simulation studies [Ogata et al. 1993]; here the principal results are summarized for the thermodynamic functions.

We consider a BIM fluid in a volume V containing N_1 and N_2 particles of species "1" and "2" with charge numbers Z_1 and Z_2, and number densities $n_1 = N_1/V$ and $n_2 = N_2/V$, respectively. The molar fraction of "i" species is denoted by

$$x_i = N_i/N, \tag{2.83}$$

where $N = N_1 + N_2$; the charge ratio R_Z has been defined in Eq. (2.4). The ion sphere radius and the Coulomb coupling parameter of the individual species are defined as

$$a_i = \left(\frac{3Z_i}{4\pi n_e} \right)^{1/3} , \qquad (2.84)$$

$$\Gamma_i = \frac{(Z_i e)^2}{a_i k_B T} , \qquad (2.85)$$

where n_e refers to the number density of the background electrons given by Eq. (1.1).

The excess internal energy of the BIM in units of $N k_B T$ is calculated in terms of the partial radial distribution functions (1.36) as

$$u_{ex}^{BIM}(R_Z, x, \Gamma_1) \equiv \sum_{i,j=1}^{2} \frac{N \sqrt{x_i x_j}}{2 V k_B T} \int d\mathbf{r} \, \frac{Z_i Z_j e^2}{r} \left[g_{ij}(r) - 1 \right] . \qquad (2.86)$$

From here on we shall write $x = x_2$ for simplicity.

Near the freezing transitions, the excess internal energy takes on a value close to the critical Γ value associated with such a transition (see Section 1.1C), that is, on the order of 200. Since the additional entropy term in the free energy, which essentially controls the outcome of phase diagrams, remains on the order of unity in the same energy units, the accuracy required in the evaluation of excess internal energy is on the order of 0.1% for a reliable determination of the phase diagrams. It may thus be concluded that the HNC scheme of Section 2.2C, if extended to the BIM cases, may not have the required accuracy.

An extensive MC simulation study of the excess internal energy (2.86) for BIM fluids has therefore been performed at 37 different combinations of the parameter values $R_Z = \{4/3, 3, 5\}$, $x = 0.01$–0.5, and $\Gamma_1 = 5$–200 [Ogata et al. 1993]. The results are compared with predictions from a *linear-mixing law*, that is,

$$u_{ex}^{LM}(R_Z, x, \Gamma_1) \equiv \sum_{i=1}^{2} x_i u_{ex}^{OCP}(\Gamma_i) , \qquad (2.87)$$

where u_{ex}^{OCP} is the OCP excess internal energy in strong coupling, as given by Eq. (2.74). The MC results for the departures,

$$\Delta u_{ex}^{BIM} = u_{ex}^{BIM} - u_{ex}^{LM} , \qquad (2.88a)$$

from the linear-mixing values are thus expressed in a parametrized form as

$$\Delta u_{\text{ex}}^{\text{BIM}}(R_Z, x, \Gamma_1) = 0.32 \frac{\sqrt{R_Z - 1}(x R_Z - 0.11)}{R_Z - 0.22} \frac{x^{0.5} + (2 \times 10^{-3}) x(1 - x)}{x^{1.7} + (5 \times 10^{-5})} \frac{}{\Gamma_1}.$$

(2.88b)

If $\Delta u_{\text{ex}}^{\text{BIM}} = 0$, then the BIM excess internal energies are said to satisfy the linear-mixing law with the constant electron-density, ion-sphere scaling [Salpeter 1954] characterized by Eq. (2.84). Since a contribution from the entropy of mixing (see Eq. 2.91a below) usually acts to lower the free energy in the mixed phase, $\Delta u_{\text{ex}}^{\text{BIM}} = 0$ would imply that a mixed phase is favored. Though small in magnitude, Eq. (2.88) does indicate that the internal energy in most cases will increase after mixing (i.e., $\Delta u_{\text{ex}}^{\text{BIM}} > 0$), implying a tendency toward phase demixing. If $x R_Z < 0.11$, however, one finds in Eq. (2.88) that $\Delta u_{\text{ex}}^{\text{BIM}} < 0$, meaning an added stability for a mixed state. Departure from the linear mixing thus plays the central part in the analysis of miscibility for BIM fluids.

The Helmholtz free energy for a BIM fluid (in units of $N k_B T$) is expressed as a sum of the linear mixing contributions and the deviations therefrom:

$$f^{\text{BIM}} = (1 - x) f^{\text{OCP}}(\Gamma_1) + x f^{\text{OCP}}(\Gamma_2)$$
$$+ \Delta f_{\text{id}}^{\text{BIM}}(R_Z, x) + \Delta f_{\text{ex}}^{\text{BIM}}(R_Z, x, \Gamma_1).$$

(2.89)

Here

$$f^{\text{OCP}}(\Gamma) = -0.898004\Gamma + 3.87144\Gamma^{1/4} - 0.882812\Gamma^{-1/4}$$
$$+ 2.13903 \ln \Gamma - 3.24222$$

(2.90)

is the free energy for the OCP fluids assuming $f^{\text{OCP}}(\Gamma = 1) = -1.1516$ [Slattery, Doolen, & DeWitt 1982],

$$\Delta f_{\text{id}}^{\text{BIM}}(R_Z, x) = \sum_{i=1}^{2} x_i \ln \left(\frac{n_i Z_i}{n_e} \right)$$
$$= (1 - x) \ln \left(\frac{1 - x}{1 - x + x R_Z} \right) + x \ln \left(\frac{x R_Z}{1 - x + x R_Z} \right)$$

(2.91a)

refers to a free-energy contribution of the ideal-gas entropy of mixing under the condition of charge neutrality at a constant density of the background electrons.

The last excess part in Eq. (2.89) can be evaluated from Eq. (2.88) via the technique of the coupling-constant integration (2.69) as

$$\Delta f_{\text{ex}}^{\text{BIM}}(R_Z, x, \Gamma_1) =$$
$$0.32 \frac{\sqrt{R_Z - 1}(x R_Z - 0.11)}{R_Z - 0.22} \frac{x^{0.5} + (2 \times 10^{-3})}{x^{1.7} + (5 \times 10^{-5})} x(1 - x)\left(1 - \frac{1}{\Gamma_1}\right)$$
$$+ \Delta f_{\text{ex}}^{\text{BIM}}(R_Z, x, \Gamma_1 = 1) . \tag{2.91b}$$

The HNC scheme of Section 2.2C with Eq. (2.51) may be used for the calculation of the last term in Eq. (2.91b), since it is known to be accurate for $\Gamma_1 \leq 1$; the result is

$$\Delta f_{\text{ex}}^{\text{BIM}}(R_Z, x, \Gamma_1 = 1) = 0.0551 \frac{(R_Z - 1)^{1.8} x(1 - x)}{1 + 1.12(R_Z - 1)x} \tag{2.92}$$

[Ogata et al. 1993].

C. Classical Coulomb Solids

Consideration of thermodynamic functions for a Coulomb solid, be it classical or quantum, begins with the concept of a Wigner sphere model consisting of charged point particles free to move in a uniform background of neutralizing charge [Wigner 1934, 1938]. For an OCP, the electrostatic energy E_{IS} of an ion sphere consisting of a uniform, negative charge sphere with radius a and charge density $-\rho_e = -3Ze/4\pi a^3$ and an ion with charge Ze located at a distance r away from the center of the sphere is given by (see Eq. 1.33 in Volume I)

$$\beta E_{\text{IS}} = -0.9\Gamma + \frac{\Gamma}{2}\left(\frac{r}{a}\right)^2 . \tag{2.93}$$

The first term on the right-hand side, -0.9Γ, is the Lieb–Nahnhofer lower bound for OCP internal energy [e.g., Lieb 1976].

The second term induces an oscillatory motion to the ion (mass M) at an angular frequency,

$$\omega_E = \sqrt{\frac{4\pi Ze\rho_e}{3M}} , \tag{2.94a}$$

called the *Einstein frequency* of the Wigner solid (see Eq. 1.17). In a classical system, the average energy, including kinetic energy, of such an oscillator contributes

$$E_{\text{osc}}^{\text{CL}} = 3k_B T \tag{2.94b}$$

to the ion sphere energy. In the quantum limit $(T \rightarrow 0)$, the zero-point oscillations at the frequency (2.94a) give rise to a contribution to energy:

$$E_{osc}^{QM} = (3/2)\hbar\omega_E . \tag{2.94c}$$

When an OCP forms a crystalline lattice with a lattice constant b, the electrostatic energy per particle E_M, called the *Madelung energy*, is expressed as

$$E_M = -\frac{1}{2}\alpha_M \frac{(Ze)^2}{b} , \tag{2.95}$$

where α_M is the *Madelung constant*. It has been assumed that an OCP in its ground state forms a bcc crystalline solid, a conclusion reached through comparison of the Madelung energies of the several cubic-lattice and other structures. Table 2.1 lists the values of the energy constant for various lattices [Foldy 1978]. Extensive MC simulations have been performed for OCP solids with cubic structures [Brush, Sahlin, & Teller 1966; Hansen, J.-P. 1973; Slattery, Doolen, & DeWitt 1982; Helfer, McCrory, & Van Horn 1984], and the bcc lattice has been shown to have the lowest free energy at finite temperatures as well.

The thermodynamic functions for classical OCP solids have been investigated through the MC simulations coupled with analytic study of anharmonic effects in the lattice vibrations [Dubin 1990]. The excess internal energy per particle in units of $k_B T$ for an OCP bcc-crystalline solid is thus given by

$$u_{ex}^{OCP}(\Gamma) = -0.895929\Gamma + 1.5 + \frac{10.84}{\Gamma} + \frac{352.8}{\Gamma^2} + \frac{1.74 \times 10^5}{\Gamma^3} . \tag{2.96}$$

The free energy in the OCP solid can be evaluated by integrating the internal energy with respect to the inverse temperature [e.g., Fetter & Walecka 1971] as

$$f(\Gamma) = f(\infty) + \int_\infty^\Gamma d\Gamma' \frac{u(\Gamma')}{\Gamma'} . \tag{2.97}$$

Table 2.1   Energy constants for various lattices

Lattice	$\alpha_M a/b$
Body centered cubic (bcc)	–1.791 858 52
Face centered cubic (fcc)	–1.791 747 23
Hexagonal close packed (hcp)	–1.791 676 90
Simple cubic (sc)	–1.760 118 90
Diamond (dmnd)	–1.670 851 41

Assuming that the ground-state free energy $f(\infty)$ is given by the harmonic lattice value [Pollock & Hansen 1973], the free energy less the ideal-gas value (2.76b) is calculated as

$$f^{\text{OCP}}(\Gamma) = -0.895929\Gamma + \frac{9}{2}\ln\Gamma - 1.885596 - 10.84\Gamma^{-1} - 176.4\Gamma^{-2}$$
$$- 5.980 \times 10^4 \Gamma^{-3}, \tag{2.98}$$

which may be referred to as the free energy for a classical OCP solid [Dubin 1990].

The free energy of a Coulomb solid consists of the sum of the bcc Madelung energy and the harmonic and anharmonic contributions:

$$f_{\text{bcc}} = -0.895929\Gamma + f_{\text{bcc}}^{\text{HM}} + f_{\text{bcc}}^{\text{AH}}. \tag{2.99a}$$

The results for the harmonic and anharmonic contributions with inclusion of the Wigner–Kirkwood quantum corrections [Hansen & Vieillefosse 1975] are

$$f_{\text{bcc}}^{\text{HM}} = -0.84588 + 3\ln Y + \frac{Y^2}{8} - 1.9038 \times 10^{-3} Y^4, \tag{2.99b}$$

$$f_{\text{bcc}}^{\text{AH}} = -\left(\frac{10.84}{\Gamma} + \frac{176.4}{\Gamma^2} + \frac{5.980 \times 10^4}{\Gamma^3}\right) - \frac{Y^4}{80}\left(\frac{1.2993}{\Gamma} + \frac{61.252}{\Gamma^2}\right). \tag{2.99c}$$

The first three terms on the right-hand side of Eq. (2.99c) stem from the inverse temperature integration of the last three terms in Eq. (2.96).

The free energies of solid mixtures, containing ions with differing charges, depend on the specific ionic configurations. We have little knowledge of these structures for BIM solids under equilibrated conditions. Outstanding issues include the following questions: Do BIMs form "disordered" solids instead of regular, crystalline solids? How does the equilibrium structure depend on the temperature or density? To elucidate some of these issues, Ogata et al. [1993] calculated the BIM Madelung energies for both crystalline and disordered solids. The excess internal energies at finite temperatures for each type of solid are then evaluated by the MC simulation method. Combining these results, one determines the free energies and thus the equilibrated structures for the BIM solids.

As in the cases of the classical BIM fluids treated in Section 2.3B, the Helmholtz free energies of the classical BIM solids are analyzed relative to the linear-mixing (LM) law. The BIM free energy per particle in units of $k_B T$ is thus expressed as

$$f^{\text{BIM}} = (1 - x)f^{\text{OCP}}(\Gamma_1) + xf^{\text{OCP}}(\Gamma_2) + \Delta f^{\text{BIM}}, \tag{2.100}$$

where we recall that $x = x_2$. In Eq. (2.100), the first two terms on the right-hand side are the LM contributions to the BIM free energy with the OCP formula (2.98).

The deviations of the total BIM free energies from the LM contributions are finally obtained as a sum of the internal energy and entropy contributions and expressed in parametrized forms for $1 < R_Z \leq 4.5$ as

$$\Delta f^{\text{BIM}} = \frac{\Delta E_M^{\text{BIM}}}{N k_B T} - \frac{\Delta S^{\text{BIM}}}{N k_B} , \tag{2.101}$$

where

$$\frac{\Delta E_M^{\text{BIM}}}{N(Z_1 e)^2 / a_1} = \frac{\dfrac{0.05(R_Z - 1)^2 x(1 - x)}{[1 + 0.64(R_Z - 1)]\left[1 + 0.5(R_Z - 1)^2\right]}}{1 + \dfrac{27(R_Z - 1)}{1 + 0.1(R_Z - 1)} \sqrt{x}(\sqrt{x} - 0.3)(\sqrt{x} - 0.7)(\sqrt{x} - 0.3)(\sqrt{x} - 1)} \tag{2.102}$$

and

$$\frac{\Delta S^{\text{BIM}}}{N k_B} = \begin{cases} 0, & \text{for crystalline solids,} \\ -\Delta f_{\text{id}}^{\text{BIM}}(R_Z, x), & \text{for random solids.} \end{cases} \tag{2.103}$$

In Eq. (2.103), $\Delta f_{\text{id}}^{\text{BIM}}(R_Z, x)$ is the entropy contribution calculated by counting the number of different configurations and is given by Eq. (2.91a). In Section 2.5E, we shall use these formulas for calculating the phase diagrams in dense BIM plasmas.

D. Quantum Mechanical Coulomb Solids

For a quantum mechanical OCP solid, the free energy per ion in units of $k_B T$ is expressed as the sum of the bcc Madelung energy term and the harmonic and anharmonic contributions:

$$f(\xi, Y) = -0.895929\Gamma + f_{\text{HM}}(Y) + f_{\text{AH}}(\xi, Y) . \tag{2.104}$$

Here

$$R_s = \left(\frac{3}{4\pi n}\right)^{1/3} \frac{M(Ze)^2}{\hbar^2} , \qquad Y = \frac{\hbar}{k_B T}\left(\frac{4\pi n(Ze)^2}{3M}\right)^{1/2} , \tag{2.105}$$

$$\xi = \Gamma \tanh\left(\frac{8.5}{Y}\right) = Y\sqrt{R_s}\,\tanh\left(\frac{8.5}{Y}\right) , \qquad (2.106)$$

and the Coulomb coupling parameter is given by $\Gamma = Y\sqrt{R_s}$.

In the semiclassical regime $Y \ll 1$, the harmonic and anharmonic contributions have been evaluated with inclusion of the quantum corrections up to $\hbar^4$ terms in Eqs. (2.99). The ground state energy ($Y \gg 1$) of a bcc Coulomb solid has been obtained by Carr, Coldwell-Horsfall, and Fein [1961] as

$$f_{QM}^{HM} = 1.3286\,\frac{\Gamma}{\sqrt{R_s}} , \qquad (2.107a)$$

$$f_{QM}^{AH} = -0.365\,\frac{\Gamma}{R_s} . \qquad (2.107b)$$

The free energies of the bcc Coulomb solids at arbitrary values of Y and R_s have been calculated by Iyetomi, Ogata, and Ichimaru [1993] using a path-integral Monte Carlo method (see Section 1.2G). The harmonic term takes the form

$$f_{HM}(Y) = 3\ln\left[2\sinh\left(\frac{Y}{2}g(Y)\right)\right] , \qquad (2.108)$$

with

$$g(Y) = \frac{0.7543 + 0.09245Y^2 + 0.003386Y^4}{1 + 0.1046Y^2 + 0.003823Y^4} . \qquad (2.109)$$

The values of $f(\xi, Y)$ computed at 48 different combinations of the parameters Y and R_s are then used for the derivation of a parametrized expression for $f_{AH}(\xi, Y)$; the result is

$$f_{AH}(\xi, Y) = -\frac{9}{4}L(\xi, Y)Y\coth^2\left(\frac{Y}{2}\right) , \qquad (2.110)$$

where

$$L(\xi, Y) = \Gamma Y\frac{P(\xi) - 0.08167P(\xi)Y^2 + Q(\xi)Y^4}{1 + 0.085Y^2 + R(\xi)Y^6} , \qquad (2.111)$$

and

$$P(\xi) = \frac{1.204}{\xi^2} + \frac{19.60}{\xi^3} + \frac{6.644 \times 10^3}{\xi^4} , \qquad (2.112a)$$

$$Q(\xi) = \frac{0.001805}{\xi^2} + \frac{0.08507}{\xi^3} + 0.009444 P(\xi) , \qquad (2.112b)$$

$$R(\xi) = 0.08532 \xi^2 Q(\xi) . \qquad (2.112c)$$

Formulas (2.108) and (2.110) reproduce both the semiclassical and ground-state results, Eqs. (2.99) and (2.107), in the respective limits. Errors in the parametrization (2.108) are confined to 0.04%. The form of parametrization (2.110) has been suggested by a model calculation of the anharmonic contribution to the free energy (Problem 2.8); it reproduces the MC data within the statistical errors inherent in the simulations. In Section 2.5D, these formulas will be used for the derivation of boundary curves of freezing transition.

The Helmholtz free energy of a dense BIM solid $F^{\mathrm{BIM}}(N_1, N_2)$ may be considered through an effective OCP-solid model for the BIM solid constructed under the assumption that the constant electron-density, ion-sphere scaling (2.84) is applicable for the local arrangements of the particles. Such an assumption is well justified for dense matters in the interiors of degenerate stars, for instance. In such an ion-sphere model, Γ and Y represent, respectively, the electrostatic energy and the Einstein-oscillator energy in units of the thermal energy $k_B T$. The BIM averages $\langle \Gamma \rangle$ and $\langle Y \rangle$ of these quantities are then obtained as

$$\langle \Gamma \rangle = \langle Z^{5/3} \rangle \frac{e^2}{a_e k_B T} , \qquad (2.113)$$

$$\langle Y \rangle = \left\langle \left(\frac{Z}{A} \right)^{1/2} \right\rangle \frac{\hbar \omega_P}{\sqrt{3} k_B T} , \qquad (2.114)$$

with

$$\langle Z^{5/3} \rangle = \frac{1}{N_1 + N_2} \left[N_1 Z_1^{5/3} + N_2 Z_2^{5/3} \right] , \qquad (2.115)$$

$$\left\langle \left(\frac{Z}{A} \right)^{1/2} \right\rangle = \frac{1}{N_1 + N_2} \left[N_1 \left(\frac{Z_1}{A_1} \right)^{1/2} + N_2 \left(\frac{Z_2}{A_2} \right)^{1/2} \right] . \qquad (2.116)$$

Here $\omega_P / \sqrt{3}$ denotes the Einstein frequency of the proton OCP solid and A means the mass number of an ion. The BIM free energy is therefore expressed as

$$\frac{\beta}{N_1 + N_2} F^{\mathrm{BIM}}(N_1, N_2) = -0.895929 \langle \Gamma \rangle + f_{\mathrm{HM}}(\langle Y \rangle) + f_{\mathrm{AM}}(\langle \xi \rangle, \langle Y \rangle) .$$

$$(2.117)$$

2.4 DYNAMIC PROPERTIES

Time-dependent correlations and transport properties of a classical OCP are studied in this section in a hydrodynamic or a viscoelastic formalism, applicable in a low-frequency regime. The frequency-moment sum rules provide a framework appropriate to a description of the dynamic properties in a high-frequency regime. Microscopic theory of dynamic correlations for the degenerate electron liquids will be considered in Chapter 3.

A. The Velocity Autocorrelation Function

The velocity autocorrelation function (VAF) represents one of the simplest physical quantities that convey information on dynamic properties of a many-particle system. Its long-time behavior is related to the coefficients of self-diffusion and friction. Its short-time behavior may reveal the extent of local field effects via the frequency-moment sum rules.

The VAF is defined by

$$Z(t) = \frac{1}{3} \langle \mathbf{v}_1(t) \cdot \mathbf{v}_1(0) \rangle = \langle v_{1x}(t) v_{1x}(0) \rangle \; ; \tag{2.118}$$

it is a correlation function associated with the self-motion of a tagged particle "1" (mass m) between its initial velocity $\mathbf{v}_1(0)$ and the velocity $\mathbf{v}_1(t)$ at time t. Isotropy of the system is assumed in the calculation of the statistical average in Eq. (2.118), so that a Cartesian component $v_{1x}(t)$ of the velocity suffices for a description of $Z(t)$. Salient features in the dynamic correlation of the system may be unveiled in the self-motion as the tagged particle interacts with the surrounding particles.

1. Self-Diffusion

The *coefficient of self-diffusion* D_s is defined and calculated as

$$D_s \equiv \lim_{t \to \infty} \frac{1}{6t} \langle |\mathbf{r}_1(t) - \mathbf{r}_1(0)|^2 \rangle = \int_0^\infty dt \; Z(t) \tag{2.119}$$

[e.g., Chandrasekhar 1943]. Defining the *friction coefficient* $1/\tau_s$ via

$$\tau_s \equiv \frac{1}{Z(0)} \int_0^\infty dt \; Z(t) \; , \tag{2.120}$$

one then finds an Einstein relation,

$$D_s = \tau_s \frac{k_B T}{m} \; , \tag{2.121}$$

since

$$Z(0) = \langle |v_{1x}(0)|^2 \rangle = \frac{k_B T}{m} .$$

The VAF for a strongly coupled OCP has been measured extensively through MD simulations over a wide range of Γ values [Hansen, McDonald, & Pollock 1975]. The VAFs so computed may be grouped into three types, as depicted in Fig. 2.14. At $\Gamma = 0.993$, the VAF exhibits the behavior of a simple decay type; at $\Gamma = 9.7$ and 19.7, the VAF turns into an oscillatory decay type; and finally at $\Gamma = 59.1$, 110.4, and 152.4, the VAF shows a damped oscillatory behavior. The oscillatory decay type and the damped oscillatory type may be distinguished in terms of the possibility of the VAF moving into negative domain [i.e., $Z(t) < 0$], indicating reversal of the tagged-particle velocity. Such a reversal is related to the *cage effect* in a dense liquid, where a single particle has a tendency to be trapped by the surrounding particles [Bengtzelius, Götze, & Sjölander 1984]. Those qualitative changes in the VAF should point to essential differences in the internal structures of strongly coupled OCPs.

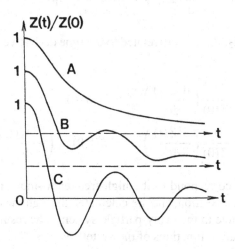

Figure 2.14 Schematic drawing of the three types of VAFs observed in the MD simulation study of the classical OCPs: (A) the simple decay type; (B) the oscillatory decay type; (C) the damped oscillatory type.

2. Memory-Function Formalism

The nature of those different VAFs may be elucidated through a *memory-function formalism* [Mori 1965] on various terms in the short-time and long-time behaviors. The memory function $M(t)$ of the VAF is introduced through the equation

$$\frac{dZ(t)}{dt} + \int_0^t ds\, M(t-s) Z(s) = 0 . \tag{2.122}$$

The Laplace transform of the memory function,

$$\tilde{M}(\omega) = \int_0^\infty dt\, M(t) \exp(i\omega t) , \tag{2.123}$$

physically describes the frequency-dependent friction coefficient, in that a *generalized friction force* acting on the tagged particle is given by

$$\tilde{F}_x(\omega) = -m\tilde{M}(\omega)\tilde{v}_x(\omega) \tag{2.124}$$

[Zwanzig & Bison 1970].

In the short-time domain, the VAF may be expanded as

$$\frac{Z(t)}{Z(0)} = 1 + \frac{A_2}{2!} t^2 + \frac{A_4}{4!} t^4 + \cdots . \tag{2.125}$$

The coefficients, A_2, A_4, ..., are related to the time derivatives of $Z(t)$ and $M(t)$, so that

$$A_2 = \frac{1}{Z(0)} \left(\frac{d^2 Z(t)}{dt^2} \right)_{t=0} = -M(0) , \tag{2.126a}$$

$$A_4 = \frac{1}{Z(0)} \left(\frac{d^4 Z(t)}{dt^4} \right)_{t=0} = M^2(0) - \left(\frac{d^2 M(t)}{dt^2} \right)_{t=0} . \tag{2.126b}$$

Those coefficients correspond to the high-frequency–moment sum rules, to be treated in the next subsection, and are calculated through the equation of motion for the tagged particle in the many-particle system. The results are expressed in terms of the correlation functions of the system as

$$A_2 = \frac{n}{m} \int dr \left[\frac{\partial \phi(r)}{\partial x} \right] \left[\frac{\partial g(r)}{\partial x} \right] , \tag{2.127a}$$

$$A_4 = \frac{2n}{mk_B T} \int d\mathbf{r} \int d\mathbf{v}\, f(\mathbf{v}) \left\{ \mathbf{v} \cdot \frac{\partial}{\partial \mathbf{r}} \left[\frac{\partial \phi(r)}{\partial x} \right] \right\}^2 g(r)$$

$$+ \frac{n^2}{mk_B T} \int d\mathbf{r} \int d\mathbf{r}' \int d\mathbf{v}\, f(\mathbf{v}) \left\{ \mathbf{v} \cdot \frac{\partial}{\partial \mathbf{r}} \left[\frac{\partial \phi(r)}{\partial x} \right] \right\} \left\{ \mathbf{v} \cdot \frac{\partial}{\partial \mathbf{r}'} \left[\frac{\partial \phi(r')}{\partial x'} \right] \right\}$$

$$\times \left[1 + h(r) + h(r') + h\left(|\mathbf{r} - \mathbf{r}'| \right) + h^{(3)}(\mathbf{r}, \mathbf{r}') \right] .$$

$$(2.127b)$$

Here $\phi(r) = (Ze)^2/r$, $h(r)$ and $h^{(3)}(\mathbf{r}, \mathbf{r}')$ refer to the pair and triple correlation functions for the homogeneous system (see Eqs. 2.46), and

$$f(\mathbf{v}) = \left(\frac{m}{2\pi k_B T} \right)^{3/2} \exp\left(-\frac{mv^2}{2k_B T} \right) \tag{2.128}$$

is the Maxwellian velocity distribution. For the OCP, one finds

$$A_2 = -\frac{4\pi n(Ze)^2}{3m} . \tag{2.129}$$

To evaluate A_4 according to Eq. (2.127b), knowledge of the triple correlation function is needed.

Theoretical studies have revealed the importance of taking into account the hydrodynamic modes in describing the long-time behavior of the VAF for a stongly coupled system such as liquids and plasmas [Alder & Wainwright 1970; Zwanzig & Bison 1970; Gould & Mazenko 1977; Ernst, Hauge, & van Leeuwen 1979]; coupling of a single particle motion with the transverse and longitudinal modes plays an essential part. In a theoretical approach to the generalized friction from a hydrodynamic point of view, one finds it instructive to consider a force exerted on a spherical body with radius R and mass m, vibrating at an angular frequency ω, in a Navier–Stokes fluid with shear viscosity η. The frequency-dependent friction coefficient $M(\omega)$ calculated with the stick boundary condition (i.e., the flow velocities relative to the spherical body vanish at the surface) reads

$$\tilde{M}(\omega) = \frac{6\pi \eta R}{m} \left[1 - i \left(i \frac{\omega}{\omega_s} \right)^{1/2} - \frac{i}{9} \frac{\omega}{\omega_s} \right] \tag{2.130}$$

[e.g., Landau & Lifshitz 1959], where

$$\omega_s \equiv \frac{\eta}{R^2 \rho_m} , \tag{2.131a}$$

with ρ_m denoting the mass density of the fluid.

The first term on the right-hand side of Eq. (2.130) is the ordinary Stokes friction. Since $\tilde{M}(0) = 1/\tau_s$ in light of Eq. (2.120), it also yields a relation,

$$\frac{1}{\tau_s} = \frac{6\pi\eta R}{m} .$$

(2.131b)

The second term is related to the penetration depth of viscous unsteady flow around the sphere, and the third term is connected with the virtual mass of the sphere.

We now extend the notion of an equivalent hard-sphere system and assume that Eq. (2.130) is applicable to the description for the long-time behavior of the VAF in a strongly coupled OCP. Similarity between those two systems has been noted in various examples: The radial distribution functions are known to exhibit quite analogous oscillatory structures. The fitting formulas proposed for the internal energy of the strongly coupled OCPs point to the applicability of an ion-sphere model; a quantitative correspondence between the Γ parameter for the OCP and the packing fraction $(R/a)^3$ for the hard-sphere system has been indicated through a variational treatment of the free energy. With the aid of these ideas, we may determine the numerical values of the coefficients in Eq. (2.130) as follows.

Among the various points already mentioned, we particularly remark on the correspondence between Γ and $(R/a)^3$ proposed by DeWitt and Rosenfeld [1979]. In light of the exact analytic solution of the Percus–Yevick equation for the hard-sphere system, a variational principle on the basis of the Gibbs–Bogoliubov inequalities for the free energy [e.g., Feynman 1972] is established, whereby the hard-sphere system is regarded as a reference system. Minimizing the OCP free energy evaluated in the reference system with respect to the packing fraction, one finds the correspondence

$$\Gamma = 2\frac{R}{a}\left[1 + 2\left(\frac{R}{a}\right)^3\right]^2\left[1 - \left(\frac{R}{a}\right)^3\right]^{-4} .$$

(2.132)

A hard-sphere system equivalent to a strongly coupled OCP with a given value of Γ may thus be determined by this relation. As the form of Eq. (2.132) implies, the value of R/a is a weak function of Γ in the strong coupling domain ($\Gamma \gg 1$); for instance, as Γ changes from 20 to 160, R/a varies only from 0.695 to 0.815. The effective hard-sphere radius R thus scales as a ; the ion-sphere model here appears sustained.

Interpolation of the memory function between those short-time and long-time evaluations has been performed by Nagano and Ichimaru [1980], resulting in

$$M(t) = \frac{4\pi n(Ze)^2}{3m}(1 - \Omega_s^2 t^2) \exp\left(-\frac{1}{2}\Omega_s^2 t^2\right)$$
$$+ \frac{\omega_s}{2\sqrt{\pi}\,\tau(\omega_s t)^{3/2}} \exp\left(-\frac{1}{4\omega_s t}\right).$$

(2.133)

Here the new parameter Ω_s is to be determined via Eq. (2.126b):

$$\Omega_s^2 = \frac{4\pi n(Ze)^2}{9m} - \frac{mA_4}{4\pi n(Ze)^2}.$$

(2.134)

With the values of A_4 estimated by Hansen [1978], the memory function (2.133) can quite accurately reproduce the features of the MD results for VAF exemplified in Fig. 2.14.

B. Frequency-Moment Sum Rules

As Eqs. (2.125)–(2.127) illustrate, short-time behaviors of the dynamic correlations are related to the short-range correlations in the many-particle system. It is useful, in these connections, to study the *frequency-moment sum rules* for the dielectric function $\varepsilon(k, \omega)$ of the plasma. In light of Eqs. (1.33) and (A.6), the retarded density-density response relation is expressed as

$$\frac{1}{\varepsilon(k, \omega)} = 1 + \frac{4\pi(Ze)^2}{k^2}\int_{-\infty}^{\infty}\frac{d\omega'}{\pi}\frac{\mathrm{Im}\,\chi(k, \omega')}{\omega' - \omega - i0},$$

(2.135)

where +0 is the positive infinitesimal,

$$\mathrm{Im}\,\chi(k, \omega) = -\frac{1}{2\hbar}\int_{-\infty}^{\infty} dt\,\langle[\rho_k(t), \rho_k^\dagger(0)]\rangle \exp(i\omega t)$$

(2.136)

and

$$\rho_k(t) = \sum_{j=1}^{n} \exp(-i\mathbf{k}\cdot\mathbf{r}_j(t)) \qquad (n = N/V),$$

(2.137)

corresponds to a time-dependent Heisenberg operator for density fluctuations with the Hamiltonian,

$$H = \frac{1}{2m}\sum_{j=1}^{n} p_j^2 + \frac{1}{2}\sum_{k(\neq 0)}\frac{4\pi(Ze)^2}{k^2}(\rho_k\rho_k^\dagger - n).$$

(2.138)

The high-frequency asymptotic expansion of Eq. (2.135) may be written as

$$\frac{1}{\varepsilon(k,\omega)} = 1 + \frac{4\pi(Ze)^2}{\omega k^2} \sum_{\nu=1} \frac{\langle \omega^\nu \rangle}{\omega^\nu} .$$ (2.139)

The νth frequency moment $\langle \omega^\nu \rangle$ defined through this expansion is then calculated as

$$\langle \omega^\nu \rangle = \frac{1}{\hbar} \left(i \frac{\partial}{\partial t} \right)^\nu \langle [\rho_{\mathbf{k}}(t), \rho_{\mathbf{k}}^\dagger(0)] \rangle \Big|_{t=0} .$$ (2.140)

For an isotropic system, $\langle \omega^\nu \rangle$ with an even value of ν identically vanishes. The sum rules with $\nu = 1$ and 3 are expressed as [Puff 1965; Ichimaru 1982]

$$\langle \omega^1 \rangle = \frac{nk^2}{m} ,$$ (2.141a)

$$\langle \omega^3 \rangle = \frac{nk^2}{m} \left\{ \frac{4}{\hbar} \omega_0(k) \langle K \rangle + \omega_0^2(k) + \omega_p^2 [1 - I(k)] \right\} .$$ (2.141b)

Here $\langle K \rangle$ represents the average kinetic energy per particle,

$$\omega_p = \sqrt{\frac{4\pi n (Ze)^2}{m}}$$ (2.142)

is the *plasma frequency*,

$$\omega_0(k) = \frac{\hbar k^2}{2m} ,$$ (2.143)

$$I(k) = -\frac{1}{n} \int \frac{d\mathbf{q}}{(2\pi)^3} K(\mathbf{k}, \mathbf{q}) \frac{\mathbf{k} \cdot \mathbf{q}}{k^2} [S(|\mathbf{k} - \mathbf{q}|) - 1] ,$$ (2.144)

and

$$K(\mathbf{k}, \mathbf{q}) = \frac{\mathbf{k} \cdot \mathbf{q}}{q^2} + \frac{\mathbf{k} \cdot (\mathbf{k} - \mathbf{q})}{|\mathbf{k} - \mathbf{q}|^2}$$ (2.145)

arising from a symmetrized Coulomb interaction.

Equation (2.141a) corresponds to the f-sum rule for an OCP [see Section 3.2B in Volume I]. The third frequency-moment sum rule (2.141b), combined with the response relation (1.53) for an OCP, yields an asymptotic expression for the dynamic local field correction,

$$\lim_{\omega \to \infty} G(k, \omega) = I(k) .$$ (2.146)

In the limit of long wavelengths, it is useful to note the relation

$$\lim_{k \to 0} I(k) = -\frac{4}{15} u_{\text{ex}} \left(\frac{k}{k_{\text{D}}}\right)^2 , \qquad (2.147)$$

where u_{ex} is the interaction energy (2.68) per particle in units of $k_{\text{B}}T$ and

$$k_{\text{D}} = \sqrt{\frac{4\pi n (Ze)^2}{k_{\text{B}}T}} \qquad (2.148)$$

is the Debye–Hückel screening parameter for the OCP.

For a classical OCP, the fifth frequency-moment sum rule is calculated in terms of the pair and triple correlation functions as [Ichimaru et al. 1975]

$$
\frac{\langle \omega^5 \rangle}{\omega_{\text{p}}^4} = \frac{4\pi n^2}{m k_{\text{B}} T}
$$

$$
\times \left(15 \left(\frac{k}{k_{\text{D}}}\right)^6 - 15 \left(\frac{k}{k_{\text{D}}}\right)^4 \int d\mathbf{r} \left[\frac{\mathbf{k}}{k} \cdot \frac{\partial \phi(r)}{\partial \mathbf{r}}\right] \frac{\mathbf{k}}{k} \cdot \frac{\partial g(r)}{\partial \mathbf{r}} \right.
$$

$$
+ 6 \left(\frac{k}{k_{\text{D}}}\right)^3 \int d\mathbf{r}\, g(r) \sin(\mathbf{k} \cdot \mathbf{r}) \left(\frac{\mathbf{k}}{k} \cdot \frac{\partial}{\partial \mathbf{r}}\right)^3 \frac{\phi(r)}{k_{\text{D}}}
$$

$$
+ \frac{2}{n} \left(\frac{k}{k_{\text{D}}}\right)^2 \int d\mathbf{r}\, g(r) \left[1 - \cos(\mathbf{k} \cdot \mathbf{r})\right] \left(\frac{\mathbf{k}}{k} \cdot \frac{\partial}{\partial \mathbf{r}} \frac{\partial \phi(r)}{\partial \mathbf{r}}\right)^2
$$

$$
+ \left(\frac{k}{k_{\text{D}}}\right)^2 \int d\mathbf{r} \int d\mathbf{r}' \left[\frac{\mathbf{k}}{k} \cdot \frac{\partial}{\partial \mathbf{r}} \frac{\mathbf{k}}{k} \cdot \frac{\partial}{\partial \mathbf{r}'} \frac{\phi(r)\phi(r')}{4\pi(Ze)^2}\right]
$$

$$
\times \frac{\partial}{\partial \mathbf{r}} \cdot \frac{\partial}{\partial \mathbf{r}'} \left[1 + \cos(\mathbf{k} \cdot (\mathbf{r} - \mathbf{r}')) - \cos(\mathbf{k} \cdot \mathbf{r}) - \cos(\mathbf{k} \cdot \mathbf{r}')\right]
$$

$$
\left. \times \left[1 + h(r) + h(r') + h\left(|\mathbf{r} - \mathbf{r}'|\right) + h^{(3)}(\mathbf{r}, \mathbf{r}')\right] \right) , \quad (2.149)
$$

where $\phi(r) = (Ze)^2/r$, $h(r)$ and $h^{(3)}(\mathbf{r}, \mathbf{r}')$ refer to the pair and triple correlation functions.

C. The Dynamic Structure Factor in a Generalized Viscoelastic Formalism

In the dielectric formulation of Section 1.2D, the dielectric function of the classical OCP is expressed as

$$\varepsilon(k, \omega) = 1 - \frac{Z^2 v(k) \chi^{(0)}(k, \omega)}{1 + Z^2 v(k) G(k, \omega) \chi^{(0)}(k, \omega)} , \qquad (2.150)$$

where $G(k, \omega)$ refers to the dynamic local field correction and

$$\chi^{(0)}(k, \omega) = -\int d\mathbf{p} \, \frac{1}{\omega - \mathbf{k} \cdot \mathbf{p}/m + i0} \mathbf{k} \cdot \frac{\partial F(\mathbf{p})}{\partial \mathbf{p}} \tag{2.151}$$

is the free-particle polarizability, with $F(\mathbf{p})$ representing the momentum distribution normalized to the average number density,

$$\int d\mathbf{p} \, F(\mathbf{p}) = n \, .$$

The dynamic structure factor can then be calculated via the fluctuation-dissipation theorem as

$$S(k, \omega) = -\frac{k_B T}{\pi Z^2 v(k)\omega} \, \text{Im} \, \frac{1}{\varepsilon(k, \omega)} \, . \tag{2.152}$$

A major physics issue on dynamic correlations in a dense plasma is how the increase in viscosity and a possible formation of a *glassy state* may be accounted for in a formalism with strong Coulomb coupling.

An explicit formulation for a dynamic local field correction capable of reflecting such an effect may be obtained with a set of linearized viscoelastic equations of motion [e.g., Frenkel 1946] for the OCP:

$$\frac{\partial}{\partial t}\delta n(\mathbf{r}, t) + n\frac{\partial}{\partial \mathbf{r}} \cdot \delta \mathbf{u}(\mathbf{r}, t) = 0 \, , \tag{2.153a}$$

$$mn\frac{\partial}{\partial t}\delta \mathbf{u}(\mathbf{r}, t) = -\frac{\partial}{\partial \mathbf{r}}\Pi(\mathbf{r}, t) \, , \tag{2.153b}$$

$$\left[1 + \tau_m\frac{\partial}{\partial t}\right]\left[\frac{\partial}{\partial \mathbf{r}}\Pi(\mathbf{r}, t) - \frac{\partial}{\partial \mathbf{r}}P(\mathbf{r}, t) + Zen\mathbf{E}(\mathbf{r}, t)\right]$$
$$= -\eta\frac{\partial}{\partial \mathbf{r}} \cdot \frac{\partial}{\partial \mathbf{r}}\delta \mathbf{u}(\mathbf{r}, t) - \left(\zeta + \frac{\eta}{3}\right)\frac{\partial}{\partial \mathbf{r}}\frac{\partial}{\partial \mathbf{r}} \cdot \delta \mathbf{u}(\mathbf{r}, t) \, , \tag{2.153c}$$

$$nk_B T\frac{\partial}{\partial t}\delta s(\mathbf{r}, t) = \kappa\frac{\partial}{\partial \mathbf{r}} \cdot \frac{\partial}{\partial \mathbf{r}}\delta T(\mathbf{r}, t) \tag{2.153d}$$

[Tanaka & Ichimaru 1987]. Here the fluctuating quantities—$\delta \mathbf{u}$, Π, P, $\mathbf{E}$, δs, and δT—represent the flow velocity, the isotropic part of the momentum flow tensor, the pressure, the electric field, the entropy, and the temperature, respectively; τ_m, η, ζ, and κ are the viscoelestic relaxation time, the shear viscosity, the bulk viscosity, and the thermal conductivity, respectively.

Equations (2.153a), (2.153b), and (2.153d) represent continuity relations of the particle number, momentum, and energy, respectively; Eq. (2.153c) then describes the effect of viscoelastic relaxation. If τ_m is set equal to zero in

Eq. (2.153c), Eq. (2.153b) would recover the Navier–Stokes equation for an OCP [Vieillefosse & Hansen 1975]. We shall find that viscoelastic relaxation plays the essential role in making the formalism compatible with the compressibility sum rule and the third frequency-moment sum rule.

We consider the isothermal (i.e., $\delta T = 0$) response of the OCP through application of a weak external potential field $\Phi^{\text{ext}}(\mathbf{k}, \omega)$. Fourier components of the electric field fluctuations are then given by

$$\mathbf{E}(\mathbf{k}, \omega) = -i\mathbf{k}\left[\Phi^{\text{ext}}(\mathbf{k}, \omega) + Zev(k)\delta n(\mathbf{k}, \omega)\right] .$$

The dielectric function calculated in the viscoelastic formalism is then expressed as

$$\varepsilon^{\text{VE}}(k, \omega) = 1 - \frac{Z^2 v(k)\chi_0^{\text{VE}}(k, \omega)}{1 + Z^2 v(k)G^{\text{VE}}(k, \omega)\chi_0^{\text{VE}}(k, \omega)} , \qquad (2.154)$$

where

$$\chi_0^{\text{VE}}(k, \omega) = \frac{n}{m}\frac{k^2}{\omega^2 - k^2/m\beta} , \qquad (2.155)$$

$$G^{\text{VE}}(k, \omega) = \frac{k^2}{k_D^2}\left[1 - \beta\left(\frac{\partial P}{\partial n}\right)_T + i\frac{\beta}{n}\omega\eta_l(\omega)\right] \qquad (2.156)$$

with

$$\eta_l(\omega) = \frac{(4/3)\eta + \zeta}{1 - i\omega\tau_m} \qquad (2.157)$$

The functions $\chi_0^{\text{VE}}(k, \omega)$ and $G^{\text{VE}}(k, \omega)$ are the free-particle polarizability and the dynamic local field correction in the viscoelastic formalism. These descriptions may be applicable in the low-frequency and long-wavelength regime.

We may now extend the viscoelastic dielectric function (2.154) into a finite k and ω regime, by making $\eta_l(\omega)$ and τ_m functions of k as well. First, a straightforward generalization of the free-particle polarizability is performed by setting

$$\chi_0^{\text{VE}}(k, \omega) \rightarrow \chi^{(0)}(\mathbf{k}, \omega) .$$

The static local field correction with the generalized compressibility sum rule (1.89) should satisfy the Ornstein–Zernike relation (2.41), so that

$$G^{\text{VE}}(k, 0) \rightarrow G(k) = 1 + \frac{k^2}{k_D^2}\left[1 - \frac{1}{S(k)}\right] . \qquad (2.158)$$

The third frequency-moment sum rule then sets the boundary condition (2.146), so that

$$G^{VE}(k, \omega) \to G(k, \omega) = \frac{G(k) - i\omega\tau_m I(k)}{1 - i\omega\tau_m} \, . \qquad (2.159)$$

The dynamic local field correction as expressed in Eq. (2.159) has a form in which the low-frequency limit $G(k)$ and the high-frequency limit $I(k)$ are interpolated through the generalized viscoelastic relaxation time; $\tau_m(k)$ plays the role of a characteristic time distinguishing between the low- and high-frequency regimes.

For a hydrodynamic mode, where only low-frequency and long-wavelength excitations are considered, the first term on the right-hand side of Eq. (2.150), unity, is negligible, since $v(k)\chi^{(0)}(k, \omega) \gg 1$. We thus find a *quasielastic peak* in the dynamic structure factor (2.152) as

$$S(k, \omega) \cong \frac{n}{\pi} \left(\frac{k}{k_D} \right)^2 [G(k) - I(k)] \frac{\tau_m(k)}{1 + [\omega\tau_m(k)]^2} \, , \qquad (2.160)$$

which in the limit of long relaxation times takes on the values

$$S(k, \omega) \underset{(\tau_m \to \infty)}{\longrightarrow} n \left(\frac{k}{k_D} \right)^2 [G(k) - I(k)] \delta(\omega) \, . \qquad (2.161)$$

This limit thus corresponds to a *frozen* state [Bengtzelius, Götze, & Sjölander 1984; Leutheusser 1984; Kirkpatrick 1985] in which the local structure does not vanish even after an infinite time or the expectation value $\langle \delta n(k, t) \delta n^*(k, 0) \rangle$ stays finite for $t \to \infty$; these features point to a dynamic transition to a glassy state [Tanaka & Ichimaru 1987].

In the limit of long wavelengths, one has

$$\tau_m(0) = \frac{\beta}{n} \frac{\eta_l(0)}{1 - \beta \left(\frac{\partial P}{\partial n} \right)_T + \frac{4}{15} u_{ex}} \, . \qquad (2.162)$$

The viscoelastic relaxation time $\tau_m(k)$ has been introduced to remove a certain rigidity inherent in the Navier–Stokes equation with regard to temporal response of the internal energy against the viscous motion of the fluid. Without the relaxation effect, the viscous motion cannot be made compatible with the third frequency-moment sum rule or with the quasielastic peak.

Tanaka and Ichimaru [1987] investigated in detail the dynamic structure factors with inclusion of evaluation for the viscosity, the possibility of *glass transition*, and comparison with MD simulation results [Hansen, McDonald, & Pollock 1975] in dense plasmas. In their investigation, the generalized visco-

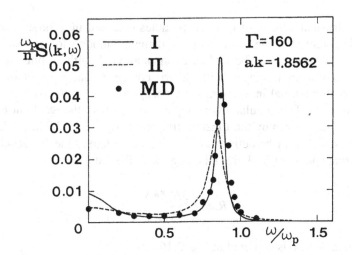

Figure 2.15 Dynamic structure factor for *ak* = 1.8562 at Γ = 160. I refers to the calculation with Eq. (2.163a) and ξ_G = 2.7 ; II, with Eq. (2.163b) and ξ_L = 1.25; MD, the MD simulation data.

elastic relaxation time was assumed to take a Gaussian

$$\tau_m(k) = \tau_m(0)\exp[-(ak/\xi_G)^2] \qquad (2.163a)$$

or a Lorentzian

$$\tau_m(k) = \frac{\tau_m(0)}{\left[1 + (ak/\xi_L)^2\right]} . \qquad (2.163b)$$

Figure 2.15 illustrates such a comparison with the MD data.

2.5   ORDERED STRUCTURES

In the limit of strong coupling, plasmas undergo phase transitions such as a freezing transition; ordered structures develop in such a plasma. Microscopic features associated with various possibilities in the resultant ordered structures are studied in this section.

A. Coulomb Clusters

Ground-state properties such as binding energies and stable structures in mesoscopic systems have been investigated in various materials, including small aggregates of rare-gas atoms [Farges et al. 1983], microclusters of simple metals [de Heer et al. 1987], and pure ions trapped in electromagnetic fields [Wuerker, Shelton, & Langmuire 1959; Diedrich et al. 1987; Bollinger et al. 1990].

The classical system of charged particles confined in a three-dimensional harmonic potential is one of the most fundamental models for mesoscopic systems. Such a system may be looked upon as an OCP with a finite size, where N particles of a single species with electric charge Ze and mass M interact via the Coulomb potential in a neutralizing background of uniform charge-density distribution $-\rho_e$. The angular frequency associated with the resultant harmonic potential is then given by the Einstein frequency ω_E in Eq. (2.94a). A Coulomb cluster of size N may be defined as a combined system of the N particles and a surrounding sphere of background charge with the radius

$$R = \left(\frac{3NZe}{4\pi\rho_e}\right)^{1/3}.$$ (2.164)

A cluster at $N = 6$ is depicted in Fig. 2.16.

Microscopic ordering in the mesoscopic systems of charged particles, such as a formation of spherical shell structures, has been observed in laboratory experiments [Wuerker, Shelton, & Langmuire 1959; Diedrich et al. 1987; Bollinger et al. 1990]. A theoretical study of the Coulomb clusters by Rafac et al. [1991] has shown that the shell structures of particles with $N \le 27$ in the ground state are distinct from the microstructures of the infinite OCP in a crystalline phase. The origin of such a distinction may be traced to the difference in the symmetry structures: In a computer simulation study of an infinite system, one may adopt the periodic boundary conditions to approach the thermodynamic limit with cubic symmetry. The Coulomb cluster, on the other hand, has its origin of spherical symmetry due to the construction of the background charge sphere.

For elucidation of these points, Tsuruta and Ichimaru [1993] performed computer simulation as well as analytic studies for binding energies and microscopic shell structures of the Coulomb clusters. The binding energy is found to take on maxima as a function of the cluster size when the particle configuration in shells

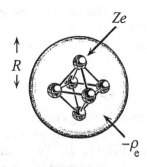

Figure 2.16 Coulomb cluster of size $N = 6$ with the octahedral configuration.

satisfies a certain local symmetry characteristic of a finite, spherical system; these maxima define the *magic numbers* of the Coulomb clusters.

The MD simulations were carried out with a standard Verlet's algorithm [Verlet 1967] and calculated the electrostatic energy of the cluster in its ground state (i.e., $T = 0$), defined as

$$E(N) = \frac{1}{2} \int d\mathbf{r} \int d\mathbf{r}' \frac{1}{|\mathbf{r} - \mathbf{r}'|} \left[\rho(\mathbf{r})\rho(\mathbf{r}') - (Ze)^2 \sum_{j=1}^{N} \delta(\mathbf{r} - r_j)\delta(\mathbf{r}' - r_j) \right].$$

(2.165)

Here

$$\rho(\mathbf{r}) = Ze \sum_{j=1}^{N} \delta(\mathbf{r} - \mathbf{r}_j) - \rho_e \theta(R - r)$$

(2.166)

refers to density distribution of the electric charges and

$$\theta(x) = \begin{cases} 1, & x \geq 0, \\ 0, & x < 0, \end{cases}$$

(2.167)

is the unit step function. The second term in the integrand of Eq. (2.165) removes the self-interaction contributions contained in the first term.

In Fig. 2.17, the ground-state values of $E(N)$ per particle calculated by the

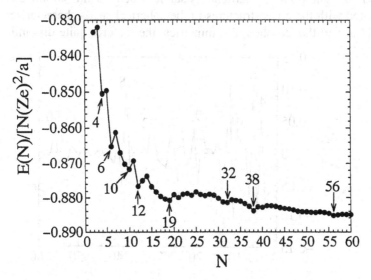

Figure 2.17 Electrostatic energy per particle of N-particle Coulomb cluster in the ground-state configuration.

MD simulations to the accuracy on the order of $10^{-4} \times (Ze)^2/a$ are shown for $2 \leq N \leq 60$. We observe that local minima of $E(N)/N$ exist at some values of N, indicating the possibility of stable cluster structures.

In characterization stability of such clusters, it is useful to introduce the *relative binding energy* defined by [de Heer 1987]

$$\Delta_2(N) = \frac{[E(N+1) + E(N-1) - 2E(N)]a}{(Ze)^2} . \qquad (2.168)$$

Physically this quantity represents the amount of binding energy gained by a formation of two N-particle clusters out of coalescence between $(N-1)$- and $(N+1)$-particle clusters. The relative binding energy is therefore a quantity more relevant for the stability analysis of an N-particle cluster than the value of $E(N)$ itself.

Figure 2.18 exhibits the values of the relative binding energy computed from the data in Fig. 2.17. Very conspicuous peaks are notable at $N = 6$, 12, and 38 (magic numbers), with relative binding energies as large as $0.1 \times (Ze)^2/a$. These, respectively, have regular structures of *octahedron, icosahedron*, and a double shell consisting of octahedron (inner) and *face-centered icosahedron* (outer) illustrated in Figs. 2.16 and 2.19. In addition, local minima in Fig. 2.17 and peaks in Fig. 2.18 coincide at $N = 4$, 10, 19, 32, and 56. Structures and binding energies of those stable clusters are listed in Table 2.2. We thus find in the magic-number configurations that particularly stable structures of Coulomb clusters are associated with the microstructures in the spherical shells characterized by the octahedral and the icosahedral symmetries, the features quite dissimilar to the

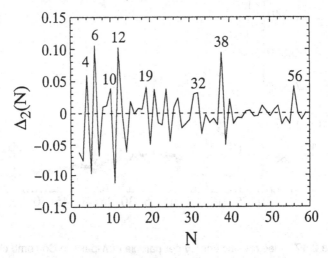

Figure 2.18 Relative binding energies (Eq. 2.168) of Coulomb clusters.

Table 2.2   Structures and binding energies [in units of $(Ze)^2/a$] of stable Coulomb clusters

N	Structure	$E(N)/N$	$\Delta_2(N)$
4	tetrahedron	–0.85042(3)	0.0596(1)
6	octahedron	–0.86518(5)	0.1052(3)
10	2-capped square skew prism	–0.87163(4)	0.0390(4)
12	icosahedron	–0.87664(1)	0.1024(1)
19	1(center) + 18-particle polyhedron	–0.88082(7)	0.0407(7)
32	tetrahedron + 28-particle polyhedron	–0.88145(7)	0.0318(9)
38	octahedron + face-centered icosahedron	–0.88391(6)	0.0947(8)
56	icosahedron + 44-particle polyhedron	–0.88527(7)	0.0416(4)

Note: Numbers in parentheses indicate the extent of errors in the last digits.

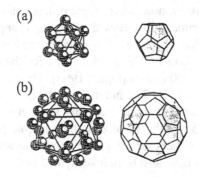

Figure 2.19   (a) Icosahedral and (b) face-centered icosahedral configurations (left) and the associated polyhedra (right) constructed by the Voronoi polygons. A Voronoi polygon is formed around each particle by the lines equally bisecting nearest-neighbor pairs on a spherical shell.

symmetries in infinite crystals listed in Table 2.1. Such large binding energies for clusters in Table 2.2 may be useful in a formulation of a cluster theory of dense Coulomb liquids.

B. The Evolution of Microstructures at Freezing Transition

Solidification such as crystallization and glass transition is one of the most interesting events in the thermal evolution of a many-body system. The cumulative effort by many investigators notwithstanding, evolution of the microscopic features associated with such a transition has remained an outstanding problem to be elucidated. Theoretical treatment is difficult, since the transition is chaotic in character; it is not easy to trace the dynamic evolution of such microscopic structures in laboratory experiments.

Computer simulation study of many-particle systems has a long history: In the late fifties [Alder & Wainwright 1957, 1959], methods of MD simulation were first applied to the hard-core systems. Later, these and MC methods were extended to other cases of the potentials, such as soft core and Lennard–Jones [e.g., Hansen & McDonald 1986]. The available size of the simulations has rapidly increased as computer capabilities have developed. Simulation studies of crystallization and glass transition for these systems have contributed much to clarification of microscopic properties in liquids and solids.

The equations of state for the OCP have been calculated accurately both for fluids and for solids in crystalline phases (see Sections 2.1 and 2.3). Comparing the Helmholtz free energies between the fluid and crystalline phases, it has been found that an OCP fluid freezes (Wigner transition) into a bcc crystal at $\Gamma = 172$–180. It is not clear, however, how microscopic orders emerge and develop when a rapid quench into a temperature below the freezing is applied to an OCP.

In this subsection, findings in the MC simulation studies of freezing processes in such a rapidly supercooled OCP are described for the case of quench from $\Gamma = 160$ (fluid) to $\Gamma = 300$ (solid) [Ogata 1992]. The number of MC particles in the simulation is $N = 1458$. The number of MC configurations generated is denoted by c; application of stepwise quenches starts at $c/N = 0$, and $\Gamma = 300$ is reached at $c/N = 2.5 \times 10^4$. The total number of configurations generated is approximately 4×10^8. Other cases of quenches, such as with $N = 432$ and/or to $\Gamma = 200, 400$, and 800, have also been investigated in detail [Ogata & Ichimaru 1989; Ogata 1992].

1. Excess Internal Energy

Excess internal energy in Eq. (2.71) is a primary quantity in characterizing the state of an OCP; evolution of the excess internal energy per particle averaged over a sequence of $\Delta c/N = 70$ is followed in Fig. 2.20 after $\Gamma = 300$ has been reached. Five subsequent stages (a)–(e) are also defined in Fig. 2.20; stage (c) corresponds to the onset of a transient in which the excess internal energy abruptly decreases.

In Fig. 2.20, the upper dashed line corresponds to an extrapolation to $\Gamma = 300$ of the fluid excess energy value due to Slattery, Doolen, and DeWitt [1982], and the lower dashed line is Eq. (2.74) due to Ogata and Ichimaru [1987]. The dot-dashed line indicates the bcc crystalline value, Eq. (2.96). It should be noted that

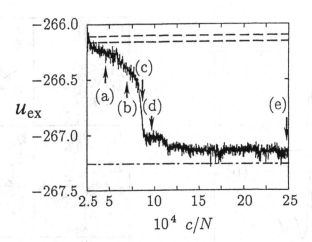

Figure 2.20 Evolution of the normalized excess internal energy for the quench to $\Gamma = 300$. Stages are specified at (a) 4.6, (b) 7.0, (c) 8.6, (d) 9.5, and (e) 24.9.

the fcc crystalline value of u_{ex} [Helfer, McCrory, & Van Horn 1984] is higher than the bcc value only by 0.01 for $\Gamma = 300$.

After the quench to $\Gamma = 300$, u_{ex} stays around the fluid extrapolation value for $(c/N) \times 10^{-4} = 2.5$–6.0. It gradually decreases as c increases for $(c/N) \times 10^{-4} = 6.0$–8.0, and decreases abruptly at $(c/N) \times 10^{-4} = 8.0$–9.0 and at 11.0–12.0; after $(c/N) \times 10^{-4} \approx 12.0$, it assumes metastable states, though jitters are observed. The deviation of u_{ex} from the bcc crystalline value is 0.08 at the metastable plateau.

2. Radial Distribution Function

Evolution of $g(r) = \langle g(r) \rangle$ averaged over $\Delta(c/N) \times 10^{-4} = 0.21$ is displayed in Fig. 2.21; $g(r)$ exhibits a smooth feature analogous to that in a fluid simulation (see Fig. 2.1) at stages (a) and (b) before the abrupt decrease in u_{ex}. At stage (c), the second and third peaks of $g(r)$ begin acquiring shoulders at the radii corresponding to bcc peaks and thus exhibit a substantial degree of bcc-like microscopic structures. No significant changes are found in $g(r)$ from (c) to (d). At stage (e), $g(r)$ appears quite similar to the one with the bcc crystalline simulation at $\Gamma = 300$, in terms of the comparison with the positions and heights of the peaks in $g(r)$. A clear difference to be remarked here, however, is the fact that the value of $g(r)$ in (e) at the first valley remains finite ($\neq 0$) while it is zero in the bcc-crystalline simulation; this points to existence of imperfections in the microscopic structures attained in the stage (e).

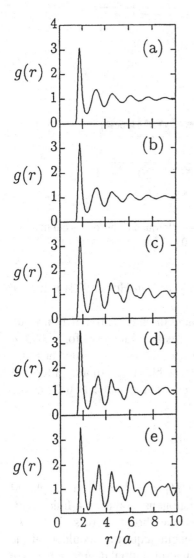

Figure 2.21 Evolution of the radial distribution function.

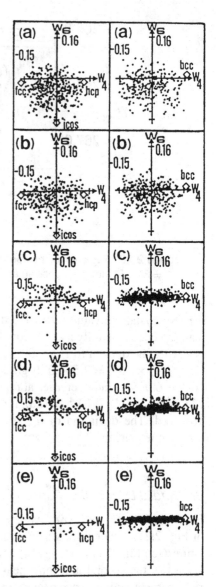

Figure 2.22 Evolution of two-dimensional (W_4, W_6) maps (left) particles with $N_c = 12$; (right) particles with $N_c = 14$.

3. *Local Bond-Orientational Order*

Microscopic orders in the particle configurations are described in terms of the orientational correlations between *bonds*, which are the connecting lines between a particle and its "neighboring particles." We define the neighboring particles as particles inside a sphere of radius $r/a = 2.3$ around a given particle; particle positions are averaged over a sequence of $\Delta(c/N) \times 10^{-4} = 0.069$ to reduce the thermal fluctuations. The radius $r/a = 2.3$ corresponds approximately to the radial position of the first valley in the radial distribution functions both for the fluid phase near the freezing condition and for the bcc crystalline phase.

Number N_c of the neighboring particles, which we shall call the *coordination number*, is 14 for the bcc cluster and 12 for the fcc, hcp, and icosahedral clusters. In a fluid simulation at $\Gamma = 160$, $N_c = 12$–14 for almost all the particles; in a bcc crystalline simulation, $N_c = 14$ for all the particles.

A set of quantities $\{Q_{lm}(\mathbf{r})\}$ is attached to each bond defined via the spherical harmonics,

$$Q_{lm}(\mathbf{r}) = Y_{lm}[\theta(\mathbf{r}), \phi(\mathbf{r})] , \qquad (2.169)$$

where $\mathbf{r}$ is the central position of a bond and $\theta(\mathbf{r})$ and $\phi(\mathbf{r})$ are its polar angles.

The local *bond-orientational order parameters* [Steinhardt, Nelson, & Ronchetti 1983; Nelson & Spaepen 1989], which are rotationally invariant combinations in the second and the third order, are introduced via

$$Q_l = \left[\frac{4\pi}{2l+1} \sum_{|m|\leq l} |\langle Q_{lm}(\mathbf{r})\rangle|^2 \right]^{1/2} , \qquad (2.170)$$

$$W_l = \sum_{m_1+m_2+m_3=0} \begin{bmatrix} l & l & l \\ m_1 & m_2 & m_3 \end{bmatrix} \frac{\overline{Q}_{lm_1}(\mathbf{r})\overline{Q}_{lm_2}(\mathbf{r})\overline{Q}_{lm_3}(\mathbf{r})}{\left[\sum_{|m|\leq l} |\overline{Q}_{lm}(\mathbf{r})|^2 \right]^{3/2}} . \qquad (2.171)$$

The coefficients in Eq. (2.171) are the Wigner $3j$ symbols [e.g., Landau & Lifshitz 1976]. The average $\overline{Q}_{lm}(\mathbf{r})$ in Eq. (2.171) is carried out with regard to all the bonds around a given particle; $\langle \cdots \rangle$ in Eq. (2.170) means an analogous average with respect to such bonds over all the MC particles. Quantities Q_l and W_l play the key roles in the cluster "shape spectroscopy" in fluids and solids [Steinhardt, Nelson, & Ronchetti 1983].

Since Q_4 assumes a first nonvanishing value (other than Q_0) in samples with cubic symmetry and Q_6 does so in icosahedral systems, those are quantities of substantial significance. The quantities (Q_4, Q_6) take on values $(0.1909, 0.5745)$ for the fcc, $(0.0972, 0.4848)$ for the hcp, $(0, 0.6633)$ for the icosahedral, and

(0.0364, 0.5107) for the bcc clusters. In the fluid simulation at $\Gamma = 160$, (Q_4, Q_6) have been found to take on small values (0.01, 0.03), while in the bcc crystalline simulation these are (0.04, 0.5).

The quantities (W_4, W_6) assume significantly different values between the clusters: $(W_4, W_6) = (-0.1593, -0.0132)$ for the fcc, (0.1341, -0.0124) for the hcp, $(-, -0.1698)$ for the icosahedral, and (0.1593, 0.0132) for the bcc clusters. We remark that W_4 is not a well-defined quantity for the icosahedron, since $Q_4 = 0$. The magnitude of W_6 is substantially larger for the icosahedron than for the other three types of clusters. The local bond-orientational symmetries around a particle can therefore be discerned through its location on the two-dimensional (W_4, W_6) map.

Evolutions of the two-dimensional (W_4, W_6) maps for clusters with $N_c = 12$ and 14 are plotted in Fig. 2.22, on p. 86 for the quench to $\Gamma = 300$. In the figure, the (W_4, W_6) values for the reference clusters are likewise displayed by the diamond markers. ($W_4 = 0$ has been arbitrarily taken for the icosahedron.)

At stages (a) and (b), (W_4, W_6) are distributed almost uniformly in the domain $|W_4| \leq 0.15$ and $|W_6| \leq 0.16$; this is a typical fluid behavior. At stage (c), a substantial fraction of clusters with $N_c = 14$ begins to exhibit local bcc symmetry; distribution of these clusters resembles that in the bcc crystalline simulation. At stage (d), distribution of clusters with $N_c = 14$ is centered at around $W_6 \approx 0.013$, the bcc value, concurrent with the decrease in the number of clusters with $N_c = 12$. At a subsequent stage (e), an increased degree of local bcc symmetry in the distribution of clusters with $N_c = 14$ is observed.

4. Extended Bond-Orientational Order

Extended bond-orientational symmetries are studied in terms of the correlation functions [Steinhardt, Nelson, & Ronchetti 1983]

$$G_l(r) = \frac{4\pi}{(2l+1)G_0(r)} \sum_{|m| \leq l} \langle Q_{lm}(\mathbf{r}) Q_{l-m}(0) \rangle \ , \qquad (2.172)$$

where $G_0(r) = 4\pi \langle Q_{00}(\mathbf{r}) Q_{00}(0) \rangle$. $G_l(r)$ take on first nonvanishing values at $l = 6$ for the bcc structure. In a fluid simulation, one finds $G_6(r) \approx 0$, indicating absence of an extended order. In a bcc crystalline simulation, $G_6(r) \approx 0.3$ extending over distances of the MC simulation cell, confirming persistence of a long-ranged bond-orientational order.

Evolution of the extended bond-orientational symmetries are displayed in terms of $G_6(r)$ in Fig. 2.23. At stage (a), no extended order is observed. At stage (b), bond correlations extend themselves approximately two-thirds of L, the lattice constant of the MC cell. At stage (c), $G_6(r) \approx 0.1$ over the entire MC cell; hence the system exhibits a degree of long-range orientational order. Final values of $G_6(r)$ are about two-thirds of the bcc value.

5. Layered Structures

One of the main objectives in these simulation studies has been a possible formation of layered structures characteristic of infinite crystalline lattices and their relation to the periodic boundary conditions imposed in the MC simulations. In connection with such an issue, two-dimensional projection maps of the system observed from various directions have been constructed. Those particles inside a sphere of radius $L/2$ are isolated and are rotated as a whole by angles around y- and z-axes; the resulting configurations are projected onto the y-z plane.

Figure 2.24 shows the projection maps so constructed at stages (a)–(e), observed from the same direction. Particle layers start to nucleate at stage (b) over a half of the sphere; nearly perfect layers are found already at stage (c).

If a formation of such layered structures had anything to do with the periodic boundary conditions in the MC cell, orientations of the emerged layers, or the angles of rotations, would have definite relations to the cubic symmetry of the cell. No such relations have been found in any simulation where quenches have been studied.

Bond angle distributions and two-dimensional radial-distribution functions in the resultant layers have been investigated in various cases of the simulated quenches [Ogata & Ichimaru 1989; Ogata 1992].

C. Shear Moduli of Coulomb Solids

A tensor of nonvanishing shear moduli is one of the principal features by which solids are distinguished from fluids. McDermott et al. [1985] analyzed nonradial oscillations of neutron stars, modeled as three-component stars consisting of fluid interior, solid crust, and fluid envelope. Novel aspects of the analysis include a prediction of the bulk and interfacial modes, associated with the nonvanishing shear moduli of Coulombic crustal solids.

Shear moduli for a Coulomb solid with cubic symmetry at $T = 0$ have been known for quite some time [e.g., Mott & Jones 1936]; these are called Fuchs [1936] values and are closely related to the Madelung energies for Coulomb crystals. For an application to neutron crust problems, however, the values of shear moduli at finite temperatures are called for. One then employs the Metropolis algorithm of MC simulation for the first principle calculations of shear moduli at finite temperatures. The free-energy increments stemming from virtual deformations of a Coulomb solid created by the MC method are evaluated by MC samplings of the relevant Ewald sums [Ogata & Ichimaru 1990].

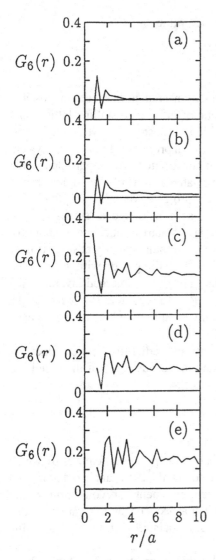

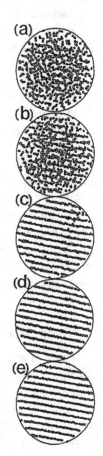

Figure 2.23 Evolution of the bond-orientational correlation function $G_6(r)$.

Figure 2.24 Evolution of layered structures.

The free-energy increment δF arising from strain u_{ij} is expressed as

$$\delta F = \frac{1}{2}S_{ij,kl}u_{ij}u_{kl} \tag{2.173}$$

[e.g., Landau & Lifshitz 1970]. Here $S_{ij,kl}$ is the elastic modulus tensor, the subscripts i, j, k, and l designate the Cartesian components x, y, and z, and the summation convention for repeated subscripts is adopted. For an isotropic body, Eq. (2.173) reduces to

$$\delta F = \frac{1}{2}\lambda u_{ii}^2 + \mu u_{ik}u_{ik} . \tag{2.174}$$

The quantities λ and μ are the *Lamé coefficients*.

The usual elastic constants [e.g., Mott & Jones 1936]—$c_{rs}(r, s = 1, 2, \ldots, 6)$—are derived from the elastic modulus tensor through the transformation $(ij, kl) \rightarrow (r, s)$ of the subscripts $(xx, yy, zz, xy, yz, zx) \rightarrow (1, 2, 3, 4, 5, 6)$, such that

$$c_{rs} = S_{ij,kl}. \tag{2.175}$$

For a solid with cubic symmetry, only three elastic constants remain: $c_{11} = c_{22} = c_{33}$, $c_{12} = c_{21} = c_{23} = c_{32} = c_{31} = c_{13}$, and $c_{44} = c_{55} = c_{66}$. When such a solid is deformed without a change in the volume (i.e., $\sum_i u_{ii} = 0$), one finds

$$\delta F = \frac{1}{2}(c_{11} - c_{12})u_{ii}^2 + c_{44}u_{ik}u_{ik} \quad (i \neq k) . \tag{2.176}$$

The first term on the right-hand side represents a differential between two compressional deformations.

If cubic symmetry is destroyed, as in the case of quenched solids, more independent elements should appear. In conjunction with the first term on the right-hand side of Eq. (2.176), it is useful to define and introduce

$$b_{11} = \frac{2c_{11} - c_{12} - c_{31}}{4} ,$$

$$b_{22} = \frac{2c_{22} - c_{23} - c_{12}}{4} , \tag{2.177}$$

$$b_{33} = \frac{2c_{33} - c_{31} - c_{23}}{4} .$$

The shear modulus tensor is then represented by the elements b_{11}, b_{22}, b_{33}, c_{44}, c_{55}, and c_{66}. For an isotropic body, all these elements take on the value that coincides with the shear modulus μ in Eq. (2.174).

Table 2.3 Elements of shear modulus tensor [in units of $n(Ze)^2/a$].

Γ	Case	b_{11}	b_{22}	b_{133}	c_{44}	c_{55}	c_{44}
∞	bcc	0.02454	0.02454	0.02454	0.1827	0.1827	0.1827
∞	fcc	0.02066	0.02066	0.02066	0.1852	0.1852	0.1852
800	bcc	0.024(2)	0.024(2)	0.024(2)	0.174(1)	0.174(1)	0.174(1)
800	Q	0.053(2)	$-0.007(1)$	0.057(1)	0.181(2)	0.171(2)	0.133(1)
400	bcc	0.025(2)	0.025(2)	0.025(2)	0.167(1)	0.167(1)	0.167(1)
400	Q	0.059(3)	$-0.009(3)$	0.062(3)	0.170(2)	0.169(1)	0.121(4)
300	bcc	0.025(2)	0.025(2)	0.025(2)	0.157(4)	0.157(4)	0.157(4)
300	Q	0.053(1)	0.025(2)	0.014(3)	0.141(3)	0.167(2)	0.149(3)
200	bcc	0.019(3)	0.019(3)	0.019(3)	0.12(1)	0.12(1)	0.12(1)

Note: In the "Case" column, bcc and fcc denote the crystalline solids and Q, a quenched solid with imperfections. Numbers in parentheses indicate the extents of errors in the last digits.

These relations can be used for explicit evaluations of the elements of shear-modulus tensor for various Coulomb solids produced by MC simulations. The results are listed in Table 2.3; significant dependence of the shear moduli on the temperature is observable. The bcc and fcc values of the shear modulus at $\Gamma = \infty$ (i.e., $T = 0$) are identical to the historic Fuchs values. Those temperature-dependent shear moduli have been applied for the analyses of nonradial oscillations of neutron stars [Strohmayer et al. 1991].

D. Fluid-Solid Phase Boundaries

As an application of the evaluations (2.78) and (2.104) for the equations of state of OCP fluids and solids, we examine the fluid–solid transition in dense carbon OCP materials—relevant to the interiors of white-dwarf stars—through comparison of the free energies. It is expected that a solid core may be formed during cooling of such a degenerate star.

The quantum effects of ions have been taken into account through the use of Eqs. (2.108) and (2.110) in Eq. (2.104) for solids. For the free energy (2.78) of a fluid OCP, on the other hand, we use the Wigner–Kirkwood expansion up to the terms on the order of $\hbar^4$; this treatment is basically semiclassical. Hence only that part of melting caused by the thermal motion of particles can be considered.

The melting curves are constructed with the OCP fluid equation of state at different degrees of quantum corrections: the zero-order expression (WK0) with $f^{(0)}$ alone, the first-order expression (WK1) with $f^{(0)} + f^{(1)}$, and the second-order expression (WK2) with $f^{(0)} + f^{(1)} + f^{(2)}$. The result so calculated for

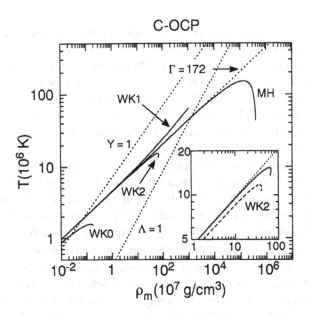

Figure 2.25 Melting curves for a carbon OCP in various schemes—WK0, WK1, WK2, and MH—explained in the text. The inset magnifies the departure of the melting curve from the classical condition (dotted line) based on WK2 (solid curve); the dashed curve corresponds to the melting curve obtained without the anharmonic term in the solid free energy.

carbon plasmas is shown in Fig. 2.25; the melting line $\Gamma = 172$ for the classical OCP and the lines, $\Lambda = 1$ and $Y = 1$, measuring the extents of quantum effects involved are also drawn for comparison. These calculations clearly demonstrate sensitive dependence of the transition curve on the treatment of ionic quantum effects.

The WK0 calculation implies that the ionic quantum effects in the solid phase suppress its stability. Inclusion of the leading quantum correction Eq. (2.80a) to the free energy acts to enlarge the domain of the solid phase to an extent exceeding a purely classical prediction, as the WK1 curve illustrates.

The second-order quantum correction Eq. (2.80b), included in the WK2 scheme, is negative definite and hence tends to destabilize the solid phase. The departure of the resultant transition curve from the classical one is magnified in the inset of Fig. 2.25. Since the leading quantum corrections are identical in both phases, the behavior of the melting curve near the classical regime is determined by a delicate balance between the quantum corrections of the second- and higher-order in the two phases. Anharmonic effects on the melting are assessed by repeating the WK2 calculations and neglecting the anharmonic term; the result is shown in the inset of Fig. 2.25 by a dashed line. Anharmonicity enhances stability of the solid phase by 10% to 20% in temperature. The WK2 calculation represents the most accurate result in these treatments. The strong bending observed in the melting curves based on the WK2 scheme implies a breakdown of the approximations in the fluid equation of state.

This result contrasts with that of Mochkovitch and Hansen (MH) [1979] based on a generalized (Y-dependent) Lindemann criterion, which led to the conclusion that the melting was completely classical below $\rho_m = 4 \times 10^{11} g/cm^3$. Figure 2.25 incorporates the transition curve calculated in the MH scheme but with fitting parameters so modified as to reproduce the classical melting condition, $\Gamma = 172$. This modification decreases the critical mass density at which the quantum effects emerges to $5 \times 10^{10} g/cm^3$, which is still considerably larger than the indication of WK2. The considerable difference between the two predictions may be ascribed to a possible improper treatment of the $\hbar^4$-correction term in the interpolation formula of the Lindemann parameter in the MH scheme. The average mass density ρ_m in carbon white dwarfs usually takes on a value in the range 10^5–$10^8 g/cm^3$. For an accreting white dwarf, which is regarded as a progenitor of type I supernova, the relevant range of the mass density is expended to $\rho_m = 10^{10} g/cm^3$ [Nomoto 1982]. According to the WK2 calculation, the quantum effects begin to change the classical melting line beyond $\rho_m = 2 \times 10^8 g/cm^3$ in the carbon plasma. Evolution of the accreting white dwarfs and a mechanism of the subsequent explosions may be influenced significantly by the quantum effects, as reflected in the melting.

E. Phase Diagrams of Dense Binary-Ionic Mixtures

Phase diagrams for the BIMs under the condition of rigid uniform background charges may be obtained by comparing the Helmholtz free energies for the fluid and solid phases derived in Sections 2.3B and 2.3C. For applications to the real physical cases, including the interiors of white dwarfs, the calculations are then extended to the cases of constant pressure that include the compressibility of the relativistically degenerate electrons [Ogata et al. 1993].

Phase diagrams for BIMs are thus determined with the Helmholtz free energies, Eqs. (2.89) and (2.100). Examples of the phase diagrams so constructed are shown in Figs. 2.26 (a)–(d) for those BIMs with $R_Z = 4/3$, $3/2$, $5/3$, and $13/3$. The quantity T_m in the figures is the freezing temperature for an OCP of the "1" particles given by

$$T_m(K) = 9.5 \times 10^6 \left(\frac{12}{A}\right)^{1/3} \left(\frac{Z}{6}\right)^2 \rho_8^{1/3}, \qquad (2.178)$$

where

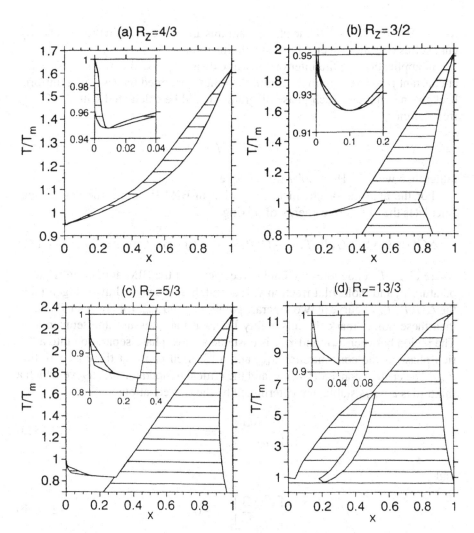

Figure 2.26 Phase diagrams for BIMs at constant, uniform background-charge density.

$$\rho_8 \equiv \frac{\rho_m}{10^8 \text{g/cm}^3} . \tag{2.179}$$

Dependence of the phase diagrams on the charge ratio R_Z can be summarized as follows: For BIMs with $R_Z < 1.4$, the diagrams are of the *azeotropic* type and no chemical separations take place at solidification except for $x < 0.01$. Phase diagrams for $R_Z \approx 1.5$ have composite structures containing features of both the azeotropic and the *eutectic* diagrams: For $x > 0.4$, a "2" rich solid with $x = 0.8$ to 0.9 is produced at solidification, while a mixed solid emerges from the fluid for $x < 0.4$. For $R_Z > 1.6$, the phase diagrams are of the eutectic type and nearly pure solids of either species form at solidification.

In applications to the interiors of white dwarfs, phase diagrams constructed at constant pressures are more useful than those obtained by assuming uniform background charges. Such phase diagrams should be calculated with the Gibbs free energies,

$$G = F + PV, \tag{2.180}$$

rather than with the Helmholtz free energies.

For the analysis of chemical separation in BIM systems, one defines and calculates the Gibbs free energy of mixing,

$$\Delta G(P, T, x) = G(P, T, x) - G(P, T, x = 0) - G(P, T, x = 1), \tag{2.181}$$

where $G(P, T, x)$ denotes the Gibbs free energy of the BIM at pressure P, temperature T, and molecular fraction x. If a straight line can be drawn tangential to the $\Delta G(P, T, x)$ curve at two separate points x_{min} and x_{max} for constant P and T, then these points mark the miscibility limits at that pressure and temperature. The region between x_{min} and x_{max} is unstable against phase separation into a "1" rich phase at the concentration x_{min} and a "2" rich phase at the concentration x_{max}. The critical concentration x_c and the critical temperature T_c, above which a mixture is always stable, are determined by the two equations:

$$\left[\frac{\partial^2 \Delta G}{\partial x^2}\right]_{P,T} = 0 \tag{2.182a}$$

and

$$\left[\frac{\partial^3 \Delta G}{\partial x^3}\right]_{P,T} = 0. \tag{2.182b}$$

The condition (2.182a) itself gives an instability (spinodal) curve, which always lies inside the coexistence curve determined from the double-tangent construction.

For the mass densities $\rho_8 \approx 10^{-1}$ to 10^2 and temperatures $T \approx 10^6$ to 10^8 K, typical parameter ranges for the white dwarfs, the electrons are in the ground state, since the ratio of the thermal energy to the electron Fermi energy (see Eq. 1.8),

$$\Theta = 1.688 \times 10^{-10} T \left[\sqrt{1 + 21.94(\rho_8/\mu_e)^{2/3}} - 1\right]^{-1}, \tag{2.183}$$

is extremely small. Here we measure the number density of electrons by the r_s parameter (see Eq. 1.2)

$$r_s = 2.991 \times 10^{-3} (\rho_8/\mu_e)^{-1/3} , \tag{2.184a}$$

and

$$\mu_e = \frac{\sum_i x_i A_i}{\sum_i x_i Z_i} , \tag{2.184b}$$

is the mean molecular weight per electron.

The dominant contribution to the pressure in a white dwarf stems from the relativistically degenerate electrons. The partial pressures coming from the ions may be treated as perturbations in comparison with the electron pressure. The Helmholtz free energy is thus expressed as the sum of the ground-state kinetic energy contribution F_0 of the electrons and the remainder F_1, which consists of the exchange-correlation contribution F_{xc} of the electrons and the ionic contribution F_i, that is, $F_1 = F_{xc} + F_i$. To the second order in F_1, the Gibbs free energy is given by

$$G(P, T) = F_0(V_0) + PV_0 + F_1(V_0) - \frac{1}{2} \left(\frac{\partial F_1}{\partial V_0} \right)^2 \left(\frac{\partial^2 F_0}{\partial V_0^2} \right)^{-1} , \tag{2.185}$$

where the volume V_0 is determined by the condition

$$P = -\frac{\partial F_0}{\partial V} \bigg]_{V=V_0} . \tag{2.186}$$

In Chapter 3, we shall study the equations of state for the electron systems in detail. For the moment it is convenient to note that Eq. (2.186) can be written in the form

$$P(\text{dynes/cm}^2) = 1.80 \times 10^{23} \left[y\sqrt{1+y^2} \left(\frac{2y^2}{3} - 1 \right) + \ln(y + \sqrt{1+y^2}) \right] , \tag{2.187}$$

where

$$y \equiv 0.01400/r_s(V_0) \tag{2.188}$$

and $r_s(V_0)$ is the r_s value in Eq. (2.184a) at $V = V_0$.

Figure 2.27 shows the phase-separation diagrams of BIM fluids into two fluids having different compositions, calculated by the double-tangent construction

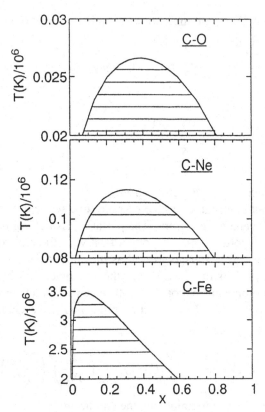

Figure 2.27 Demixing curves for BIM fluids at $P = 10^{22}$ dynes/cm^2.

with Eq. (2.181) for dense C-O ($R_Z = 4/3$), C-Ne ($R_Z = 5/3$), and C-Fe ($R_Z = 13/3$) fluids at $P = 10^{22}$ dynes/cm^2, which corresponds to $r_s(V_0) = 0.02145$. The critical temperature T_c for demixing increases with R_Z, concurrent with a decrease in the critical concentration x_c: Specifically, $(T_c/10^6 \text{ K}, x_c) = (0.027, 0.39)$ for the C-O fluid, $(0.11, 0.31)$ for C-Ne, and $(3.5, 0.082)$ for C-Fe. Since the values of T_c for the C-O and C-Ne fluids are far below the freezing temperature, $T_m = 1.7 \times 10^6$ K, for a pure carbon OCP, solidification takes place prior to demixing for those fluids. For C-Fe fluid, however, T_c is comparable to T_m, leaving open for the moment the possibility that demixing may occur prior to solidification of this fluid.

We now finally consider the phase diagrams for fluid-solid mixtures at constant pressure. Figures 2.28 and 2.29 are the phase diagrams for dense C-Ne ($R_Z = 5/3$) and C-Fe ($R_Z = 13/3$) matter at $P = 10^{22}$ dynes/cm^2. The compressibility of the electrons changes the diagrams only slightly for $R_Z \leq 5/3$; for instance, the eutectic concentration and temperature decrease only by 8%

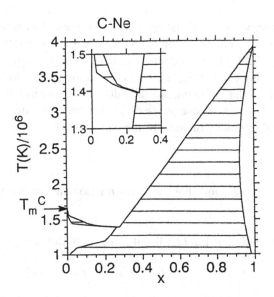

Figure 2.28   Phase diagram for C-Ne BIM at $P = 10^{22}$ dynes/cm^2.

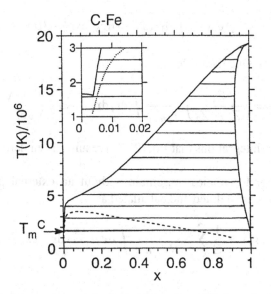

Figure 2.29   Phase diagram for C-Fe BIM at $P = 10^{22}$ dynes/cm^2. The dotted curve represents the fluid demixing curve from Fig. 2.27.

and 5%, respectively, for C-Ne matter in Fig. 2.28. For C-Fe matter, however, the
eutectic concentration changes rather significantly, from $x = 0.04$ to $x = 0.004$,
as shown in Fig. 2.29. The fluid demixing curve for C-Fe matter, determined in
Fig. 2.27, is superimposed in Fig. 2.29, where it is shown as a dotted curve. It can
be seen in Fig. 2.29 that demixing actually does not take place before solidification in C-Fe BIM matter. Those phase diagrams carry significant consequences
for evolution and internal structures of the white dwarfs.

PROBLEMS

2.1   Derive Eq. (2.12), and show how this relation can be extended to the case
of a multi-ionic plasma.

2.2   In the calculation of Eq. (2.13b), it is useful to define a potential energy,

$$U(\mathbf{r}_1, \mathbf{r}_3, \ldots, \mathbf{r}_N) \equiv W(\mathbf{r}_1, \mathbf{r}_1, \mathbf{r}_3, \ldots, \mathbf{r}_N) ,$$

and the associated configuration integral,

$$Q_{N-1} = \int d\mathbf{r}_1 d\mathbf{r}_3 \cdots d\mathbf{r}_N \exp(-\beta U)$$

[Jancovici 1977]. One then calculates

$$\left\langle \left(\frac{\partial W}{\partial \mathbf{r}_1} \right)^2 \right\rangle = \left(\frac{Z_1}{Z_1 + Z_2} \right)^2 \frac{1}{Q_{N-1}} \int d\mathbf{r}_1 d\mathbf{r}_3 \cdots d\mathbf{r}_N \left(\frac{\partial U}{\partial \mathbf{r}_1} \right)^2 \exp(-\beta U) .$$

Show by partial integrations that Eq. (2.13b) results. Then derive Eq. (2.14b).

2.3   In the grand canonical ensemble without an external potential, the ν-
particle densities are defined and calculated as

$$\rho^{(n)}(\mathbf{r}_1, \ldots, \mathbf{r}_n) = \frac{1}{\Xi} \sum_{N \geq n}^{\infty} \frac{z^N}{(N - n)!} \int d\mathbf{r}_{n+1} \cdots d\mathbf{r}_N \exp \left[-\beta V_N(\mathbf{r}^N) \right] ,$$

where

$$V_N(\mathbf{r}^N) = \frac{1}{2} \sum_{j \neq k}^{N} \frac{(Ze)^2}{|\mathbf{r}_j - \mathbf{r}_k|}$$

and z is the activity (2.34) without an external potential. Using the functional
derivatives (see Appendix C), show Eq. (2.35) where $n(\mathbf{r}) = \rho^{(1)}(\mathbf{r})$.

2.4 Derive Eqs. (2.41) and (2.42).

2.5 For OCP, the excess chemical potential is expressed as

$$\mu_{ex} = n \int_0^1 d\lambda \int d\mathbf{r}\, \phi(r)h(r;\lambda)\,, \tag{A}$$

where $\phi(r) = (Ze)^2/r$ and λ is the coupling constant explained in conjunction with Eq. (2.50). Writing

$$h(r;\lambda) = \exp\{-\lambda\beta\phi(r) + \gamma(r;\lambda)\} - 1\,, \tag{B}$$

one performs partial integrations with respect to λ in (A), to find

$$\mu_{ex} = n \int d\mathbf{r}\,[\gamma(r) - h(r) - \beta v(r)] + n \int d\lambda \int d\mathbf{r}\, h(r;\lambda)\frac{\partial}{\partial\lambda}\gamma(r;\lambda)\,, \tag{C}$$

where $h(r) = h(r;\lambda = 1)$ and $\gamma(r) = \gamma(r;\lambda = 1)$. Show that the λ integration in the last term of (C) can be performed in the HNC approximation, $\gamma(r;\lambda) = h(r;\lambda) - c(r;\lambda)$, and with the aid of the Ornstein–Zernike relation (2.41) adopted to the present case,

$$\gamma(r;\lambda) = n \int d\mathbf{r}'\, h(|\mathbf{r} - \mathbf{r}'|;\lambda)\left[h(r') - \gamma(r')\right]\,, \tag{D}$$

resulting in a derivation of Eq. (2.51) [Baus & Hansen 1980].

2.6 Derive Eq. (2.72).

2.7 It has been shown [Totsuji & Ichimaru 1973] that, in the long-wavelength domain such that $k \ll k_B T/(Ze)^2$, the structure factor for a classical OCP is expressed as

$$S^1(k) - 1 = -\frac{k_D^2}{k^2 + k_D^2} + \frac{\varepsilon}{2}\frac{k^3 k_D}{(k^2 + k_D^2)^2}\left[\frac{\pi}{2} - \tan^{-1}\left(2\frac{k_D}{k}\right)\right]$$

to the first order in the plasma-parameter expansion, where

$$\varepsilon = (4\pi n)^{1/2}(Ze)^3(k_B T)^{-3/2}$$

and $k_D = (4\pi n)^{1/2}|Ze|(k_B T)^{-1/2}$ is the Debye wave number parameter. In the short-wavelength domain such that $k \gg k_D$, one likewise finds that

$$S^s(k) - 1 = \varepsilon \frac{k_D}{k} \int_0^\infty dx\, x \left\{ \exp\left[-\frac{1}{x} \exp(-\varepsilon x) \right] - 1 \right\} \sin\left(\frac{\varepsilon k x}{k_D} \right)$$

is the appropriate expression when $\varepsilon \ll 1$. In such a weak coupling case, it is then possible to select a wave number k_1 such that

$$\varepsilon \ll k_D/k_1 \ll 1 \,.$$

Show that Eq. (2.73) can be obtained through a calculation of

$$u_{ex} = \frac{\varepsilon}{\pi} \int_0^{k_1} \frac{dk}{k_D} \left[S^l(k) - 1 \right] + \frac{\varepsilon}{\pi} \int_{k_1}^\infty \frac{dk}{k_D} \left[S^s(k) - 1 \right] \,,$$

which stems from Eq. (2.68).

2.8   For an Einstein oscillator with the Hamiltonian,

$$H_E = \frac{p^2}{2M} + \frac{M}{2} \omega_E^2 r^2 - \delta(x^4 + y^4 + z^4) \,,$$

calculate the Helmholtz free energy F_E to the first order in the anharmonicity parameter δ and show that

$$\beta F_E = 3 \ln \left[2 \sinh\left(\frac{\beta \hbar \omega_E}{2} \right) \right] - \frac{9}{4} \beta \delta \left(\frac{\hbar}{M \omega_E} \right)^2 \coth^2\left(\frac{\beta \hbar \omega_E}{2} \right) \,.$$

2.9   Show the relation (2.119).

2.10   Derive Eqs. (2.127a) and (2.129).

2.11   Derive Eqs. (2.141).

2.12   Show the relation (2.147).

2.13   Consider a Coulomb cluster of size $N = 6$ as depicted in Fig. 2.16, in which the six particles form a regular octahedron. Calculate the electrostatic binding energy (2.165) of the cluster, minimize it with respect the interparticle distance, and compare the result with the value in Table 2.2.

2.14   Show that the coordination number N_c of neighboring particles, defined in Section 2.5B3, is 14 for the bcc cluster and is 12 for the fcc, hcp, and icosahedral clusters.

CHAPTER 3

Quantum Electron Liquids

3.1 THERMODYNAMIC PROPERTIES

A statistical ensemble of electrons as an OCP constitutes a typical example of a charged Fermi liquid. Conduction electrons in metals are relatively low in density so that the exchange and Coulomb correlation effects tend to dominate the system properties over those arising from the kinetic energies in the Fermi distribution; such may therefore be looked upon as a *strongly coupled quantum plasma*. At an ambient temperature, the metallic electrons can be regarded as in the ground state, since their thermal effects are negligible. When the temperature is elevated as in the cases of liquid metals and the solar interior, the thermal effects become significant. Those electrons in the interiors of degenerate stars, on the other hand, are so dense that the Fermi energies take on relativistic values; the electrons form a *relativistic degenerate plasma*. In this section, we shall study thermodynamic properties for those various quantum OCPs of electrons, including the relativistic and thermal effects. Strong coupling effects in the static local field correction approximation will be considered in the next section.

A. Free Energy: Ideal-Gas Contribution

For an electron OCP, in which the average charge density is fixed at a value specified by the density of uniform neutralizing charges, the Helmholtz free energy is a thermodynamic potential appropriate for a description of such a system [e.g., Landau & Lifshitz 1969]. Let the Helmholtz free energy per unit volume be expressed as a sum of the ideal-gas contribution and the *exchange-correlation* part:

$$F = F_0 + F_{xc} . \tag{3.1}$$

The quantity F_{xc} is the same as what has been called the "excess free energy" in the previous chapter; for the electron systems the present nomenclature has been customary due to the fermion nature of the electrons. The free energy (in energy units) is expressible as a function of a pair of dimensionless quantities, r_s and Θ, given by Eqs. (1.2) and (1.8). In this subsection, we consider F_0 for various cases of the electron OCPs.

The value of the free energy at $T = 0$ is the *ground-state energy*. Kinetic energy of an electron with momentum $\mathbf{p}$ is given by

$$\varepsilon_\mathbf{p} = mc^2 \left[\sqrt{1 + \left(\frac{p}{mc} \right)^2} - 1 \right] . \tag{3.2}$$

In the ground state, each momentum state for $p \le p_F = \hbar(3\pi^2 n)^{1/3}$ is occupied by a pair of electrons with opposite spins, so that the ideal-gas contribution to the free energy is calculated as (Problem 3.1)

$$F_0 = \frac{nmc^2}{x_F^3} \left[\frac{3}{8} x_F (2x_F^2 + 1) \sqrt{x_F^2 + 1} - x_F^3 - \frac{3}{8} \sinh^{-1} x_F \right] , \tag{3.3}$$

where n is the number density of the electrons and x_F is the *relativity parameter* defined by Eq. (1.6) [Salpeter 1961; Jancovici 1962]. In a nonrelativistic case where $x_F \ll 1$,

$$F_0 = \frac{3n\hbar^2}{10m} (3\pi^2 n)^{2/3} . \tag{3.4a}$$

In an extreme relativistic case where $x_F \gg 1$,

$$F_0 = \frac{3n\hbar c}{4} (3\pi^2 n)^{1/3} . \tag{3.4b}$$

Relativistic effects are negligible in most of the electron gases at finite temperatures, that is, those with $\Theta \ge 0.1$. The quantum states, designated by the momentum $\mathbf{p}$ and spin σ, are occupied by electrons according to the *Fermi distribution*:

$$f_\sigma(\mathbf{p}) = \frac{1}{\exp\left\{ \beta \left[(p^2/2m) - \mu_\sigma \right] \right\} + 1} . \tag{3.5}$$

Here $\beta = 1/k_B T$ and μ_σ is the chemical potential for the ideal-gas system of electrons with spin σ, which is determined from the normalization (1.44). Analytic formulas for the ideal-gas contribution to the thermodynamic functions in various cases of electron gases are derived in Appendix B.

B. The Hartree–Fock Approximation

The exchange and correlation free energy F_{xc} in Eq. (3.1) is split further into two parts: the Hartree–Fock exchange contribution F_x and the correlation energy F_c. That is,

$$F_{xc} = F_x + F_c . \tag{3.6}$$

In the dielectric formulation of Section 1.2B, the structure factors are calculated in the Hartree–Fock approximation as

$$S_{\sigma\tau}^{HF}(\mathbf{k}, \omega) = -\frac{\hbar}{2\pi} \coth(\hbar\beta\omega/2) \, \mathrm{Im} \, \chi_{\sigma\tau}^{HF}(\mathbf{k}, \omega) , \tag{3.7}$$

$$S_{\sigma\tau}^{HF}(\mathbf{k}) = \frac{1}{\sqrt{n_\sigma n_\tau}} \int_{-\infty}^{\infty} d\omega \, S_{\sigma\tau}^{HF}(\mathbf{k}, \omega) , \tag{3.8a}$$

where $\chi_{\sigma\tau}^{HF}(\mathbf{k}, \omega)$ are the density-density response functions in the Hartree–Fock approximation given by Eq. (1.40). Substitution of Eqs. (1.40) and (1.41) in Eq. (3.7) yields explicit evaluation for the Hartree–Fock static structure factors,

$$S_{\sigma\tau}^{HF}(\mathbf{k}) = \frac{2}{\hbar\beta\sqrt{n_\sigma n_\tau}} \sum_\rho \delta_{\sigma\rho}\delta_{\tau\rho} \sum_{\mathbf{p}} \sum_{\nu=-\infty}^{\infty} \frac{f_\rho(\mathbf{p}) - f_\rho(\mathbf{p}+\hbar\mathbf{k})}{\omega_{\mathbf{pk}} - z_\nu} , \tag{3.8b}$$

where

$$z_\nu = i2\pi\nu/\hbar\beta , \qquad \nu = 0, \pm 1, \pm 2, \dots , \tag{3.9}$$

represents the poles of $\coth(\hbar\beta z/2)$ along the imaginary axis on the complex z plane (see Section 3.1C of Volume I).

For paramagnetic electrons in the ground state, one thus obtains

$$S_{\uparrow\uparrow}^{HF}(\mathbf{k}) = S_{\downarrow\downarrow}^{HF}(\mathbf{k}) = \begin{cases} \dfrac{3}{4}\dfrac{k}{k_F} - \dfrac{1}{16}\dfrac{k^3}{k_F^3} , & k < 2k_F , \\ 1 , & k \geq 2k_F , \end{cases} \tag{3.10a}$$

$$S_{\uparrow\downarrow}^{HF}(\mathbf{k}) = S_{\downarrow\uparrow}^{HF}(\mathbf{k}) = 0 . \tag{3.10b}$$

The Hartree–Fock approximation accounts for the first-order exchange processes and induces spin-dependent correlations, as Eqs. (3.10) illustrate.

The Hartree–Fock pair distribution functions are calculated as

$$g_{\sigma\tau}^{HF}(\mathbf{r}) = 1 + \frac{1}{\sqrt{n_\sigma n_\tau}} \int_{-\infty}^{\infty} \frac{d\mathbf{k}}{(2\pi)^3} \left[S_{\sigma\tau}^{HF}(\mathbf{k}) - \delta_{\sigma\tau} \right] \exp(i\mathbf{k} \cdot \mathbf{r}) \,. \tag{3.11}$$

Corresponding to Eqs. (3.10), one finds

$$g_{\uparrow\uparrow}^{HF}(\mathbf{r}) = g_{\downarrow\downarrow}^{HF}(\mathbf{r}) = 1 - 9 \left[\frac{\sin k_F r - k_F r \cos k_F r}{(k_F r)^3} \right]^2 \,, \tag{3.12a}$$

$$g_{\uparrow\downarrow}^{HF}(\mathbf{r}) = g_{\downarrow\uparrow}^{HF}(\mathbf{r}) = 1 \,. \tag{3.12b}$$

Again, spin-dependent correlations are manifested in these formulas.

The Hartree–Fock exchange energies are evaluated with those correlation functions through the coupling-constant integrations in accordance with Eqs. (2.68) and (2.69). From Eqs. (3.10) or (3.12) one thus finds that

$$F_x = -\frac{3}{4} \left(\frac{3}{2\pi} \right)^{2/3} \frac{ne^2}{a_e} \approx -0.4582 \frac{ne^2}{a_e} \,, \tag{3.13}$$

where a_e refers to the Wigner–Seitz radius (1.3) for the electrons. For the relativistic electron gases, Jancovici [1962] derived a dielectric function, whence

$$F_x = \frac{e^2 m^4 c^4}{(2\pi)^3 \hbar^5} \left[\left(\sinh^{-1} x_F - x_F \sqrt{x_F^2 + 1} \right)^2 - \frac{4}{3} x_F^3 \sqrt{x_F^2 + 1} \sinh^{-1} x_F \right.$$

$$\left. + \frac{2}{3} (x_F^2 + 1)^2 \ln(x_F^2 + 1) - \frac{5}{3} x_F^4 - \frac{2}{3} x_F^2 \right] \,. \tag{3.14}$$

In the nonrelativistic limit ($x_F \ll 1$), this evaluation reproduces Eq. (3.13).

The Hartree–Fock energies for nonrelativistic electrons at finite temperatures have been accurately evaluated and parametrized by Perrot and Dharma-wardana [1984]. The interaction energy u_x per unit volume in units of $k_B T$ is expressed as a function of Θ and Γ_e in Eqs. (1.8) and (1.9):

$$u_x(\Gamma_e, \Theta) = -a^{HF}(\Theta)\Gamma_e \,, \tag{3.15}$$

with the coefficient given by

$$a^{HF}(\Theta) = \left(\frac{3}{2\pi} \right)^{2/3} \frac{a_0}{a_1} \tanh \frac{1}{\Theta} \,, \tag{3.16}$$

$$a_0 = 0.75 + 3.04363\Theta^2 - 0.09227\Theta^3 + 1.7035\Theta^4 \,, \tag{3.17a}$$

$$a_1 = 1 + 8.31051\Theta^2 + 5.1105\Theta^4 \,. \tag{3.17b}$$

The exchange contribution to the free energy is then evaluated through the coupling-constant integration as

$$F_x \equiv \frac{n}{\beta} f_x(\Gamma_e, \Theta) = \frac{n}{\beta} \int_0^{\Gamma_e} dx \, \frac{u_x(x, \Theta)}{x} . \tag{3.18}$$

The excess part of the pressure is thus given by

$$P_x \equiv \frac{n}{\beta} p_x(\Gamma_e, \Theta) = \frac{n}{\beta} \left\{ \frac{u_x(\Gamma_e, \Theta)}{3} - \frac{2\Theta}{3} \frac{\partial f_x(\Gamma_e, \Theta)}{\partial \Theta} \bigg|_{\Gamma_e} \right\} \tag{3.19}$$

[e.g., Ichimaru, Iyetomi, & Tanaka 1987].

Asymptotic formulas for the exchange contribution to the thermodynamic potential of a partially degenerate, semirelativistic electron gas have been obtained and evaluated by Kovetz, Lamb, and Van Horn [1972].

C. Correlation Energy in the Ground State

Correlation energy is defined as the difference in the free energy between the Hartree–Fock evaluation and any better evaluation. It therefore corresponds to the quantity F_c introduced via Eq. (3.6).

For the paramagnetic electrons, the RPA dielectric function (1.47) may be rewritten as

$$\frac{1}{\varepsilon_0(k, \omega)} - 1 = v(k)\chi_0(k, \omega) + \sum_{\nu=2}^{\infty} [v(k)\chi_0(k, \omega)]^\nu , \tag{3.20a}$$

where $v(k) = 4\pi e^2/k^2$ and $\chi_0(k, \omega)$ denotes the free-electron polarizability defined by Eq. (1.59). The first term on the right-hand side amounts to the Hartree–Fock evaluation, and the remainder represents the summation of the *ring diagram* contributions [Gell-Mann & Brueckner 1957]. (See Fig. 3.1.)

The RPA correlation energy in the ground state is calculated as

$$\begin{aligned} F_c^{\mathrm{RPA}} &= -\frac{\hbar}{(2\pi)^4} \int dk \int_0^1 \frac{d\eta}{\eta} \int_0^\infty d\omega \, \mathrm{Im} \left\{ v(k)\chi_0(k, \omega) \left[\frac{1}{\varepsilon_0(k, \omega)} - 1 \right] \right\} \\ &= \frac{\hbar}{(2\pi)^4} \int dk \int_0^\infty d\omega \, \mathrm{Im} \left\{ \ln [1 - v(k)\chi_0(k, \omega)] + v(k)\chi_0(k, \omega) \right\} , \end{aligned} \tag{3.20b}$$

where η refers to the coupling constant in the sense of Eq. (2.69). The η integration has been carried out explicitly in Eq. (3.20b), since $v(k)$ is proportional to and $\chi_0(k, \omega)$ is independent of η.

(a)

(b)

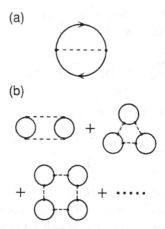

Figure 3.1 (a) Hartree–Fock diagram and (b) summation of the ring diagrams. The solid circles represent the polarizations $\chi_0(\mathbf{k}, \omega)$ and the dotted lines, Coulomb interaction $v(k)$.

The RPA contribution to the correlation energy for the nonrelativistic electrons in the ground state in the limit of high density ($r_s \ll 1$) is thus obtained as

$$F_c = \frac{ne^4 m}{\pi^2 \hbar^2}(1 - \ln 2)\ln r_s \tag{3.21}$$

[Macke 1950; Gell-Mann & Brueckner 1957]. The RPA correlation energy for the relativistic electrons has been likewise calculated by Jancovici [1962] in the dielectric formulation, with the result

$$F_c = \frac{e^4 m^4 c^3}{3\pi^4 \hbar^5}(1 - \ln 2).x_F^3 \sqrt{x_F^2 + 1}\ln r_s . \tag{3.22}$$

This result again recovers the evaluation (3.21) in the nonrelativistic cases where $x_F \ll 1$.

The correlation energy per electron is usually expressed in units of rydbergs:

$$Ry = \frac{e^4 m}{2\hbar^2} = 13.6058 \text{ eV} . \tag{3.23}$$

Hence in such units, the correlation energy is

$$E_c(r_s) = \frac{2\hbar^2}{ne^4 m}F_c . \tag{3.24}$$

For ground-state energy in nonrelativistic electron gases, the most accurate evaluations thus far have been those of Ceperley and Alder [1980] using the GFMC method (see Section 1.2G); the correlation energy was calculated at $r_s = 1, 2, 5, 10, 20, 50,$ and 100. These values were then interpolated by Vosko, Wilk, and Nusair [1980] through a Padé approximant technique in the formula

$$r_s \frac{dE_c(r_s)}{dr_s} = y_0 \frac{1 + y_1 x}{1 + y_1 x + y_2 x^2 + y_3 x^3} . \tag{3.25}$$

Here $x = \sqrt{r_s}$, and

$$y_0 = 0.0621814, \qquad y_1 = 9.81379,$$
$$y_2 = 2.82224, \qquad y_3 = 0.736411.$$

The differential equation (3.25) may be integrated with the boundary condition in the small r_s (i.e., high-density) regime:

$$E_c^{GB}(r_s) = y_0 \ln r_s - 0.09329 . \tag{3.26}$$

The first term on the right-hand side corresponds to the leading RPA contribution (3.22), and the second term represents the sum of the next RPA term in the r_s expansion and the second-order exchange contribution (Fig. 3.2) [Gell-Mann & Brueckner 1957; Onsager, Mittag, & Stephen 1966]. The latter contribution, the first non-RPA term in the r_s expansion, is calculated as [e.g., Pines 1963]

$$E_c^{(2x)}(r_s) = \frac{2\hbar}{ne^4} \sum_{p,q,k} v(k)v\left(|k + (p+q)/\hbar|\right)$$
$$\times \frac{n(p)\,[1 - n(p + \hbar k)]\,n(q)\,[1 - n(q + \hbar k)]}{k \cdot (p + q + \hbar k)} , \tag{3.27}$$

where

$$n(p) = \begin{cases} 1 , & |p| \le p_F , \\ 0 , & |p| > p_F \end{cases} \tag{3.28}$$

Figure 3.2 Second-order exchange diagram.

(see Problem 3.6). As Eq. (3.27) indicates, $E_c^{(2x)}(r_s)$ is a constant independent of r_s.

The result of integrating Eq. (3.25) with the boundary condition (3.26) takes the form

$$
E_c(r_s) = y_0 \left\{ \ln \frac{x^2}{X(x)} + \frac{2b}{Q} \tan^{-1} \frac{Q}{2x+b} \right.
$$
$$
\left. - \frac{bx_0}{X(x_0)} \left[\ln \frac{(x-x_0)^2}{X(x)} + \frac{2(b+2x_0)}{Q} \tan^{-1} \frac{Q}{2x+b} \right] \right\} , \quad (3.29)
$$

where

$$
X(x) = x^2 + bx + c, \qquad Q = \sqrt{(4c - b^2)} . \quad (3.30)
$$

The best fitting parameters, x_0, b, and c, were found to be -0.409286, 13.0270, and 42.7198 for the paramagnetic case, and -0.743294, 20.1231, and 101.578 for the ferromagnetic case (i.e., $n_\uparrow = n$ and $n_\downarrow = 0$), respectively [Vosko, Wilk, & Nusair 1980].

3.2   STATIC LOCAL FIELD CORRECTIONS

The effects of exchange and Coulomb correlations beyond the RPA may be taken into account through the local field corrections $G_{\sigma\tau}(k, \omega)$ in the dielectric formulation of Section 1.2D. The density-density response function for the paramagnetic electrons is thus written

$$
\chi(k, \omega) = \frac{\chi_0(k, \omega)}{1 - v(k) [1 - G(k, \omega)] \chi_0(k, \omega)} , \quad (3.31)
$$

where $v(k) = 4\pi e^2/k^2$, $\chi_0(k, \omega)$ denotes the free-electron polarizability, and

$$
G(k, \omega) = \frac{1}{2} \left[G_{\uparrow\uparrow}(k, \omega) + G_{\uparrow\downarrow}(k, \omega) \right] . \quad (3.32)
$$

The structure factor is then calculated via the fluctuation-dissipation theorem, (1.34) and (1.35):

$$
S(k) = -\frac{1}{n\beta} \sum_{\nu=-\infty}^{\infty} \chi(k, z_\nu) , \quad (3.33)
$$

where $z_\nu = i2\pi\nu/\hbar\beta$ refers to the poles of $\coth(\hbar\beta z/2)$ along the imaginary axis on the complex z plane [e.g., Fetter & Walecka 1971; Ichimaru 1992].

A. Static Local Field Correction Approximation

Formulas (3.31) and (3.33) imply that for a rigorous formulation of correlations and thermodynamic functions in a many-particle system, knowledge on frequency dependence of $G(k, \omega)$ is essential. When the classical limit $\hbar \to 0$ is approached, however, only the term with $\nu = 0$ remains in the summation, so that

$$S(k) \approx -\chi(k, 0)/n\beta \ . \tag{3.34}$$

In such a classical approximation, strong coupling effects beyond the RPA are represented solely by the static values,

$$G(k) \equiv G(k, \omega = 0) \ , \tag{3.35}$$

of the local field corrections. Because of the relation (1.89), the exchange-correlation potentials (1.84) of the density-functional theory are closely related to a static local field correction approximation, in which the dynamic local field corrections are replaced by the static values such as Eq. (3.35). Any strong coupling theory that depends on such a static local field description involves a classical approximation.

In such a static local field correction approximation, the static structure factor is calculated as

$$
\begin{aligned}
S(k) &= -\frac{\hbar}{2\pi n} \int_{-\infty}^{\infty} d\omega \ \coth\left(\frac{\hbar\omega}{2k_B T}\right) \operatorname{Im} \frac{\chi_0(k, \omega)}{1 - v(k)\left[1 - G(k)\right]\chi_0(k, \omega)} \\
&= \frac{3}{2}\Theta \sum_{\nu=-\infty}^{\infty} \frac{\varphi(x, \nu)}{1 + (2\Gamma_e\Theta/\pi\lambda x^2)\left[1 - G(k)\right]\varphi(x, \nu)} \ .
\end{aligned}
\tag{3.36}
$$

Here $x = k/k_F$, $\lambda = (4/9\pi)^{1/3}$,

$$
\begin{aligned}
\varphi(x, \nu) &= -\frac{2E_F}{3n}\chi_0(k, z_\nu) \\
&= \frac{1}{2x} \int_0^{\infty} dy \ \frac{y}{\exp\left[(y^2/\Theta) - \alpha\right] + 1} \ln\left|\frac{(2\pi\nu\Theta)^2 + (x^2 + 2xy)^2}{(2\pi\nu\Theta)^2 - (x^2 + 2xy)^2}\right|
\end{aligned}
\tag{3.37}
$$

is a dimensionless free-electron polarizability, and $\alpha = \beta\mu_\sigma$. Once the static local field correction $G(k)$ is known by some means, Eq. (3.36) makes it possible to calculate the static structure factor $S(k)$. It is therefore an approximate generalization of the classical Ornstein–Zernike relation (2.41) to the cases of quantum electron liquids.

B. Self-Consistent Schemes

In the polarization potential approach to the theory of condensed plasmas as elucidated in Section 1.2D, the effective interaction arising from correlated behavior of the particles plays an important part. Such a potential differs from the bare potential of binary interaction, owing to many-body correlations in the system. The difference between the effective and bare potentials therefore measures the extent to which the correlation effects are involved in the description of plasma properties.

In this subsection, we formulate the static local field corrections produced by exchange and Coulomb correlations in strongly coupled plasmas. Since Eq. (3.36) has set a relation between the two unknown functions—the local field corrections and the static structure factors—a self-consistent scheme to determine the correlations and thermodynamic functions will be established if another independent relation is formulated.

A number of approximation schemes have been advanced for the calculation of local field corrections in a strongly coupled plasma. Notable among them are those proposed by Singwi and his collaborators [1968] and by Ichimaru and Totsuji [Ichimaru 1970; Totsuji & Ichimaru 1973, 1974], called the *STLS scheme* and the *CA scheme*, respectively; the latter applies the convolution approximation (CA), introduced in Section 2.2C, to the triple correlation functions.

In the STLS scheme, the local field correction is expressed in a linear functional of the structure factor as

$$G^{\text{STLS}}(k) = -\frac{1}{n} \int \frac{d\mathbf{q}}{(2\pi)^3} \frac{\mathbf{k} \cdot \mathbf{q}}{q^2} [S(|\mathbf{k} - \mathbf{q}|) - 1] . \qquad (3.38)$$

This scheme corresponds to approximating the nonlinear term (1.28c) by

$$-\frac{1}{2n} \sum_{\mathbf{q}(\neq \mathbf{k})}' v(q) \frac{\mathbf{k} \cdot \mathbf{q}}{k^2} [S(|\mathbf{k} - \mathbf{q}|) - 1] \left\{ \rho_\mathbf{k}, n_{\mathbf{p}\sigma} - n_{\mathbf{p}+\hbar\mathbf{k},\sigma} \right\} \qquad (3.39)$$

in the equation of motion (1.28) for the induced density fluctuations at an application of the external potential field. Such a procedure is equivalent to assuming, in the calculation of the density-density response, that the two-particle distribution may be factored as a product of two single-particle distributions and an *unperturbed* radial distribution (see Problem 3.23 in Volume I).

The STLS equation (3.38) makes it possible to truncate the hierarchical structure of Eq. (1.28) and to establish a set of self-consistent integral equations when combined with Eq. (3.36). The STLS scheme provides significantly improved predictions over the RPA scheme. STLS accurately reproduces the GFMC values of ground-state energy for electron liquids at the metallic densities $2 \leq r_s \leq 6$, as Table 3.2 (p. 125) illustrates. It exhibits a systematic departure from the exact values (2.74) for the interaction energy, however, when it is applied to a classical OCP.

Another problem in the STLS scheme concerns the *compressibility sum rule*, which we introduced in Section 1.2D and shall revisit in Section 3.2C. This sum rule requires that the isothermal compressibility calculated directly from the equation of state agree with that obtained from the long-wavelength behavior of $S(k)$. The STLS theory exhibits a substantial discrepancy between these two sets of values for compressibility.

A modification of the STLS scheme to correct such a flaw has been advanced [Schneider 1971; Vashishta & Singwi 1972]. The Vashishta–Singwi (VS) scheme proposes to use the local field correction in the form

$$G^{\mathrm{VS}}(k) = \left(1 + \xi n \frac{\partial}{\partial n}\right) G^{\mathrm{STLS}}(k) . \tag{3.40}$$

The parameter ξ is chosen in such a way that the resultant scheme satisfies the compressibility sum rule. For a classical OCP, it can be proven that the choice of $\xi = 1/2$ exactly fulfills the sum rule requirement (see Problem 3.7). For an electron liquid in the ground state, Vashishta and Singwi [1972] have chosen $\xi = 2/3$, by which the sum rule can be satisfied almost identically.

The static local field correction $G(k)$ can be formulated in terms of the pair and triple correlation functions without resort to a perturbation calculation. Tago, Utsumi, and Ichimaru [1981] have shown, through an analysis of the density-fluctuation excitations (cf. Section 7.3D in Volume I), that

$$G(k) = -\frac{1}{nS(k)} \int \frac{d\mathbf{q}}{(2\pi)^3} K(\mathbf{k}, \mathbf{q}) \left[S\left(|\mathbf{k} - \mathbf{q}|\right) - 1 + \frac{1}{2}\tilde{h}^{(3)}(\mathbf{k} - \mathbf{q}, \mathbf{q}) \right] .$$
$$\tag{3.41}$$

Here $K(\mathbf{k}, \mathbf{q})$, defined by Eq. (2.145), stems from a symmetrization of Coulomb interaction, and

$$\tilde{h}^{(3)}(\mathbf{k}, \mathbf{q}) = n^2 \int d\mathbf{r}_{13} \int d\mathbf{r}_{23}\, h^{(3)}(r_{12}, r_{23}, r_{31}) \exp(-i\mathbf{k} \cdot \mathbf{r}_{13} - i\mathbf{q} \cdot \mathbf{r}_{23}) \tag{3.42}$$

is the Fourier transform of the triple correlation function $h^{(3)}(r_{12}, r_{23}, r_{31})$ for a homogeneous system. This function is defined in terms of the two- and three-particle radial-distribution functions, $g(r)$ and $g^{(3)}(r_{12}, r_{23}, r_{31})$, as

$$h^{(3)}(r_{12}, r_{23}, r_{31}) = g^{(3)}(r_{12}, r_{23}, r_{31}) - g(r_{12}) - g(r_{23}) - g(r_{31}) + 2 \tag{3.43}$$

[see Eq. (2.46)]. Equation (3.41) gives an exact formula for the local field correction in terms of the pair and triple correlation functions. When it is combined with the classical fluctuation-dissipation theorem (3.34), the result corresponds to the Born–Green equation between the two- and three-particle radial-distribution functions (see Problem 7.9 in Volume I).

To proceed further, one adopts an approximation to the triple correlation function in Eq. (3.41) and thereby evaluates $G(k)$. Truncation of the hierarchical structure of the correlations is thus completed. Combination of such a formulation with Eq. (3.36) makes a set of self-consistent equations for determining $S(k)$ and $G(k)$.

An ansatz for the triple correlation function based on the CA reads

$$\tilde{h}^{(3)}(\mathbf{k}, \mathbf{q}) = S(k)S(q)S(|\mathbf{k} + \mathbf{q}|) - S(k) - S(q) - S(|\mathbf{k} + \mathbf{q}|) + 2 \quad (3.44)$$

(see Problem 3.8). Substitution of this expression in Eq. (3.41) yields

$$G^{\text{CA}}(k) = -\frac{1}{n} \int \frac{d\mathbf{q}}{(2\pi)^3} K(\mathbf{k}, \mathbf{q}) \frac{1 + S(q)}{2} [S(|\mathbf{k} - \mathbf{q}|) - 1] \; . \quad (3.45)$$

This is the local field correction in the CA scheme. It takes a nonlinear form with respect to the structure factor $S(k)$, which is an unknown function for an iterative solution to the resultant self-consistent equations. Such a nonlinear structure in the local field correction may sometimes impair numerical accuracy of the integrations required for the solution to such integral equations. Hence a modification to the CA scheme in this regard has been considered.

For such a purpose, it is instructive to express the local field correction in a general form:

$$G(k) = -\frac{1}{n} \int \frac{d\mathbf{q}}{(2\pi)^3} K(\mathbf{k}, \mathbf{q}) R(q) [S(|\mathbf{k} - \mathbf{q}|) - 1] \; . \quad (3.46)$$

The function $R(q)$ introduced here may be interpreted as a screening factor to the generalized Coulomb interaction $K(\mathbf{k}, \mathbf{q})$ in the local field correction. The STLS local field correction (3.38) may be recovered if $R(q) = 1$ in Eq. (3.46).

An approximation scheme that retains the main features of the CA scheme (3.45) and attains the desired linearization may be offered by the *modified convolution approximation* (MCA) [Tago, Utsumi, & Ichimaru 1981]; its screening function takes the form

$$R^{\text{MCA}}(q) = \frac{1}{2} \left[S^{\text{IS}}(q) + 1 \right] \; , \quad (3.47a)$$

with

$$S^{\text{IS}}(k) = \frac{k^2}{k^2 + k_{\text{IS}}^2} \; . \quad (3.47b)$$

Here the parameter k_{IS} is determined from the self-consistency condition

$$\int_0^\infty dk \left[S^{IS}(k) - 1 \right] = \int_0^\infty dk \left[S(k) - 1 \right] \qquad (3.48a)$$

or, in light of Eq. (2.68),

$$k_{IS} = -2E_{int}/ne^2 . \qquad (3.48b)$$

It has been shown [Yan & Ichimaru 1987] that for strongly coupled classical OCPs (i.e., $\Gamma \geq 1$) the MCA scheme predicts the values of the interaction energy as accurately as the HNC scheme and satisfies the compressibility sum rule almost exactly. The MCA can likewise reproduce GFMC values of the correlation energy closely for the strongly coupled electron liquids (i.e., $r_s \gg 1$) in the ground state, as we shall see in Section 3.2G. The MCA scheme has also been used for analyzing various multicomponent plasmas in strong coupling [Tanaka & Ichimaru 1989; Tanaka, Yan, & Ichimaru 1990].

C. Compressibility and Spin-Susceptibility Sum Rules

The thermodynamic sum rules set by Eq. (A.13) in Appendix A establish rigorous boundary conditions to the static local-field corrections, as Eqs. (1.66) and (1.67) have exemplified. For an electron gas in an external magnetic field **H** and an external potential ϕ_{ext}, the increments of thermodynamic potential per unit volume [e.g., Landau & Lifshitz 1960],

$$\tilde{\Omega} \equiv \Omega - \frac{\mathbf{H} \cdot \mathbf{B}}{4\pi} ,$$

may be expressed as

$$d\tilde{\Omega} = -s\, dT - \frac{P}{V} dV - n(d\mu + d\phi_{ext}) - \frac{\mathbf{B} \cdot d\mathbf{H}}{4\pi} . \qquad (3.49)$$

Here s is the entropy per unit volume and

$$\mathbf{B} = \mathbf{H} + 4\pi \mathbf{M}$$

is the total magnetic field, which defines the magnetization **M**.

On the basis of the increments (3.49), the compressibility sum rule (1.66) or equivalently

$$\lim_{k \to 0} v(k)G(k) = -\left(\frac{\partial \mu_{xc}}{\partial n} \right)_{T,V} \qquad (3.50)$$

may be derived, connecting the long-wavelength values of the local-field correction $G(k)$ for the density-density response and the density derivative of the exchange-correlation chemical potential, $\mu_{xc} = \mu - \mu_0$, the balance between the chemical potential and its ideal-gas part.

Analogously, one expresses the spin-susceptibility sum rule (1.67) in the form

$$\lim_{k \to 0} J(k) = -\left(\frac{g\mu_B}{2}\right)^2 \left(\frac{1}{\chi_P} - \frac{1}{\chi_{P0}}\right). \tag{3.51}$$

This formula thus connects the long-wavelength values of the local-field correction $J(k)$ for the magnetic spin response, the spin susceptibility χ_P, and its ideal-gas value,

$$\chi_{P0} = \left(\frac{g\mu_B}{2}\right)^2 \frac{3n}{2E_F}, \tag{3.52}$$

which is the *Pauli susceptibility*.

D. Screening Lengths

The field of an external test charge $Z_0 e$ (located at the origin) is screened by dielectric polarization of the electron liquid. In terms of the Fourier components, the potential field is calculated as

$$\tilde{V}_{scr}(k) = \frac{4\pi Z_0 e}{k^2 \varepsilon(k, 0)}, \tag{3.53}$$

where $\varepsilon(k, 0)$ denotes the static dielectric function of the electrons. Evaluation of such a potential constitutes a basis for constructing pseudopotentials of ions in metals [Singwi & Tosi 1981; Hafner 1987].

In light of Eq. (1.60) and the compressibility sum rule (3.50), one finds in the limit of long wavelengths

$$\varepsilon(k, 0) \to 1 + \frac{4\pi e^2}{k^2} \left(\frac{\partial n}{\partial \mu}\right)_{T,V}. \tag{3.54}$$

This limiting behavior then leads to a definition of the *long-range screening distance* D_L stemming from the compressibility:

$$D_L^2 = \frac{1}{4\pi e^2} \left(\frac{\partial \mu}{\partial n}\right)_{T,V}. \tag{3.55}$$

The Thomas–Fermi screening parameter k_{TF} in Eq. (1.50) and the Debye–Hückel screening parameter k_{DH} in Eq. (1.52) may be obtained as $1/D_L$ in this formula when the ideal-gas values of the chemical potential are substituted in the respective quantum and classical limits.

The formula (3.53) leads also to a definition of the *short-range screening distance* D_S through the short-range expansion of the screened potential,

$$V_{scr}(r) = \frac{Z_0 e}{2\pi^2} \int d\mathbf{k}\, \frac{\exp(-i\mathbf{k}\cdot\mathbf{r})}{k^2\varepsilon(k,0)} \rightarrow \frac{Z_0 e}{r}\left[1 - \frac{r}{D_S}\cdots\right],$$

such that

$$\frac{a_e}{D_S} = \left(\frac{18}{\pi^2}\right)^{1/3} \int_0^\infty dt\,\left[1 - \frac{1}{\varepsilon(k_F t, 0)}\right]. \tag{3.56}$$

Since D_S characterizes short-range behavior of the screened Coulomb forces, it plays an essential role in the treatment of nuclear reaction rates in dense plasmas, which we shall consider in Chapter 5.

Of significance is an essential difference between the two screening lengths defined through Eqs. (3.55) and (3.56). Since D_L in Eq. (3.55) has been defined in terms of the isothermal compressibility of the electrons, the latter quantity may take on a negative value at low densities (i.e., in the strong Coulomb coupling), when D_L would become an ill-defined quantity. On the other hand, D_S remains a well-defined quantity, since one generally proves

$$\frac{1}{\varepsilon(k,0)} < 1$$

from the causality condition for the density-density response function [Martin 1967; Dolgov, Kirzhnits, & Meksimov 1981; Ichimaru 1982].

E. Screening by Relativistic Electrons

In an ultradense plasma with $r_s \leq 0.01$, appropriate to the interiors of degenerate stars, the Fermi energy of electrons is relativistically high (see Eq. 1.7), so that their coupling with ions is indeed weak. The kinematic effects of relativistic degenerate electrons [e.g., Landau & Lifshitz 1969] soften the electrons against compression, however, and thus act to enhance their polarizations.

The RPA static dielectric function for the relativistic electrons in the ground state has been obtained by Jancovici [1962] as

$$
\begin{aligned}
\varepsilon(k, 0) = 1 + \left(\frac{k_{\text{TF}}}{k}\right)^2 & \left[\frac{2}{3}\sqrt{1 + x_{\text{F}}^2} - \frac{2x_{\text{F}}x^2}{3}\sinh^{-1}x_{\text{F}}\right. \\
& + \sqrt{1 + x_{\text{F}}^2}\frac{1 + x_{\text{F}}^2 - 3x_{\text{F}}^2 x^2}{6x_{\text{F}}^2 x}\ln\left|\frac{1 + x}{1 - x}\right| \\
& \left. - \frac{1 - 2x_{\text{F}}^2 x^2}{6x_{\text{F}}^2 x}\sqrt{1 + x_{\text{F}}^2 x^2}\ln\left|\frac{\sqrt{1 + x_{\text{F}}^2 x^2} + x\sqrt{1 + x_{\text{F}}^2}}{\sqrt{1 + x_{\text{F}}^2 x^2} - x\sqrt{1 + x_{\text{F}}^2}}\right|\right].
\end{aligned}
\tag{3.57}
$$

Here $x = k/2k_{\text{F}}$ and k_{TF} is the (nonrelativistic) Thomas–Fermi screening parameter as defined in Eq. (1.50). In the nonrelativistic limit $x_{\text{F}} \to 0$, the expression (3.57) reduces to the static ($\omega = 0$) evaluation of the Lindhard dielectric function (1.47).

The short-range screening length (3.56) has been computed with this screening function, and the result has been parametrized as

$$
a_{\text{e}}/D_{\text{S}} = 0.1718 + 0.09283R + 1.591R^2 - 3.800R^3 + 3.706R^4 - 1.311R^5,
\tag{3.58}
$$

where $R \equiv 10r_{\text{s}}$ [Ichimaru & Utsumi 1983]. This formula reproduces the computed values over the domain, $0 \leq r_{\text{s}} \leq 0.1$, with digressions of less than 0.7%.

It is noteworthy that the screening length (in units of a_{e}) takes on the finite value 5.8 in the limit of high densities ($r_{\text{s}} \to 0$), while the nonrelativistic Thomas–Fermi length $1/(k_{\text{TF}}a_{\text{e}})$ vanishes in the same limit. Effects of the electron screening may thus remain considerable in dense stellar materials.

F. Parametrized Spin-Dependent Local Field Corrections

In the self-consistent formulation of Section 3.1D2, essential quantities such as the local field corrections are obtained through a numerical solution to a complex set of nonlinear integral equations. The result is usually presented in the form of a numerical table at discrete values of r_{s}. For numerical studies of strong coupling effects in degenerate electron liquids, it proves useful to derive a parametrized expression for a local field correction that accurately fits the results of self-consistent analyses as well as those of variational calculations [Ceperley 1978; Ceperley & Alder 1980].

To achieve such an end, Taylor [1978] proposed a simple formula,

$$
G(k) = \gamma_0(r_{\text{s}})\left(\frac{k}{k_{\text{F}}}\right)^2,
\tag{3.59}
$$

emphasizing the long-wavelength behavior, Eq. (3.50). The coefficient $\gamma_0(r_s)$ is related to the compressibility and can be parametrized in terms of the correlation energy $E_c(r_s)$ of Section 3.1C as

$$\gamma_0(r_s) = \frac{1}{4} - \frac{\pi}{24}\left(\frac{4}{9\pi}\right)^{1/3}\left[r_s^3\frac{d^2E_c(r_s)}{dr_s^2} - 2r_s^2\frac{dE_c(r_s)}{dr_s}\right] . \qquad (3.60)$$

The first term on the right-hand side (1/4) stems from the Hartree–Fock contribution. It has been argued that apparent inaccuracy of the setting (3.59) in the short-wavelength domain may be inconsequential, since the polarizability $\chi_0(k, 0)$ quickly vanishes there. A closer numerical examination has revealed, however, that the divergence of Eq. (3.59) at large k leads to seriously incorrect predictions.

In the static local field correction approximation, it can be proven that

$$\lim_{k\to\infty} G(k) = 1 - g(0) \qquad (3.61)$$

(see Problem 3.9). Since

$$0 \le g(0) \le 1 , \qquad (3.62)$$

the ansatz (3.59) obviously violates the condition (3.61). Through a resummation of the electron-electron ladder diagrams, Yasuhara [1972] has obtained the expression

$$g(0) = \frac{1}{8}\left[\frac{z}{I_1(z)}\right]^2 , \qquad (3.63)$$

where $z = 4\sqrt{\lambda r_s/\pi}$, $\lambda = (4/9\pi)^{1/3}$, and $I_1(z)$ is the modified Bessel function of the first kind and of the first order.

Kimball [1973, 1976] analyzed some of the exact properties for the two-particle distribution function $g(r)$. It has been noted particularly that the short-range contributions to $g(r)$ arise from the s-wave scattering acts between a pair of electrons. This observation stems from the fact that the wave function $\psi_l(\mathbf{r})$ of scattering in a Coulomb potential with the azimuthal quantum number l is proportional to r^l in short ranges. An s-wave scattering is described by a solution to the two-particle Schrödinger equation:

$$\left(-\frac{\hbar^2}{m}\frac{\partial^2}{\partial r^2} - \frac{2\hbar^2}{mr}\frac{\partial}{\partial r} + \frac{e^2}{r}\right)\psi_0(r) = E\psi_0(r) . \qquad (3.64)$$

A short-range solution to Eq. (3.64) yields the *cusp condition*,

$$\frac{\partial g(r)}{\partial r}\bigg|_{r=0} = \frac{1}{a_B} g(0) \qquad (3.65)$$

[a_B = the Bohr radius (1.4)], since $g(r)$ is proportional to $|\psi_0(r)|^2$. Since Eq. (3.65) sets an exact boundary condition for a Coulombic $g(r)$ in the limit of short interparticle separations, it plays an essential role in the analysis of nuclear reaction rates, to be considered in Chapter 5.

The local field corrections describe the effects of exchange and Coulomb correlations, which the RPA does not take into account. In the high-density r_s expansion of the ground-state energy, the first of such non-RPA contributions corresponds to the second-order exchange energy (3.27). Comparing this contribution with that stemming from the RPA second-order direct processes (i.e., the first diagram in Fig. 3.1b),

$$E_c^{(2d)}(r_s) = -\frac{4\hbar}{ne^4} \sum_{\mathbf{p,q,k}} [v(k)]^2 \frac{n(\mathbf{p})\,[1 - n(\mathbf{p}+\hbar\mathbf{k})]\,n(\mathbf{q})\,[1 - n(\mathbf{q}+\hbar\mathbf{k})]}{\mathbf{k}\cdot(\mathbf{p}+\mathbf{q}+\hbar\mathbf{k})} ,$$

$$(3.66)$$

we find the local field correction for the density response, arising from the second-order exchange processes, as

$$G_{2x}(k) = \frac{X(k)}{D(k)} , \qquad (3.67)$$

where

$$X(k) = \sum_{\mathbf{p,q}} v\left(\left|\mathbf{k} + \frac{\mathbf{p}+\mathbf{q}}{\hbar}\right|\right) \frac{n(\mathbf{p})\,[1 - n(\mathbf{p}+\hbar\mathbf{k})]\,n(\mathbf{q})\,[1 - n(\mathbf{q}+\hbar\mathbf{k})]}{\mathbf{k}\cdot(\mathbf{p}+\mathbf{q}+\hbar\mathbf{k})} ,$$

$$(3.68a)$$

$$D(k) = 2 \sum_{\mathbf{p,q}} v(k) \frac{n(\mathbf{p})\,[1 - n(\mathbf{p}+\hbar\mathbf{k})]\,n(\mathbf{q})\,[1 - n(\mathbf{q}+\hbar\mathbf{k})]}{\mathbf{k}\cdot(\mathbf{p}+\mathbf{q}+\hbar\mathbf{k})} . \qquad (3.68b)$$

[Sato & Ichimaru 1989]. Since Eq. (3.67) originates from exchange processes, which act only between electrons with parallel spins as in Eq. (1.57), the local field correction associated with the spin-magnetic response is also formulated as

$$J_{2x}(k) = \frac{v(k)X(k)}{D(k)} . \qquad (3.69)$$

The multidimensional integrations in Eqs. (3.68) can be evaluated numerically by the MC method. The numerical results are then parametrized in the formula

$$
G_{2x}(k) = \frac{J_{2x}(k)}{v(k)}
$$

$$
= \begin{cases} \left(\dfrac{x^2}{4} + 0.0057x^4\right)\left[0.79 + 0.21 \tanh\left(\dfrac{4 - x^2}{0.45}\right)\right], & 0 \le x \le 2, \\ \dfrac{1}{2} + \dfrac{1}{2(x^2 - 2.46)} + \dfrac{0.222}{x^4 - 10}, & 2 < x, \end{cases} \quad (3.70)
$$

with $x = k/k_F$. This parametrized expression, substituted in Eq. (1.60), reproduces the second-order exchange energy to 0.08% accuracy and exactly satisfies the compressibility and spin-susceptibility sum rules as well as the short-range Kimball relation (3.61) in the limit $r_s \to 0$.

Parametrized expressions for the local field corrections, such as $G(k)$ and $J(k)$, which can accurately predict the thermodynamic properties at finite values of r_s consistently with those boundary conditions, have been obtained by Ichimaru and Utsumi [1981; Utsumi & Ichimaru 1982, 1983]. The formula proposed for $G(k)$ takes the form

$$
G(k) = Ax^4 + Bx^2 + C
$$
$$
+ \left[Ax^4 + \left(B + \frac{8}{3}A\right)x^2 - C\right]\frac{4 - x^2}{4x}\ln\left|\frac{2 + x}{2 - x}\right|, \quad (3.71)
$$

with $x = k/k_F$. Here

$$
A = 0.029, \qquad 0 \le r_s \le 15, \tag{3.72a}
$$
$$
B = \frac{9}{16}\gamma_0(r_s) - \frac{3}{64}[1 - g(0)] - \frac{16}{15}A, \tag{3.72b}
$$
$$
C = -\frac{3}{4}\gamma_0(r_s) + \frac{9}{16}[1 - g(0)] - \frac{16}{5}A, \tag{3.72c}
$$

and the values of $\gamma_0(r_s)$ and $g(0)$ should be those given by Eqs. (3.60) and (3.63) with the parametrized correlation energy (3.29). The function (3.71) in the limit of $r_s \to 0$ closely resembles Eq. (3.70).

Table 3.1 compares the values of the correlation energy and the two-particle distribution function at zero separation between the RPA calculations and those with the local field correction (3.71). The RPA description clearly overemphasizes Coulomb repulsion between nearby electrons to such an extent that $g(0)$ takes on unphysical, negative values at metallic densities, violating the condition (3.62); consequently, the RPA values of $E_c(r_s)$ always stay underestimated. With the local field correction, such a flaw has been cured, and a self-consistent description of correlations in the electron systems has been obtained.

Table 3.1 Values of $-E_c(r_s)$ in mRy (= $10^{-3}Ry$) and $g(0)$ calculated in the RPA and in IU with the parametrized local field correction (3.71)

r_s		1	2	4	6	10
$-E_c(r_s)$	RPA	157	124	94	78	
	IU	117.4	86.9	61.0	48.3	35.0
$g(0)$	RPA	-0.12	-0.65	-1.57	-2.60	
	IU	0.279	0.181	0.094	0.052	0.011

Parametrized expression of the local field correction $J(k)$ for the spin-magnetic response has been analogously obtained in the formula [Utsumi & Ichimaru 1983]

$$J(k) = J_0(k) - J(k; r_s) + \eta(r_s)J(k; r_s)\exp\left[-\left(\frac{k}{k_0}\right)^2\right], \qquad (3.73)$$

with

$$J_0(k) = \lim_{r_s \to 0} v(k)G(k), \qquad (3.74a)$$

$$J(k; r_s) = v(k)G(k) - J_0(k), \qquad (3.74b)$$

$$k_0/k_F = 2.559 + 0.1319r_s + 0.01881r_s^2, \qquad r_s \le 7 \qquad (3.74c)$$

$$\eta(r_s) = 1 + \frac{\gamma_a(r_s) - 1/4}{\gamma_0(r_s) - 1/4}. \qquad (3.74d)$$

In this expression, $\gamma_a(r_s)$ is the spin-susceptibility coefficient, analogous to the compressibility coefficient $\gamma_0(r_s)$ defined by Eq. (3.60), given by

$$\gamma_a(r_s) = \frac{\pi}{4\lambda r_s}\left(\frac{3}{2}\lambda^2 r_s^2 \left.\frac{\partial^2 E(r_s, \zeta)}{\partial \zeta^2}\right|_{\zeta=0} - 1\right), \qquad (3.75)$$

where $\lambda = (4/9\pi)^{1/3}$ and ζ denotes the spin polarization

$$\zeta \equiv \frac{n_\uparrow - n_\downarrow}{n_\uparrow + n_\downarrow}. \qquad (3.76)$$

The spin stiffness in Eq. (3.75) has been parametrized by Vosko, Wilk, and Nusair

[1980] as

$$
\frac{\partial^2 E(r_s, \zeta)}{\partial \zeta^2}\bigg|_{\zeta=0} = \frac{2}{3\lambda^2 r_s^2} - \frac{2}{3\pi\lambda r_s} - \frac{1}{3\pi^2}\left[\ln\frac{x^2}{X(x)} + \frac{2b}{Q}\tan^{-1}\frac{Q}{2x+b}\right.
$$
$$
\left. - \frac{bx_0}{X(x_0)}\left(\ln\frac{(x-x_0)^2}{X(x)} + \frac{2(b+2x_0)}{Q}\tan^{-1}\frac{Q}{2x+b}\right)\right] \quad (3.77)
$$

on the basis of the GFMC data obtained by Ceperley and Alder [1980], where $x = \sqrt{r_s}$, the functions $X(x)$ and Q have been defined by Eq. (3.30), and the values $x_0 = -0.00475840$, $b = 1.13107$, and $c = 13.0045$ fit the data.

G. Thermodynamic Functions at Finite Temperatures

Various investigators have evaluated interaction energies at finite temperatures in the RPA [Montroll & Ward 1958; Englert & Brout 1960; Gupta & Rajagopal 1980; Dharma-wardana & Taylor 1981]. For those strongly coupled electron liquids at metallic densities with $1 \leq r_s$, however, a treatment based on the RPA may not be justified. It becomes necessary to take account of the strong electron-electron correlations represented by the local field corrections in the dielectric formulation and thereby to calculate the thermodynamic functions applicable to such an electron liquid.

The exchange and correlation parts of the thermodynamic functions for electron liquids may be derived when the interaction energy u_{xc} per particle in units of $k_B T$, which is a statistical average of the interaction part of the Hamiltonian (2.138), is expressed as a function of Γ_e and Θ. The exchange-correlation free energy per particle and the excess pressure, both in the same energy units, are then calculated through the coupling-constant integrations as in Eqs. (3.18) and (3.19):

$$
F_{xc} \equiv \frac{n}{\beta} f_{xc}(\Gamma_e, \Theta) = \frac{n}{\beta}\int_0^{\Gamma_e} dx\, \frac{u_{xc}(x, \Theta)}{x}, \quad (3.78a)
$$
$$
P_{xc} \equiv \frac{n}{\beta} p_{xc}(\Gamma_e, \Theta) = \frac{n}{\beta}\left[\frac{u_{xc}(\Gamma_e, \Theta)}{3} - \frac{2\Theta}{3}\frac{\partial f_{xc}(\Gamma_e, \Theta)}{\partial\Theta}\bigg|_{\Gamma_e}\right]. \quad (3.78b)
$$

The exchange-correlation energies of electron gases at finite temperatures were evaluated in the static local-field approximation through solution to the set of STLS equations (3.36) and (3.38) [Tanaka & Ichimaru 1986]. The results were then parametrized in the analytic formulas as [Ichimaru, Iyetomi, & Tanaka 1987]

$$
u_{xc}^{STLS}(\Gamma_e, \Theta) = -\Gamma_e \frac{a(\Theta) + b(\Theta)\sqrt{\Gamma_e} + c(\Theta)\Gamma_e}{1 + d(\Theta)\sqrt{\Gamma_e} + e(\Theta)\Gamma_e}, \quad (3.79)
$$

with

$$a(\Theta) = a^{HF}(\Theta), \tag{3.80a}$$

$$b(\Theta) = \frac{0.341308 + 12.070873\Theta^2 + 1.148889\Theta^4}{1 + 10.495346\Theta^2 + 1.326623\Theta^4} \sqrt{\Theta} \tanh\left(\frac{1}{\sqrt{\Theta}}\right), \tag{3.80b}$$

$$c(\Theta) = \left[0.872756 + 0.025248 \exp\left(-\frac{1}{\Theta}\right)\right] e(\Theta), \tag{3.80c}$$

$$d(\Theta) = \frac{0.614925 + 16.996055\Theta^2 + 1.489056\Theta^4}{1 + 10.10935\Theta^2 + 1.22184\Theta^4} \sqrt{\Theta} \tanh\left(\frac{1}{\sqrt{\Theta}}\right), \tag{3.80d}$$

$$e(\Theta) = \frac{0.539409 + 2.522206\Theta^2 + 0.178484\Theta^4}{1 + 2.555501\Theta^2 + 0.146319\Theta^4} \Theta \tanh\left(\frac{1}{\Theta}\right). \tag{3.80e}$$

In the classical limit ($\Theta \gg 1$), the ratio $c(\Theta)/e(\Theta)$ approaches 0.898004, the coefficient of the first term in the liquid excess free-energy formula (2.74). Formulas (3.79)–(3.80) in fact reproduce the HNC values for $\Gamma \leq 1$ with digressions of less than 1% and agree with the interaction energy formula (2.74) within 0.5% for $1 \leq \Gamma \leq 200$ in the classical limit. The functions (3.80) vanish at $\Theta = 0$ in such a way that Eq. (3.79) becomes a function of r_s only. Formulas (3.79)–(3.80) are therefore applicable to the electron liquids in the ground state as well, and the interaction energy (3.79) agrees with that derived from Eq. (3.29) for $r_s \leq 100$ within 0.4%.

The coupling-constant integration of Eq. (3.78a) is performed with Eq. (3.79) to yield

$$\begin{aligned}
f_{xc}^{STLS}(\Gamma_e, \Theta) = &-\frac{c}{e}\Gamma_e - \frac{2}{e}\left[b - \frac{cd}{e}\right]\sqrt{\Gamma_e} \\
&- \frac{1}{e}\left[\left(a - \frac{c}{e}\right) - \frac{d}{e}\left(b - \frac{cd}{e}\right)\right] \ln\left|e\Gamma_e + d\sqrt{\Gamma_e} + 1\right| \\
&+ \frac{2}{e\sqrt{4e - d^2}}\left[d\left(a - \frac{c}{e}\right) + \left(2 - \frac{d^2}{e}\right)\left(b - \frac{cd}{e}\right)\right] \\
&\times \left[\tan^{-1}\left(\frac{2e\sqrt{\Gamma_e} + d}{\sqrt{4e - d^2}}\right) - \tan^{-1}\left(\frac{d}{\sqrt{4e - d^2}}\right)\right].
\end{aligned} \tag{3.81}$$

Condition that $4e - d^2 > 0$ is satisfied for any Θ.

The ground-state energy of the ferromagnetic electrons (i.e., $\zeta = 1$), as well as of the paramagnetic electrons (i.e., $\zeta = 0$), has been calculated by the GFMC method [Ceperley & Alder 1980] and by the solution to the STLS integral equations [Tanaka & Ichimaru 1989]. Spin-dependent correlations and thermodynamic functions for electron liquids at arbitrary degeneracy and spin polarization have been investigated through a solution to another self-consistent

Table 3.2 Ground-state energy per particle in mRy for the paramagnetic ($\zeta = 0$) and ferromagnetic ($\zeta = 1$) cases calculated in the GFMC [Ceperley & Alder 1980], STLS, and MCA methods [Tanaka & Ichimaru 1989]

r_s	ζ	GFMC	STLS	MCA
2	0	4.1	3.5	-0.3
	1	251.7	248.3	244.3
5	0	-151.2	-151.2	-154.4
	1	-121.4	-123.4	-126.7
10	0	-106.75	-105.86	-108.36
	1	-101.3	-102.0	-104.6
20	0	-63.29	-62.33	-64.18
	1	-62.51	-62.33	-64.21
50	0	-28.84	-27.97	-29.06
	1	-28.78	-28.21	-29.31
100	0	-15.321	-14.699	-15.365
	1	-15.340	-14.812	-15.480

set of the MCA integral equations [Tanaka & Ichimaru 1989].

In the STLS scheme, the spin-dependent local field corrections are given by [Singwi & Tosi 1981]

$$G_{\sigma\tau}(k) = -\frac{1}{n} \int \frac{d\mathbf{q}}{(2\pi)^3} \frac{\mathbf{k} \cdot \mathbf{q}}{q^2} \gamma_{\sigma\tau}(|\mathbf{k} - \mathbf{q}|) \ . \tag{3.82a}$$

The partial structure factors $\gamma_{\sigma\tau}(k)$ are related to the corresponding pair distributions by

$$g_{\sigma\tau}(r) = 1 + \frac{1}{n} \int \frac{d\mathbf{k}}{(2\pi)^3} \gamma_{\sigma\tau}(k) \exp(-i\mathbf{k} \cdot \mathbf{r}) \tag{3.82b}$$

In the MCA scheme, the spin-dependent local field corrections are formulated at an arbitrary value of the spin polarization ζ [Tanaka & Ichimaru 1989]. For a paramagnetic electron liquid,

$$G(k) = -\frac{1}{n} \int \frac{d\mathbf{q}}{(2\pi)^3} K(\mathbf{k}, q) R(q) [S(|\mathbf{k} - \mathbf{q}|) - 1] \ , \tag{3.83a}$$

Table 3.3 Values of $-F_{xc}/n(e^2/a_e)$ calculated in the MCA scheme for the paramagnetic (P: $\zeta = 0$) and ferromagnetic (F: $\zeta = 1$) cases at $\Theta = 0.2$ and 0.5 for various values of r_s

r_s	$\Theta = 0.2$		$\Theta = 0.5$	
	P	F	P	F
1	0.51223	0.60394	0.47115	0.56120
2	0.54860	0.62668	0.52207	0.59653
3	0.57415	0.64308	0.55615	0.62056
5	0.60739	0.66523	0.59686	0.65006
10	0.65412	0.69758	0.65142	0.69049
20	0.69912	0.72996	0.70010	0.72750
30	0.72316	0.74777	0.72511	0.74688
50	0.75017	0.76826	0.75248	0.76845
75	0.76878	0.78271	0.77098	0.78327
100	0.78046	0.79193	0.78246	0.79260

$$J(k) = -\frac{v(k)}{n} \int \frac{dq}{(2\pi)^3} \left\{ \frac{\mathbf{k} \cdot \mathbf{q}}{q^2} R(q) \left[S_- (|\mathbf{k} - \mathbf{q}|) - 1 \right] \right.$$
$$\left. + \frac{\mathbf{k} \cdot (\mathbf{k} - \mathbf{q})}{|\mathbf{k} - \mathbf{q}|^2} R_-(q) \left[S (|\mathbf{k} - \mathbf{q}|) - 1 \right] \right\} . \tag{3.83b}$$

Here

$$S_-(k) = S_{\uparrow\uparrow}(k) - S_{\uparrow\downarrow}(k) \tag{3.84a}$$

is the spin structure factor,

$$R(k) = \frac{1}{2}\left[1 + \overline{S}(k)\right] = \frac{1}{2}\left[1 + \frac{k^2}{k^2 + k_0^2}\right], \tag{3.84b}$$

$$R_-(k) = \frac{1}{2}\left[1 + \overline{S}_-(k)\right] = \frac{1}{2}\left[1 + \frac{k^2 + k_1^2}{k^2 + k_2^2}\right], \tag{3.84c}$$

and the parameters, k_0, k_1, and k_2, are to be determined from the internal energy relation (3.47a) as well as the compressibility and the spin-susceptibility sum rules. For a ferromagnetic electron liquid, only Eqs. (3.83a) and (3.84a) remain.

The results of the ground-state energy calculations in those three methods are compared in Table 3.2 for the paramagnetic and ferromagnetic cases. The following observations are in order: At metallic densities ($r_s = 2$ and 5), agreement

Table 3.4 Values of $-F_{xc}/n(e^2/a_e)$ calculated in the MCA scheme for the paramagnetic (P: $\zeta = 0$) and ferromagnetic (F: $\zeta = 1$) cases at $\Theta = 1$ and 2 for various values of Γ_e

| Γ_e | $\Theta = 1$ | | $\Theta = 2$ | |
	P	F	P	F
0.5	0.39243	0.47417	0.36938	0.42200
1	0.45654	0.52476	0.44739	0.49075
2	0.52474	0.57936	0.52748	0.56155
3	0.56871	0.61476	0.57793	0.60621
5	0.61859	0.65538	0.63284	0.65499
8	0.66056	0.68981	0.67743	0.69471
10	0.67889	0.70492	0.69646	0.71169
15	0.70922	0.73001	0.72733	0.73930
20	0.72834	0.74590	0.74636	0.75636
30	0.75190	0.75659	0.76931	0.77698

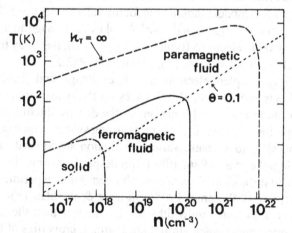

Figure 3.3 Phase diagram of the electron liquid on the number density versus temperature plane, based on the result of the MCA calculations. The dashed curve corresponds to the condition at which the isothermal compressibility κ_T diverges to infinity in the paramagnetic fluid. The solid curve describes the condition at which the spin susceptibility χ_s diverges to infinity in the paramagnetic fluid; it corresponds to the boundary between the paramagnetic and ferromagnetic fluid phases. The dot-dashed curve represents an interpolation between the solidification conditions, $\Gamma_e = 178$ [Ogata & Ichimaru 1987] in the classical limit and $r_s = 100$ [Ceperley & Alder 1980] in the quantum limit.

between the GFMC and STLS values is excellent. The MCA values always stay below the GFMC values and closely reproduce the latter in the strong coupling regime, $r_s \geq 20$.

Some selected values of excess free energies at finite temperatures, calculated in the MCA scheme [Tanaka & Ichimaru 1989], are listed in Table 3.3 for $\Theta = 0.2$ and 0.5 and in Table 3.4 for $\Theta = 1$ and 2, both for the paramagnetic and ferromagnetic cases. Analytic expressions for the thermodynamic functions, such as the excess free energy $f_{ex}(r_s, \Theta, \zeta)$ analogous to Eq. (3.81), have been derived from those computed results obtained at numerous combinations of r_s, Θ, and ζ. Phase boundary curves, arising from divergence of the isothermal compressibility κ_T and of the spin susceptibility,

$$\chi_s \equiv \lim_{k \to 0} \chi_s(k, 0) , \tag{3.85}$$

(see Eq. 1.39) were thereby obtained (Fig. 3.3).

3.3   DYNAMIC PROPERTIES

In simple metals such as Na and Al, effects of the crystal potentials are weak so that an electron liquid model is applicable for the valence electrons. The Coulomb coupling parameter, $r_s = (3/4\pi n)^{1/3} m e^2/\hbar$, takes on a magnitude greater than unity for such metallic electrons (typically, $2 \leq r_s \leq 6$); they may thus be looked upon as strongly coupled systems [e.g., Ichimaru 1982].

In a strongly coupled electron liquid, exchange and Coulomb-induced many-body effects exert essential influences on the quasiparticle properties of those electrons and the spectral functions of the density-fluctuation excitations. Progress in experimental techniques such as strong synchrotron radiation sources has made it possible to measure various excitation spectra with improved accuracy [e.g., Schülke et al. 1986, 1987]. Single-particle excitation spectra have been investigated through angle-resolved photoemission experiments [Jensen & Plummer 1985; Lyo & Plummer 1988; Lindgren & Walldén 1988], resulting in observation of conduction bandwidths much narrower than those predicted in a perturbation-theoretical calculation. Quasiparticle properties of metallic electrons have been monitored also through their mean free paths [Kanter 1970; Tracy 1974; Powell 1974; Kammerer et al. 1982]. The dynamic structure factor $S(\mathbf{k}, \omega)$ has been measured in inelastic scattering experiments of x-ray or electron beams with fine resolutions [Höhberger, Otto, & Petri 1975; Zacharias 1975; Gibbons et al. 1976; Batson, Chen, & Silcox 1976; Raether 1980; Schülke et al. 1986, 1987]. The double peak structures observed in the intermediate wave number regime were attributed primarily to the lattice structure effects, rather than to universal properties of a strongly correlated electron liquid; a possibility was also pointed out that some of the fine structures observed might be correlation-induced [Schülke et al. 1986, 1987]. In this section, we elucidate some of those issues related to the dynamic properties of electron liquids in strong coupling.

A. Basic Formalisms

Many of the theoretical treatments on the static and dynamic correlations have been either perturbation-theoretical or dependent on intuitive model considerations. Since the coupling parameter for the exchange and Coulomb effects is larger than unity for metallic electrons, a perturbative method usually fails to describe the properties of such a system correctly. To overcome such a difficulty, one may resort to a self-consistent theory, whereby the many-body effects are taken into account in a nonperturbative manner. In Section 3.2, we studied such a theory for a description of static correlations and thermodynamic properties. In this subsection, we consider a self-consistent microscopic theory for dynamic correlations in strongly coupled, degenerate electron liquids.

We thus analyze the dynamic evolutions of Green's functions in the grand canonical formalism in the presence of an external perturbation. Such a formalism was introduced in Section 1.2F, and we shall henceforth follow the notation there unless otherwise specified; in particular we recall the shorthand convention adopted in Eqs. (1.92), (1.96), and (1.99).

The multiparticle response functions (1.100) have been formulated as the functional derivatives of the single-particle Green's function with respect to the external perturbation. The multiparticle correlation potentials, analogous to the classical counterparts (2.28), may likewise be defined and introduced as functional derivatives of the *self-energy*, the interaction part of the quasiparticle energy, with respect to Green's functions.

The response functions are expressible in terms of Green's functions and correlation potentials and thus form a chain of equations. The first of such hierarchical equations represent *Dyson equations* for single-particle Green's functions, in which the self-energies depend on multiparticle correlations. The hierarchy of the multiparticle responses and correlations may be truncated if an approximation is found by which an independent expression for the correlation potentials may be established in terms of Green's functions and the response functions. A *dynamic convolution approximation* (DCA) may be introduced for a truncation of the three-body correlation potentials, analogously to the classical convolution approximation in Section 2.2C, which may lead to a *dynamic hypernetted-chain* (DHNC) *scheme* for the degenerate electron liquids. In this section, we confine ourselves to the consideration of dynamics of charge-density variations in a spin-paramagnetic electron liquid; the sum over the spin indices is appropriately performed. A spin-dependent formulation of Green's functions may be found in the literature [e.g., Nozières 1964; Nakano & Ichimaru 1989a].

1. Single-Particle Green's Functions

In the absence of the external field, the creation and annihilation operators, $c_{p\sigma}^{\dagger}$ and $c_{p\sigma}$, obey the usual equation of motion with the Hamiltonian (1.91), so that

$$c_{p\sigma}(t) = \exp\left(\frac{i}{\hbar}Ht\right) c_{p\sigma} \exp\left(-\frac{i}{\hbar}Ht\right)$$

is a Heisenberg operator. A Fourier-transformed, single-particle Green's function is expressed as

$$G_{\sigma}(\mathbf{k}, t) = \begin{cases} -i\left(\Psi, \exp\left(\frac{i}{\hbar}Ht\right) c_{\hbar\mathbf{k}\sigma} \exp\left(-\frac{i}{\hbar}Ht\right) c_{\hbar\mathbf{k}\sigma}^{\dagger} \Psi\right), & t > 0 \\ i\left(\Psi, c_{\hbar\mathbf{k}\sigma}^{\dagger} \exp\left(\frac{i}{\hbar}Ht\right) c_{\hbar\mathbf{k}\sigma} \exp\left(-\frac{i}{\hbar}Ht\right) \Psi\right), & t < 0. \end{cases}$$

$$\text{(3.86)}$$

Since

$$\left(\Psi, c_{p\sigma}^{\dagger} c_{p\sigma} \Psi\right) = \tilde{n}_{p\sigma} \tag{3.87a}$$

measures the occupation number of a single-particle state $(\mathbf{p}, \sigma)$, Eq. (3.86) implies

$$\begin{cases} G_{\sigma}(\mathbf{k}, t = +0) = -i(1 - \tilde{n}_{\hbar\mathbf{k}\sigma}) \\ G_{\sigma}(\mathbf{k}, t = -0) = i\tilde{n}_{\hbar\mathbf{k}\sigma}, \end{cases} \tag{3.87b}$$

where 0 is a positive infinitesimal.

The spectral representation $G_{\sigma}(\mathbf{k}, \omega)$ of the single-particle Green's function is thus calculated as

$$G_{\sigma}(\mathbf{k}, \omega) = \int_{-\infty}^{\infty} dt\, G_{\sigma}(\mathbf{k}, t) \exp(i\omega t)$$
$$= [\omega + \mu_{\sigma}/\hbar - \hbar k^2/2m - \Sigma_{\sigma}(\mathbf{k}, \omega)]^{-1}. \tag{3.88}$$

Here the self-energy $\Sigma_{\sigma}(\mathbf{k}, \omega)$ is formulated exactly in terms of the electron-hole pair response functions as

$$\Sigma_{\sigma}(\mathbf{k}, \omega) = -\int \frac{d\mathbf{q}}{(2\pi)^3} \int \frac{dx}{2\pi} \exp(ix0) v(k) \frac{\chi_{\mathbf{k}+\mathbf{q}/2, \omega+x/2}(\mathbf{q}, x)}{G_{\sigma}(\mathbf{k}, \omega)}, \tag{3.89}$$

with $v(k) = 4\pi e^2 / k^2$. In Eq. (3.89), the exponential with a positive infinitesimal 0 sets the boundary conditions so that the x integration, from $-\infty$ to ∞, may be closed by an infinite semicircle in the upper half of the complex x plane. The electron-hole pair response function $\chi_{q,x}(\mathbf{k}, \omega)$ is related to the two-body response function (1.100) as

$$\chi_{q,x}(\mathbf{k}, \omega) =$$

$$\sum_{\sigma_2} \int d(\mathbf{r}_1 - \mathbf{r}_1') \int d(t_1 - t_1') \int d\left[\frac{\mathbf{r}_1 + \mathbf{r}_1'}{2} - \mathbf{r}_2\right] \int d\left[\frac{t_1 + t_1'}{2} - t_2\right]$$

$$\times \chi^{(2)}(1', 1; 2^+, 2) \exp\left[-i\mathbf{q} \cdot (\mathbf{r}_1 - \mathbf{r}_1') + ix(t_1 - t_1')\right.$$

$$\left. - i\mathbf{k} \cdot \left(\frac{\mathbf{r}_1 + \mathbf{r}_1'}{2} - \mathbf{r}_2\right) + i\omega \left(\frac{t_1 + t_1'}{2} - t_2\right)\right], \tag{3.90}$$

where $\sigma_1 = \sigma$. In Eq. (3.88), the chemical potential μ_σ is given by

$$\frac{\mu_\sigma}{\hbar} = \frac{\hbar k_F^2}{2m} + \Sigma_\sigma(\mathbf{k}, 0)|_{k=k_F}. \tag{3.91}$$

The poles of the single-particle Green's function (3.88) on the complex ω plane determine the *quasiparticle energy* measured from the Fermi surface (3.91), which is calculated from a solution to the equation:

$$\omega = \frac{\hbar k^2}{2m} + \Sigma_\sigma(\mathbf{k}, \omega) - \frac{\mu_\sigma}{\hbar}. \tag{3.92}$$

Equation (3.88) has been referred to as the Dyson equation.

In an ideal-gas system where the exchange and Coulomb correlations are absent, the self-energy vanishes identically. The free-electron Green's function is

$$G_0(\mathbf{k}, \omega) = \left[\omega + \mu_0/\hbar - \hbar k^2/2m\right]^{-1}, \tag{3.93}$$

where

$$\mu_0 = \frac{(\hbar k_F)^2}{2m} \tag{3.94}$$

corresponds to the chemical potential for the ideal-gas system.

2. Ground-State Properties

The spectral representation $G_\sigma(\mathbf{k}, \omega)$ of Green's function is useful in describing the properties of electron liquids in the ground state. In light of Eq. (3.87b), the occupation number of a state $(\hbar\mathbf{k}, \sigma)$ is given by

$$n_{\hbar\mathbf{k}\sigma} = \frac{1}{2\pi i} \int d\omega \, \exp(i\omega 0) G_\sigma(\mathbf{k}, \omega) , \qquad (3.95a)$$

the total number of particles is

$$N = \frac{1}{2\pi i} \sum_{\mathbf{k},\sigma} \int d\omega \, \exp(i\omega 0) G_\sigma(\mathbf{k}, \omega) . \qquad (3.95b)$$

We write the Hamiltonian (1.91) as a sum of contributions from the kinetic and interaction energies,

$$H = K + H_{\text{int}} , \qquad (3.96)$$

and then calculate the average kinetic and interaction energies per particle in the ground state (see Problem 3.11):

$$E_K = \frac{1}{N}(\Psi, K\Psi) = \frac{1}{2\pi N i} \sum_{\mathbf{k},\sigma} \int d\omega \, \exp(i\omega 0) \left[\frac{(\hbar\mathbf{k})^2}{2m} - \mu_\sigma \right] G_\sigma(\mathbf{k}, \omega) , \qquad (3.97a)$$

$$\begin{aligned} E_{\text{int}} &= \frac{1}{N}(\Psi, H_{\text{int}}\Psi) \\ &= \frac{1}{4\pi N i} \sum_{\mathbf{k},\sigma} \int d\omega \, \exp(i\omega 0) \left[\hbar\omega - \frac{(\hbar\mathbf{k})^2}{2m} + \mu_\sigma \right] G_\sigma(\mathbf{k}, \omega) . \end{aligned} \qquad (3.97b)$$

3. Density-Density Response

The dielectric response function $\varepsilon(\mathbf{k}, \omega)$ is related to the density-density response function $\chi(\mathbf{k}, \omega)$ and the electron-hole pair response function through the relations

$$\begin{aligned} \frac{1}{\varepsilon(\mathbf{k}, \omega)} &= 1 + v(k)\chi(\mathbf{k}, \omega) \\ &= 1 + v(k) \int \frac{d\mathbf{q}}{(2\pi)^3} \int \frac{dx}{2\pi} \exp(ix0) \chi_{\mathbf{q},x}(\mathbf{k}, \omega) . \end{aligned} \qquad (3.98)$$

The dynamic local field correction, $G(\mathbf{k}, \omega)$, defined via

$$\chi(\mathbf{k}, \omega) = \frac{\chi_0(\mathbf{k}, \omega)}{1 - v(k)\,[1 - G(\mathbf{k}, \omega)]\,\chi_0(\mathbf{k}, \omega)} \,, \tag{3.99}$$

where

$$\chi_0(\mathbf{k}, \omega) = -\frac{1}{\hbar} \int \frac{d\mathbf{p}}{(2\pi\hbar)^3} \frac{F(\mathbf{p} + \hbar\mathbf{k}/2) - F(\mathbf{p} - \hbar\mathbf{k}/2)}{\omega - \mathbf{k}\cdot\mathbf{p}/m + i0\,\mathrm{sgn}(\omega)} \tag{3.100}$$

refers to the free-electron polarizability, may then be calculated in terms of the three-body response function as [Nakano & Ichimaru 1989a]

$$G(\mathbf{k}, \omega) =$$
$$-\frac{ih}{2n\chi(\mathbf{k}, \omega)} \int \frac{d\mathbf{q}}{(2\pi)^3} \int \frac{dx}{2\pi} \exp(ix0) \frac{\mathbf{k}\cdot\mathbf{q}}{q^2} \chi^{(3)}(\mathbf{k} - \mathbf{q}, \omega - x; \mathbf{q}, x).$$
$$\tag{3.101}$$

In Eq. (3.100), $F(\mathbf{p})$ represents the momentum distribution function; in Eq. (3.101),

$$\chi^{(3)}(\mathbf{k} - \mathbf{q}, \omega - x; \mathbf{q}, x) =$$
$$\int d(1 - 2) \int d(3 - 2)\, \chi^{(3)}(1^+, 1; 2^+, 2; 3^+, 3)$$
$$\times \exp\left[-i\mathbf{k}\cdot(\mathbf{r}_1 - \mathbf{r}_2) + i\omega(t_1 - t_2) - i\mathbf{q}\cdot(\mathbf{r}_3 - \mathbf{r}_2) + ix(t_3 - t_2)\right]\,.$$
$$\tag{3.102}$$

With Eq. (3.101), it is possible to prove the following two lemmas [Nakano & Ichimaru 1989c] with regard to the asymptotic behavior $G(\mathbf{k}, \omega)$:
Lemma I:

$$\lim_{k \to \infty} G(\mathbf{k}, \omega) = \frac{2}{3}[1 - g(0)] \tag{3.103}$$

in the large k regime satisfying $k^2 \gg mE_K/\hbar^2$, where E_K is the average kinetic energy per electron.
Lemma II: If the classical limit ($\hbar \to 0$) is approached first, then one has

$$\lim_{k \to \infty} \lim_{\hbar \to 0} G(\mathbf{k}, \omega) = 1 - g(0)\,. \tag{3.104}$$

Lemma I is an obvious generalization to finite-temperature cases of the Niklasson [1974] relation proven in the ground state. Lemma II is an exact relation, resembling Eq. (3.61) applicable in the static local field correction approximation. Lemmas I and II imply that the large k limit and the classical limit do not commute for $G(\mathbf{k}, \omega)$.

It is illuminating to note a certain similarity between Eqs. (3.89) and (3.101): The parts that $\chi_{\mathbf{q},x}(\mathbf{k}, \omega)$ and $G_\sigma(\mathbf{k}, \omega)$ play in $\Sigma_\sigma(\mathbf{k}, \omega)$ are now replaced by $\chi^{(3)}(\mathbf{k}, \omega; \mathbf{q}, x)$ and $\chi(\mathbf{k}, \omega)$ in $G(\mathbf{k}, \omega)$. The dynamic local field correction is therefore a counterpart in a two-particle Green's function to the self-energy in a single-particle Green's function.

The functions $G_\sigma(\mathbf{k}, \omega)$ and $\chi(\mathbf{k}, \omega)$ retain many-body effects in $\Sigma_\sigma(\mathbf{k}, \omega)$ and $G(\mathbf{k}, \omega)$, which in turn depend on two- and three-body response functions. If the latter functions are expressed in terms of $G_\sigma(\mathbf{k}, \omega)$ and $\chi(\mathbf{k}, \omega)$, a truncation is completed and a closed set of equations may be obtained.

To achieve such an end, we adopt a dynamic version of the CA, as illustrated in Fig. 2.11. Introducing the two- and three-body response functions in a DCA via

$$
\chi_0^{(2)}(\mathbf{k}, \omega; \mathbf{q}, x) =
$$
$$
-\frac{i}{\hbar} \sum_\sigma \int \frac{d\mathbf{q}}{(2\pi)^3} \int \frac{dx}{2\pi} \, G_\sigma\left(\mathbf{q} - \frac{\mathbf{k}}{2}, x - \frac{\omega}{2}\right) G_\sigma\left(\mathbf{q} + \frac{\mathbf{k}}{2}, x + \frac{\omega}{2}\right) ,
$$
$$(3.105a)$$

$$
\chi_0^{(3)}(\mathbf{k}, \omega; \mathbf{q}, x) =
$$
$$
-\frac{i}{\hbar^2} \sum_\sigma \int \frac{d\mathbf{q}'}{(2\pi)^3} \int \frac{dx'}{2\pi} \left[G_\sigma(\mathbf{q}', x') G_\sigma(\mathbf{q}' + \mathbf{k}, x' + \omega) G_\sigma(\mathbf{q}' - \mathbf{q}, x' - x) \right.
$$
$$
\left. + G_\sigma(\mathbf{q}', x') G_\sigma(\mathbf{q}' + \mathbf{q}, x' + x) G_\sigma(\mathbf{q}' - \mathbf{k}, x' - \omega) \right] ,
$$
$$(3.105b)$$

and

$$
\frac{1}{\tilde{\varepsilon}(\mathbf{k}, \omega)} = \frac{\chi(\mathbf{k}, \omega)}{\chi_0^{(2)}(\mathbf{k}, \omega)} ,
$$
$$(3.105c)$$

we finally find expressions for the self-energy and the dynamic local field correction in the CA [Nakano & Ichimaru 1989a],

$$
\Sigma^{CA}(\mathbf{k}, \omega) = \frac{i}{\hbar} \int \frac{d\mathbf{q}}{(2\pi)^3} \int \frac{dx}{2\pi} \, \exp(ix0) \frac{v(q)}{\tilde{\varepsilon}(\mathbf{q}, x)} G(\mathbf{k} + \mathbf{q}, \omega + x) ,
$$
$$(3.106)$$

$$
G^{CA}(\mathbf{k}, \omega) =
$$
$$
-\frac{i\hbar}{n\chi_0^{(2)}(\mathbf{k}, \omega)} \int \frac{d\mathbf{q}}{(2\pi)^3} \int \frac{dx}{2\pi} \frac{\mathbf{k} \cdot \mathbf{q}}{q^2} \exp(ix0) \frac{\chi_0^{(3)}(\mathbf{k} - \mathbf{q}, \omega - x; \mathbf{q}, x)}{\tilde{\varepsilon}(\mathbf{q}, x)\tilde{\varepsilon}(\mathbf{k} - \mathbf{q}, \omega - x)} ,
$$
$$(3.107)$$

which constitute the DHNC scheme.

B. Quasiparticle Properties

Quasiparticle properties are analyzed through a solution to the Dyson equation (3.88) in terms of self-energy (3.89).

1. Self-Energy

In his pioneering work, Hedin [1965] expressed Eq. (3.89) approximately as

$$\Sigma_\sigma(\mathbf{k}, \omega) = \frac{i}{\hbar} \int \frac{d\mathbf{q}}{(2\pi)^3} \int \frac{dx}{2\pi} W(\mathbf{q}, x) G_\sigma(\mathbf{k} - \mathbf{q}, \omega - x) , \qquad (3.108)$$

which may be depicted as in Fig. 3.4. Equation (3.88) was then solved by adopting for the W-function a Coulomb interaction screened by the RPA dielectric function;

$$W^{RPA}(\mathbf{k}, \omega) = v(k)/\varepsilon_0(k, \omega) , \qquad (3.109)$$

and by assuming that the ideal-gas values (3.93) may be substituted for the Green's function in Eq. (3.108); it therefore constitutes a lowest-order perturbative solution.

Rietschel and Sham [1983] obtained a self-consistent iterative solution to Eqs. (3.88) and (3.108) for $G_\sigma(k, \omega)$ while retaining Eq. (3.109). Static local field effects beyond the RPA were subsequently treated by Northrup, Hybertsen, and Louie [1987] and by Lyo and Plummer [1988] in the calculations of self-energy (3.108).

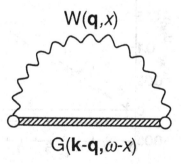

$$W(\mathbf{q}, x)$$

$$G(\mathbf{k}-\mathbf{q}, \omega-x)$$

Figure 3.4 Self-energy diagram.

More rigorously, self-energy (3.89) may be approximated by the formula (3.108) with the W-function expressed as

$$W(k, \omega) = \frac{v(k)}{\varepsilon(k, \omega)} \Gamma(k, \omega) . \qquad (3.110)$$

Here $\varepsilon(k, \omega)$ is an exact dielectric function given in terms of the dynamic structure factor $S(k, \omega)$ as

$$\frac{1}{\varepsilon(k, \omega)} = 1 + \frac{v(k)}{3\pi^2} \int_0^\infty dx \, \frac{S(k, x) - S(k, -x)}{\omega - x + i0 \, \mathrm{sgn}(\omega)} , \qquad (3.111)$$

and $\Gamma(k, \omega)$ is a vertex correction,

$$\Gamma(k, \omega) = \chi_0^{-1}(k, \omega) \big[v(k) + \chi^{-1}(k, \omega) \big]^{-1} \qquad (3.112)$$

[Nakano & Ichimaru 1990]. Physically this function describes the effects of short-range interparticle scattering beyond the Born approximation.

Once the W-function (3.110) is determined, the Dyson equation coupled with Eq. (3.108) may be solved self-consistently for $G_\sigma(k, \omega)$ and $\Sigma_\sigma(k, \omega)$. In so doing, it is useful to note a short-range boundary condition for the electron-hole pair response function,

$$\lim_{k \to \infty} \frac{3\pi k^4}{8\lambda r_s} \left[1 + 3\pi \int_0^\infty d\omega \, \mathrm{Im} \int \frac{dq}{(2\pi)^3} \int \frac{dx}{2\pi} \chi_{qr}(k, \omega) \right] = g(0) , \qquad (3.113)$$

analogous to the cusp condition (3.65), where $\lambda = (4/9\pi)^{1/3}$. In light of this boundary condition applied to Eq. (3.98), an approximate expression for the

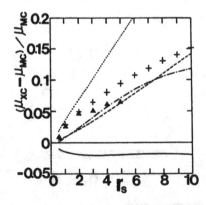

Figure 3.5 Exchange-correlation chemical potentials.

vertex correction may be obtained as

$$\Gamma(k, \omega) = \varepsilon(k, \omega) + \frac{k^2 g(0) + k_s^2}{k^2 + k_s^2}[1 - \varepsilon(k, \omega)] , \qquad (3.114)$$

where k_s is a parameter to be determined consistently with the parametrization of $S(k, \omega)$ [Nakano & Ichimaru 1989a].

In Fig. 3.5, values of the exchange-correlation chemical potential,

$$\mu_{xc} = \mu_\sigma - \mu_0 , \qquad (3.115)$$

calculated in these different schemes are compared with those values μ_{MC} obtained by the GFMC method [Ceperley & Alder 1980; see Section 3.2F]. The solid curve is the result of a self-consistent calculation with the W-function given by Eq. (3.110). The dot-dashed curve represents a similar self-consistent solution, however, with the vertex correction Γ set equal to unity in Eq. (3.110). The dashed curve uses the full W-function but the free-electron values G_0 of Eq. (3.93) substituted in Eq. (3.108). The dotted curve results from the use of the W-function with $\Gamma = 1$ and G_0 in Eq. (3.108). The crosses are Hedin's [1965] perturbative calculation, and the solid triangles, the results of Rietschel–Sham [1983] calculation. The importance of the vertex correction as well as of securing self-consistency in the dielectric function and Green's function is clearly demonstrated in the figure.

2. Fermi Surface Parameters

When k crosses k_F, the excitations change from quasiparticles to quasiholes: The occupation number $\tilde{n}_{\hbar k \sigma}$ must have a discontinuity $Z(k_F)$ at the Fermi surface, called a *quasiparticle renormalization constant*. It is related to the residue of Green's function (3.88) at the quasiparticle energy (3.92),

$$\omega = \omega_k - i\gamma_k ; \qquad (3.116)$$

that is

$$Z(k_F) = \left[1 - \frac{\partial \Sigma_\sigma(k, \omega_k)}{\partial \omega_k}\bigg|_{k=k_F}\right]^{-1} . \qquad (3.117)$$

Figure 3.6 shows the values the renormalization constant calculated in various schemes. The self-consistent calculation with Eqs. (3.110), (3.112), and (3.114) agrees well with the results of variational calculation based on Fermi hypernetted-chain approximation [Lantto 1980].

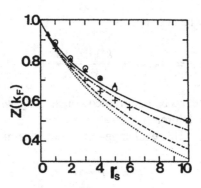

Figure 3.6 Renormalization constant at the Fermi surface. The open circles denote the values obtained by Lantto [1980]; other symbols correspond to the cases in Fig. 3.5.

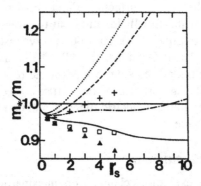

Figure 3.7 Effective-mass ratio. The open squares denote the values obtained by Yasuhara and Ousaka [1987]; other symbols correspond to the cases in Fig. 3.5.

Effective mass m_* of a quasiparticle at the Fermi surface is given by

$$\frac{m_*}{m} = \frac{\hbar k_F}{m} \frac{\partial k}{\partial \omega_k}\bigg|_{k=k_F} = \frac{1 - \dfrac{\partial \Sigma_\sigma(k, \omega_k)}{\partial \omega_k}\bigg|_{k=k_F}}{1 + \dfrac{m}{\hbar k_F}\dfrac{\partial \Sigma_\sigma(k, \omega_k)}{\partial k}\bigg|_{k=k_F}} . \quad (3.118)$$

Values of this effective mass ratio calculated in the various schemes are plotted and compared in Fig. 3.7.

3. Bandwidths and Mean Free Paths

In simple metals such as Na and Al, quasiparticle properties observable experimentally in photoemission bandwidths and electron mean free paths convey crucial information on the many-body effects in the valence and core electrons.

The many-body effects on the valence-band structures are represented in the self-energy correction,

$$\Delta \operatorname{Re} \Sigma(k) = \omega_k - \frac{\hbar}{2m}(k^2 - k_F^2) , \tag{3.119}$$

where ω_k given by Eq. (3.116) denotes the peak position of Green's function in its spectral representation [Mahan 1970 ; Shung & Mahan 1988]. The experimental data obtained by Lyo and Plummer [1988] and by Lindgren and Walldén [1988] indicated a substantial narrowing of the bandwidths observed universally in simple metals, while computed results with Eq. (3.119) implied a widening; such a theoretical bandwidth widening is related to the decrease of effective mass in $m_*/m < 1$ (see Fig. 3.7). These discords have been resolved subsequently in the framework of a Green's function formalism [Northrup, Hybertsen, & Louie 1987; Lyo & Plummer 1988; Nakano & Ichimaru 1989b].

The mean-free path of a quasiparticle (electron) with energy $\hbar\omega_k$ and momentum $\hbar\mathbf{k}$ is calculated as $\hbar k/m_*|\gamma_k|$. A Green's function formulation of the mean free path with inclusion of the effects of the core electrons for Na and Al [Nakano & Ichimaru 1989b] has shown that the experimental data [Kanter 1970; Tracy 1974; Powell 1974; Kammerer et al. 1982] can be well accounted for by such a theory, to the extent of predicting and explaining kinks in the values of the mean free path at an energy near 50 eV (130 eV) for Na (Al) associated with an onset of the core electron excitations. Predicted also are separate kinks at an energy near 6 eV (20 eV) arising from the plasmon excitations.

C. Spectra of Elementary Excitations

The dynamic structure factor $S(\mathbf{k}, \omega)$, representing a spectral function of the density-fluctuation excitations, is a quantity that can be measured directly by inelastic scattering experiments with beams of x-ray or of electrons [e.g., Pines 1963; Raether 1980; March & Parrinello 1982]. The elementary excitations may consist of *plasmons* (i.e., quantized plasma oscillations), *single-pair* and *multipair excitations* formed by quasiparticles of electrons and holes [Pines & Nozières 1966]. Dynamic local field corrections describe those effects of exchange and Coulomb correlations in the dynamic structure factors that are not accounted for in the RPA calculations. Various theoretical approaches have been advanced in the treatment of such spectra of elementary excitations, some of which are studied in this subsection.

1. Cross Sections of Inelastic Scattering

Consider a monochromatic beam of particles with mass M and momentum $\mathbf{P}_I$, incident into and scattered by a system of electrons. The extra Hamiltonian due to the introduction of a beam is

$$H_{ext} = \sum_k{}' \Phi_k \rho_k \exp(-i\mathbf{k} \cdot \mathbf{R}) + P^2/2M , \qquad (3.120)$$

where $\mathbf{P}$ and $\mathbf{R}$ refer to the momentum and spatial coordinates of an incident particle, ρ_k is the density fluctuations (1.20b), and Φ_k is the Fourier transform of the interaction potential between the electron system and the beam. When the beam particle is an electron, one has $M = m$ and

$$\Phi_k = v(k) = 4\pi e^2/k^2 . \qquad (3.121)$$

To calculate the differential cross section for those incident particles to be scattered into final states with momentum $\mathbf{P}_F$, we evoke the second-order perturbation theory of the quantum mechanics [e.g., Schiff 1968; Landau & Lifshitz 1976] and write the transition probability from an initial state where the electrons are in the ground state Ψ_0 with energy E_0 to a final state where the electrons are in an excited state Ψ_m with energy E_m as

$$W\,(\mathrm{I} \to \mathrm{F}) = \frac{2\pi}{\hbar} \left| \Phi_k (\rho_{-k})_{m0} \right|^2 \delta(E_F - E_I) . \qquad (3.122)$$

Here the matrix elements are

$$(\rho_{-k})_{m0} = (\Psi_m, \rho_{-k}\Psi_0) , \qquad (3.123a)$$

and conservation of energy and momentum sets the relations:

$$E_F - E_I = E_m - E_0 + P_F^2/2M - P_I^2/2M , \qquad (3.123b)$$
$$\hbar k = \mathbf{P}_I - \mathbf{P}_F . \qquad (3.123c)$$

The differential cross section dQ of the electron system with volume V for scattering of the beam into a solid angle do and an energy interval $d(\hbar\omega)$ is thus calculated as

$$\frac{d^2 Q}{do\,d\omega} = \frac{M^2 P_F V}{4\pi^2 \hbar^4 P_I} \Phi_k^2 S(\mathbf{k}, \omega) , \qquad (3.124)$$

where

$$S(\mathbf{k}, \omega) = \sum_m \left| (\rho_{-k})_{m0} \right|^2 \delta(\omega_{m0} - \omega) \qquad (3.125)$$

is the dynamic structure factor, and

$$\hbar\omega \equiv P_F^2/2M - P_I^2/2M \ , \tag{3.126a}$$
$$\hbar\omega_{m0} \equiv E_m - E_0 \ . \tag{3.126b}$$

If the incident energy $P_I^2/2M$ is far greater than any of the excitation energies of the electron system such as a plasmon energy $\hbar\omega_p$ and the Fermi energy E_F, then we may ignore the difference between P_I and P_F in the forefactor of Eq. (3.124); we thus obtain

$$\frac{d^2Q}{do\,d\omega} = \frac{M^2V}{4\pi^2\hbar^4}\Phi_k^2 S(\mathbf{k}, \omega) \ . \tag{3.127}$$

Features in the spectral function of the density-fluctuation excitations are thus revealed explicitly in the differential cross section for the inelastic scattering of incident particles.

Inelastic scattering of the electromagnetic radiation such as x-ray may be analyzed quite analogously (see Section 7.5D in Volume I). The differential cross section of the electron system for scattering of electromagnetic radiation with an angular frequency ω_1, a wave vector $\mathbf{k}_1$, and a unit vector of polarization $\hat{\boldsymbol{\eta}}_1$ into a plane wave state with an angular frequency ω_2, a wave vector $\mathbf{k}_2$, and a unit vector of polarization $\hat{\boldsymbol{\eta}}_2$ is thus obtained as

$$\frac{d^2Q}{do\,d\omega} = \frac{3V}{8\pi}\sigma_T(\hat{\boldsymbol{\eta}}_1 \cdot \hat{\boldsymbol{\eta}}_2)^2 S(\mathbf{k}, \omega) \ , \tag{3.128}$$

where

$$\sigma_T = \frac{8\pi}{3}\left(\frac{e^2}{mc^2}\right)^2 = 6.553 \times 10^{-25} \ \text{cm}^2 \tag{3.129}$$

is the cross section of *Thomson scattering*, and

$$\omega = \omega_2 - \omega_1 \ , \tag{3.130a}$$
$$\mathbf{k} = \mathbf{k}_2 - \mathbf{k}_1 \ . \tag{3.130b}$$

In the derivation of the cross section (3.128), it has been assumed that the energy $\hbar\omega_1$ of an incident photon is far greater than any of the excitation energies of the electron system.

If no attention is paid to the polarization directions in the scattering experiment, then the cross section (3.128) may be averaged over the two states of $\hat{\eta}_1$ and summed over $\hat{\eta}_2$ to yield

$$\frac{d^2 Q}{d o\, d\omega} = \frac{3V}{8\pi}\sigma_T \left(1 - \frac{1}{2}\sin^2\theta\right) S(\mathbf{k}, \omega) , \qquad (3.131)$$

where θ is the angle of scattering between $\mathbf{k}_1$ and $\mathbf{k}_2$. X-ray scattering spectroscopy with high resolution can be a powerful tool for probing microscopic structures of the elementary excitations in an electron liquid with strong correlation.

2. Dynamic Local Field Corrections

The dynamic structure factor of an electron liquid, measurable in such scattering experiments, may be formulated through combination of Eqs. (1.33), (1.34), (3.98), and (3.99); for a degenerate electron liquid, it reads

$$S(\mathbf{k}, \omega) = -\frac{\hbar}{2\pi v(k)} \operatorname{Im} \frac{1}{\varepsilon(\mathbf{k}, \omega)} , \qquad (3.132)$$

with

$$\varepsilon(\mathbf{k}, \omega) = 1 - \frac{v(k)\chi_0(\mathbf{k}, \omega)}{1 + v(k)G(\mathbf{k}, \omega)\chi_0(\mathbf{k}, \omega)} . \qquad (3.133)$$

Spectral function (3.132) consists of three kinds of elementary excitations [Pines & Nozières 1966]: (1) plasmons determined by the dispersion relation,

$$\varepsilon(\mathbf{k}, \omega) = 0 ; \qquad (3.134)$$

(2) single-pair excitations corresponding to free propagation of quasiparticle pairs of electrons and holes as represented by Eq. (1.19); and (3) multipair excitations representing dynamically correlated configurations arising from interaction between quasiparticles.

For an accuracy of such a dielectric formulation, it has been regarded as essential that the following exact sum rules and boundary conditions be properly taken into consideration:

1. Compressibility sum rule (3.50), that is,

$$\lim_{k \to 0} G(\mathbf{k}, 0) = \gamma_0(r_s)(k/k_F)^2 , \qquad (3.135a)$$

where $\gamma_0(r_s)$ is the compressibility coefficient given by Eq. (3.60).

2. Frequency-moment sum rules (2.142) in the high-frequency limit. In particular, from Eqs. (2.146) and (2.147), one derives (the third frequency-moment sum rule)

$$\lim_{k \to 0} G(\mathbf{k}, \infty) = \gamma_\infty(r_s)(k/k_F)^2 , \tag{3.135b}$$

where

$$\gamma_\infty(r_s) = \frac{3}{20} - \frac{\pi}{10}\left(\frac{4}{9\pi}\right)^{1/3}\left[r_s^2\frac{dE_c(r_s)}{dr_s} + 2r_sE_c(r_s)\right] . \tag{3.135c}$$

The first term on the right-hand side (3/20) stems from the Hartree–Fock contribution.

3. The cusp condition (3.65), or equivalently,

$$\lim_{k \to \infty} k^4\left[1 - \frac{1}{n}\int_0^\infty d\omega\, S(\mathbf{k}, \omega)\right] = \frac{8}{3\pi}\left(\frac{4}{9\pi}\right)^{1/3} r_s g(0) . \tag{3.135d}$$

4. Niklasson condition (3.103).

Various theories have been proposed for taking account of those strong coupling effects that go beyond the RPA description. For example, Toigo and Woodruff [1970, 1971] calculated the dynamic local field correction arising from exchange contributions contained in the terms (1.28c), with an approximation that would conserve a frequency moment of the nonlinear density response. Dynamic exchange effects were considered through a perturbation-theoretical method, and detailed numerical computations were also conducted by Devreese, Brosens, and Lemmens [1980; Brosens, Devreese, & Lemmens 1980].

Niklasson [1974] formulated a dielectric response of the electron system with inclusion of the non-RPA terms (1.28c), through a study of the equation of motion for the two-particle distribution in the presence of an external field. Limiting behavior of the resulting $G(\mathbf{k}, \omega)$ at large k or ω was investigated explicitly in connection with relations such as Eqs. (3.103) and (3.135d). The equation-of-motion approach to the dielectric function was pursued further by Dharma-wardana [1976]. The self-energy operator and the dynamic local field correction were formulated with successive approximations in the interaction potential; contacts with the compressibility sum rule and the frequency-moment sum rules were made. Subsequently, Tripathy and Mandal [1977] and Mandal, Rao, and Tripathy [1978] treated a similar problem with a perturbation-theoretical approximation. A first-order perturbation theory of dynamic correlations was developed by Holas, Aravind, and Singwi [1979], wherin the first-order diagrams for the proper polarizability were evaluated and their properties examined.

Utsumi and Ichimaru [1980] started with the exact equation of motion (1.28) and projected it onto a model equation, reflecting the memory-function formalism of Eq. (2.122), as

$$
i\hbar\frac{\partial}{\partial t}\rho_{pk}(t) =
$$

$$
\hbar\omega_{pk}\rho_{pk}(t) + v(k)\left[1 - G(k)\right]\left(n_p - n_{p+\hbar k}\right)\rho_k(t) - i\hbar\left(\frac{2}{\pi}\right)^{1/2}\frac{\Omega(k)}{\tau(k)}
$$

$$
\times\int_0^t ds\,\exp\left[-\frac{1}{2}\Omega^2(k)s^2\right]\left[\rho_{pk}(t-s) + \frac{2(n_p - n_{p+\hbar k})}{\hbar\omega_{pk}\tilde\chi_0(k,0)}\rho_{pk}(t-s)\right]
$$

$$
+ e\Phi_\sigma^{ext}(\mathbf{k},\omega)\left(n_{p\sigma} - n_{p+\hbar k,\sigma}\right)\exp(-i\omega t + 0t)\,, \tag{3.136}
$$

where

$$
\rho_{pk} = \sum_\sigma \rho_{pk\sigma}\,, \qquad n_p = \left\langle\sum_\sigma n_{p\sigma}\right\rangle\,, \tag{3.137a}
$$

$$
\tilde\chi_0(\mathbf{k},\omega) = \frac{1}{\hbar}\sum_p \frac{n_p - n_{p+\hbar k}}{\omega - \omega_{pk} + i0}\,. \tag{3.137b}
$$

Function (3.137b) reduces to the Lindhard polarizability $\chi_0(\mathbf{k},\omega)$ if the expectation value n_p is approximated by $2n(\mathbf{p})$ defined by Eq. (3.28). Mukhopadhyay and Sjölander [1978] evaluated $\tilde\chi_0(\mathbf{k},\omega)$ on the basis of a lowest-order perturbation calculation of Eq. (3.95a) [Hedin 1965; Lundqvist 1968] and thereby concluded that the differences between $\tilde\chi_0(\mathbf{k},\omega)$ and $\chi_0(\mathbf{k},\omega)$ may be negligible in practice.

By the construction of its memory-functional term, the model equation (3.136) conserves the local number of particles, as it readily reduces to

$$
i\hbar\frac{\partial}{\partial t}\rho_k(t) = \sum_P \hbar\omega_{pk}\rho_{pk}(t)\,, \tag{3.138}
$$

which is a continuity equation for the number of particles. Mermin [1970] emphasized the importance of securing such a conservation law within a given relaxation-time approximation.

The RPA contributions in Eq. (3.136) are contained in the first term on the right-hand side and that part of the second term independent of $G(k)$; the rest of the terms are an approximate representation of the non-RPA terms. The exchange and Coulomb coupling effects are taken into account through three functions: the static local field correction $G(k)$, the relaxation rate in the long-time response $1/\tau(q)$, and the relaxation frequency in the short-time response $\Omega(k)$.

The functions $G(k)$ and $1/\tau(q)$ in Eq. (3.136) are calculated from the first two terms in the low-frequency expansion of the nonlinear density response stemming from Eq. (1.28). The static local field correction $G(k)$ has been thus formulated and obtained in Section 3.2. The long-time relaxation rate $1/\tau(q)$ can be calculated analogously [Utsumi & Ichimaru 1981] and takes a form,

$$\frac{1}{\omega_p \tau(k)} = -\frac{\pi m \omega_p}{2\hbar k_F n} \int \frac{d\mathbf{q}}{(2\pi)^3} K(\mathbf{k}, \mathbf{q}) \frac{\theta(2k_F - q)}{k} \left[S^{IS}(q) \right]^2$$
$$\times \left[S(|\mathbf{k} - \mathbf{q}|) - S^{HF}(|\mathbf{k} - \mathbf{q}|) \right], \qquad (3.139)$$

where $\theta(x)$ is the unit step function and Eqs. (2.195), (3.10a), and (3.47b) have been used. It is clear that $1/\tau(k)$ is proportional to k^2 in the long-wavelength regime and vanishes as $k \to 0$, reflecting the conservation of the total momentum in the system of mutually interacting electrons [e.g., Pines & Nozières 1966]. We may recall an analogous feature observed in the calculation, Eq. (9.32) in Volume I, for the rate of nonlinear damping of the plasma oscillations in a classical electron gas. Quantities related to $1/\tau(k)$ were also formulated by DuBois and Kivelson [1969] and by Hasegawa and Watabe [1969].

The characteristic frequency $\Omega(k)$ is then determined from the third frequency-moment sum rule (2.142) as

$$\frac{\Omega(k)}{\omega_p} = \left(\frac{\pi}{2} \right)^{1/2} \omega_p \tau(k) \left[G(k) - I(k) \right]. \qquad (3.140)$$

It is therefore proportional to the difference between the low- and high-frequency limits of the dynamic local field correction. This frequency provides a characteristic frequency for the multipair excitations; it remains finite in the long-wavelength limit, as the limiting behaviors, (3.135a) and (3.135b), and those of $1/\tau(k)$ indicate.

With Eq. (3.136), the dielectric response function can be calculated as [Utsumi & Ichimaru 1980]

$$\varepsilon(k, \omega) = 1 - \frac{v(k)(\tilde{\omega}/\omega)\tilde{\chi}_0(k, \tilde{\omega})}{1 + \left\{ v(k)(\tilde{\omega}/\omega)G(k) + \left[(\tilde{\omega} - \omega)/\omega \tilde{\chi}_0(k, 0) \right] \right\} \tilde{\chi}_0(k, \tilde{\omega})}, \qquad (3.141)$$

where

$$\tilde{\omega} = \omega + \left(\frac{2}{\pi} \right)^{1/2} \frac{\Omega(k)}{\omega \tau(k)} \left[W \left(\frac{\omega}{\Omega(k)} \right) - 1 \right] \qquad (3.142)$$

and

$$W(z) \equiv 1 - z \exp\left(-\frac{z^2}{2}\right) \int_0^z dy \, \exp\left(\frac{y^2}{2}\right) + i\left(\frac{\pi}{2}\right)^{1/2} z \exp\left(-\frac{z^2}{2}\right) \quad (3.143)$$

(see Section 4.1A in Volume I for details on the properties of this function).

Explicit expression for the dynamic local field correction in this formalism may be obtained by comparing Eqs. (3.113) and (3.141). In the sequel we investigate the nature of elementary excitations predicted in those theories.

3. Long-Wavelength Excitations

Substantial effort has been expended on the experimental study of the dynamic structure factors associated with the valence electrons in metals, through the techniques of x-ray scattering spectroscopy and electron energy-loss spectroscopy [e.g., Raether 1980]. The experiments have revealed the frequency dispersion, the linewidth, and the spectral shape of the plasmon excitations, as well as the detailed features of the contributions coming from other elementary excitations.

Platzman and Eisenberger [1974] reported an earlier experimental observation of double-peak or peak-shoulder structures in the excitation spectra of electrons in Be, Al, and C over the wave number domain of $1.13 < k/k_F < 2.10$. Theoretical efforts were directed toward accounting for such fine structures in the excitation spectra [e.g., Mukhopadhyay, Kalia, & Singwi 1975; De Raedt & De Raedt 1978; Barnea 1979]. The basic idea evoked in some of those theories was to consider the lifetime effects arising from higher-order coupling between plasmons and single-pair excitations. More recent experiments [Schülke et al. 1986, 1987] performed with improved resolutions have attributed those double-peak structures primarily to the lattice structure effects, rather than to universal properties of a strongly correlated electron liquid. A possibility remains, however, that some of the fine structures observed—for example, in polycrystalline samples—may be correlation-induced.

It is instructive in these connections to investigate the dynamic structure factor (3.132) calculated from the dielectric function (3.141) in light of the long-wavelength sum rules [Pines & Nozières 1966] for various elementary excitations with inclusion of the multipair excitations. Let the dynamic structure factor be expressed therefore as a sum of the three contributions,

$$S(\mathbf{k}, \omega) \approx S_{pl}(\mathbf{k}, \omega) + S_{sp}(\mathbf{k}, \omega) + S_{mp}(\mathbf{k}, \omega) , \quad (3.144)$$

in such a long-wavelength regime (see Fig. 3.8).

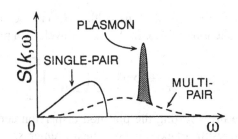

Figure 3.8 A rough sketch of the various long-wavelength excitations in electron liquids [Pines & Nozières 1966].

The plasmon contribution $S_{pl}(\mathbf{k}, \omega)$ is given by

$$S_{pl}(\mathbf{k}, \omega) \approx \frac{n\hbar k^2}{2\pi m\omega(k)} \frac{\gamma(k)}{[\omega - \omega(k)]^2 + \gamma(k)^2} \,. \tag{3.145}$$

The plasmon pole,

$$\omega = \omega(k) - i\gamma(k) \,, \tag{3.146}$$

of the dynamic structure factor may be determined from a solution to the dispersion equation (3.134). The quantities $\omega(k)$ and $\gamma(k)$ describe the frequency dispersion and decay rate of the plasmon. In the long-wavelength regime, the dispersion relation may be written as

$$\omega(k) = \omega_p + 2\alpha\omega_F(k/k_F)^2 \,, \tag{3.147}$$

with $\omega_F = E_F/\hbar$; this equation defines the coefficient α of the plasmon dispersion. The dispersion coefficient is calculated from Eq. (3.141) as

$$\alpha = \alpha_{RPA} - \frac{\omega_p}{\omega_F} \left\{ \gamma_\infty(r_s) + [\gamma_0(r_s) - \gamma_\infty(r_s)] \operatorname{Re} W\left(\frac{\omega_p}{\Omega(0)}\right) \right\} \,, \tag{3.148}$$

where

$$\alpha_{RPA} = 3\omega_F/5\omega_p \tag{3.149}$$

is the dispersion coefficient evaluated in the RPA. Formula (3.148) thus takes account of both static and dynamic strong coupling effects in the dispersion coefficient, arising from the exchange and Coulomb correlations. As Eqs. (3.135a) and (3.135b) indicate, the quantities $\gamma_0(r_s)$ and $\gamma_\infty(r_s)$ stem from the static and high-frequency limits of the local field correction; Eq. (3.148) then mixes those two contributions in the way determined by the frequency ratio $\omega_p/\Omega(0)$.

The plasmon decay rate or the linewidth $\gamma(k)$ can be calculated likewise with the dielectric function (3.141). In the long-wavelength regime, we find

$$\gamma(k) = \frac{1}{2\tau(k)} \exp\left[-\frac{1}{2}\left(\frac{\omega_p}{\Omega(0)}\right)^2\right].$$ (3.150)

We shall shortly be investigating the plasmon dispersion and lifetime in more detail, including the results of experimental measurements.

The spectrum of single-pair excitations is a contribution well known in RPA theory. Their characteristic frequencies are those ω_{pk} in Eq. (1.29) associated with the electron-hole pair excitations. Effects of Coulomb collisions do not show up in the long-wavelength regime, where the leading term of the spectrum is

$$S_{sp}(\mathbf{k}, \omega) \approx \frac{3n\hbar}{\pi^2 E_F}\left(\frac{k}{k_F}\right)^4 \frac{\omega/kv_F}{(2/\pi)^2 + (\omega/kv_F)^2}\theta(kv_F - \omega),$$ (3.151)

with v_F referring to the velocity of an electron on the Fermi surface. A step function is involved here due to restrictions arising from the Fermi distribution.

An extra non-RPA effect is reflected in the emergence of a multipair spectrum from the imaginary part of $W(\omega/\Omega(k))$; the frequency variables here scale as $\Omega(k)$, which remains finite in the limit of long wavelengths. No Pauli principle restrictions are involved in these parts of the frequency spectrum. The leading contribution for the multiparticle excitations is given in the long-wavelength regime by

$$S_{mp}(\mathbf{k}, \omega) \approx \begin{cases} \dfrac{n\hbar\omega}{2\pi E_F\omega_p^2\tau(k)}\left(\dfrac{kv_F}{\omega_p}\right)^2 \exp\left[-\dfrac{1}{2}\left(\dfrac{\omega}{\Omega(k)}\right)^2\right], & \omega < \omega_p, \\[4mm] \dfrac{n\hbar}{2\pi E_F\omega\tau(k)}\left(\dfrac{kv_F}{\omega}\right)^2 \exp\left[-\dfrac{1}{2}\left(\dfrac{\omega}{\Omega(k)}\right)^2\right], & \omega > \omega_p. \end{cases}$$ (3.152)

Salient dynamic features being correctly taken into account, each of those spectral functions—(3.145), (3.151), and (3.152)—satisfy the frequency-moment sum rules set out by Pines and Nozières [1966]. Multipair excitations peaked at $\omega \approx \Omega(k)$ are the direct consequences of strong coupling effects, which are observable in the metallic electrons.

4. Plasmon Dispersion

In various instances, a theory on the basis of the RPA has failed to account for salient features observed in experiments. Such a discord has been anticipated, however, since the RPA is basically a weak-coupling theory; the electrons in metals are, on the contrary, a strongly coupled system, for which the coupling constant $r_s = 2$–6. We consider certain aspects of such comparison on the basis of the theoretical model (3.141), since it encompasses other existing theories as its limiting cases.

Assuming that the electron liquid model may provide a workable description of the electrons in metals, we resume consideration of the dispersion coefficient α given by Eq. (3.148). The expressions obtained in various theories may be looked upon as certain limiting cases of Eq. (3.148). For instance, if one disregards the relaxation effect in the short-time domain and thereby lets $\Omega(0) \to \infty$, then Eq. (3.148) reduces to

$$\alpha_0 = \alpha_{\mathrm{RPA}} - (\omega_p/4\omega_F)\gamma_0(r_s) \ . \tag{3.153}$$

This expression corresponds to the one obtained by Singwi et al. [1968], by Vashishta and Singwi [1972], and by Lowy and Brown [1975], although the schemes of evaluation for $\gamma_0(r_s)$ differed from one theory to the other. Basically, they are static theories and thus involve the static local-field correction represented by $\gamma_0(r_s)$.

In the weak coupling limit of $r_s \to 0$, only the lowest-order Hartree–Fock contribution may be retained, and one finds $\gamma_0(r_s) = 1/4$ (see Eq. 3.60). In this limit, Eq. (3.153) becomes

$$\alpha_{\mathrm{TW}} = \alpha_{\mathrm{RPA}} - \omega_p/16\omega_F \ , \tag{3.154}$$

the expression derived by Toigo and Woodruff [1970, 1971].

If, on the other hand, the plasmon is assumed to have a sufficiently high energy so that an approximation $\omega_p/\Omega(0) \to \infty$ is applicable, then Eq. (3.148) reduces to

$$\alpha_\infty = \alpha_{\mathrm{RPA}} - \left(\frac{\omega_p}{4\omega_F}\right)\gamma_\infty(r_s) \ . \tag{3.155}$$

This is the expression obtained by Pathak and Vashishta [1973] and by Jindal, Singh, and Pathak [1977]. These authors took account of the third frequency-moment sum rule, which is reflected in the involvement of $\gamma_\infty(r_s)$.

Finally, one notes that the lowest-order Hartree–Fock contributions to $\gamma_\infty(r_s)$ may be evaluated by going over to the weak coupling limit, $r_s \to 0$, so that $\gamma_\infty(0) = 3/20$ (see Eq. 3.135c). In this limit, one finds from Eq. (3.155)

$$\alpha_{\mathrm{NP}} = \alpha_{\mathrm{RPA}} - 3\omega_p/80\omega_F \ . \tag{3.156}$$

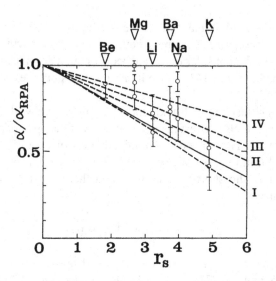

Figure 3.9 Plasmon dispersion coefficient α divided by its RPA value. The solid curve corresponds to Eq. (3.148). Dashed curves I–IV correspond to the α values in Eqs. (3.153)–(3.156), respectively. The experimental data for Be, Mg, Li, Ba, Na, and K are taken from Raether [1980].

This expression was obtained by Nozières and Pines [1958] and by DuBois [1959].

Figure 3.9 compares the values of α/α_{RPA} as computed according to Eqs. (3.148) and (3.153)–(3.156). Figure 3.9 also plots the experimental data for various metals as compiled by Raether [1980]. We may interpret those data as indicating collectively that the experimental values for α are significantly different from the RPA prediction (3.149) and that the discrepancies widen as r_s increases. Since the measured values are rather widely scattered and are attached with large error bars, however, it does not appear feasible to draw a quantitative conclusion from a comparison such as the one in Fig. 3.9.

5. Plasmon Linewidth

A critical wave number k_c is defined as that wave number at which the plasmon dispersion merges into the continuum of the single-pair excitations. Approximately, it is estimated as

$$k_c \approx \frac{\omega_p}{v_F} = 1.7 r_s^{-1/2} (\text{Å}^{-1}) . \tag{3.157}$$

In the long-wavelength regime such that $k < k_c$, plasmons do not couple with single-pair excitations; no plasmon damping results. This is therefore a prediction of the RPA theory.

When collisional processes are considered, plasmons decay at the rate given by Eq. (3.150); their lifetimes become finite. In the limit of long wavelengths, the decay rate would vanish, since $1/\tau(k)$ is proportional to k^2. In actual metals, however, linewidths of the plasmon spectra observed in scattering experiments [e.g., Kloos 1973; Zacharias 1975; Gibbons et al. 1976; Krane 1978] take on nonvanishing, finite values in the limit of $k \to 0$. These observations clearly indicate the necessity of considering those additional scattering processes of electrons that would not conserve the total momentum of electrons. Examples of such metallic effects are the interband transitions, pseudopotentials, and scattering with phonons, impurities, and so on [e.g., Nozières & Pines 1959; Hasegawa 1971; Sturm 1976, 1977; Raether 1980]; they are outside the scope of this volume.

3.4 TWO-DIMENSIONAL LAYERS OF ELECTRONS

Electrons or holes trapped in interfaces of semiconductors, metals, and insulators may be regarded as forming quasi-two-dimensional layers of charged particles [e.g., Ando, Fowler, & Stern 1982; Ando 1990]. It will be shown that spin-dependent correlations due to exchange processes are particularly strong in such a two-dimensional (2-D) system of electrons. The magnetic-order structures in layered electrons may be related to a microscopic account for the superconductivity [e.g., Anderson 1987; Schrieffer, Wen, & Zhang 1988; Monthoux, Balatsky, & Pines 1991] in a 2-D model of cuprate-oxide high-temperature superconductors [Bednorz & Mueller 1986]. In this section we study the ground-state properties and the possible inhomogeneous spin-density wave (SDW) states for such a 2-D layered system of degenerate electrons.

A. Ground-State Energy

Consider a 2-D layer of electrons with the areal number density n. The r_s parameter is given by the ratio between the circular radius $a(\pi n)^{-1/2}$ of a specific area $1/n$ and the Bohr radius a_B of Eq. (1.4):

$$r_s = \frac{me^2}{\hbar^2(\pi n)^{1/2}} \cdot \tag{3.158}$$

As with the three-dimensional (3-D) cases, the ground-state energy $E(r_s; \zeta)$ in units of rydbergs is expressed as a sum of the ideal-gas, Hartree–Fock exchange, and correlation parts:

$$E(r_s; \zeta) = E_0(r_s; \zeta) + E_x(r_s; \zeta) + E_c(r_s; \zeta), \tag{3.159}$$

where ζ denotes the spin polarization defined by Eq. (3.76). The sum of ideal-gas and Hartree–Fock exchange contributions is calculated as [e.g., Isihara 1989; Tanatar & Ceperley 1989]

$$E_0(r_s; \zeta) + E_x(r_s; \zeta) = \frac{1}{r_s^2}(1 + \zeta^2) + \frac{4\sqrt{2}}{3\pi r_s}\left[(1 + \zeta^2)^{3/2} + (1 - \zeta^2)^{3/2}\right].$$

(3.160)

This is the Hartree–Fock evaluation of the ground-state energy.

The most accurate evaluations for the ground-state energy in the 2-D system of electrons thus far have been those made by Tanatar and Ceperley [1989] using variational and fixed-node Green's-function Monte Carlo (MC) methods. For the electron liquids, the MC data were obtained at the density values $r_s = 1, 5, 10, 15, 20, 30, 50$ for the paramagnetic state ($\zeta = 0$) and $r_s = 1, 5, 10, 15, 20, 30, 40, 50, 75$ for the spin-polarized state ($\zeta = 1$). These values for the correlation energy have then been fitted by a Padé approximant formula,

$$E_c(r_s; \zeta) = a_0 \frac{1 + a_1 x}{1 + a_1 x + a_2 x^2 + a_3 x^3},$$

(3.161)

where $x = \sqrt{r_s}$. The parameters giving the best fit for the computed values are listed in Table 3.5 [Tanatar & Ceperley 1989].

Leading terms in the r_s expansion of the RPA correlation energy are

$$E_c^{RPA}(r_s) = -0.6137 - 0.1726r_s \ln r_s + 0.8653r_s.$$

(3.162a)

The second-order exchange energy stemming from electrons with parallel spins represents the first term in the r_s expansion of non-RPA corrections to the correlation energy, taking on the value

$$E_c^{2x}(r_s) = 0.2287$$

(3.162b)

[Ioriatti & Isihara 1981].

It is instructive to compare this value of the second-order exchange energy with the 3-D counterpart of Eq. (3.27), which has been evaluated to be 0.04836.

Table 3.5 Parameters of the Padé approximants (3.161) for the paramagnetic ($\zeta = 0$) and spin-polarized ($\zeta = 1$) fluids

ζ	a_0	a_1	a_2	a_3
0	-0.3568	1.1300	0.9052	0.4165
1	-0.0515	340.5813	75.2293	37.0170

The magnitude of $E_c^{2x}(r_s)$ is greater for 2-D than for 3-D by almost an order of magnitude. Such a comparison points therefore to the primary importance of the exchange effects in a 2-D system. Furthermore, in the high-density (small r_s) limit, we observe in 2-D that the second-order exchange energy (3.162b) takes on a magnitude similar to the RPA value (3.162a); in 2-D the RPA terms do *not* represent the leading contribution to the correlation energy in the limit of high densities. This is to be contrasted with the 3-D case, where the constant term $E_c^{2x}(r_s)$ is masked in high densities by the $\ln r_s$ term in Eq. (3.26), which is the leading RPA contribution. Hence it may be reasonable to conclude that spin-dependent features in the electron liquids should be more pronounced in 2-D than in 3-D.

The MC simulations were also performed for the 2-D systems of electrons in a crystalline-solid phase [Tanatar & Ceperley 1989]. The energy per electron in rydbergs is expected [Ceperley 1978] to have the form

$$E(r_s) = \frac{c_1}{r_s} + \frac{c_{3/2}}{r_s^{3/2}} + \frac{c_2}{r_s^2} + \cdots . \tag{3.163}$$

The coefficient of the first Madelung term is known to be $c_1 = -2.2122$ [Bonsall & Maradudin 1977]. The coefficients of the harmonic and anharmonic terms have been extracted from the MC data at $r_s = 30, 40, 50, 75, 100$ to be $c_{3/2} = 1.6284$ and $c_2 = 0.0508$ [Tanatar & Ceperley 1989].

B. Dielectric Formulation

Correlation energy in 2-D can be approached with dielectric formulation as well. In the static local field correction approximation of Section 1.2D, one writes the density and spin-density responses as

$$\chi(k, \omega) = \frac{\chi_0(k, \omega)}{1 - v(k)\left[1 - G(k)\right]\chi_0(k, \omega)} , \tag{3.164}$$

$$\chi^s(k, \omega) = -\left(\frac{g\mu_B}{2}\right)^2 \frac{\chi_0(k, \omega)}{1 + J(k)\chi_0(k, \omega)} . \tag{3.165}$$

Here g ($= 2.0023$) is Lande's g-factor, μ_B is the Bohr magneton,

$$v(k) = 2\pi e^2 / k , \tag{3.166}$$

$$\chi_0(k, \omega) = \frac{2}{\hbar} \sum_p \frac{f(|\mathbf{p}|) - f(|\mathbf{p} + \hbar\mathbf{k}|)}{\omega - \omega_{pk} + i0} , \tag{3.167}$$

$$f(p) = \begin{cases} 1 , & p \le \hbar k_F, \\ 0 , & p > \hbar k_F, \end{cases} \tag{3.168}$$

$$k_F = \sqrt{2\pi n} , \tag{3.169}$$

and the excitation frequency ω_{pk} has been given by Eq. (1.29); all vector quantities, summation, and integration are two-dimensional.

The functions $G(k)$ and $J(k)$ are related to the spin-dependent local field corrections $G_{\sigma\tau}(k)$ through Eqs. (1.62) and (1.63). Because of the aforementioned importance of the second-order exchange processes in spin-dependent correlations for 2-D electron systems, Sato and Ichimaru [1989] carried out explicit calculations of the 2-D counterparts of $G_{2x}(k)$ and $J_{2x}(k)$ in Eqs. (3.67) and (3.69). The numerical results are then parametrized in the formula

$$G_{2x}(k) =$$
$$\begin{cases} \left(\dfrac{x}{\pi} + 0.009x^4 - 0.00038x^6\right)\left[0.965 + 0.035\tanh\left(\dfrac{4 - x^2}{0.078}\right)\right], & 0 \le x \le 2 \\ \dfrac{1}{2} + \dfrac{0.0475}{x - 1.66} + \dfrac{0.09}{x^2 - 3.0}, & 2 < x, \end{cases}$$
$$(3.170a)$$
$$J_{2x}(k) = v(k)G_{2x}(k) , \qquad\qquad\qquad\qquad\qquad\qquad\qquad (3.170b)$$

with $x = k/k_F$. These functions are positive definite and peak at $k \cong 2k_F$. Moroni, Ceperley, and Senatore [1992] evaluated the density-density static response by the quantum MC method and obtained a result corroborating Eq. (3.170a) near $r_s = 1$.

Values of the correlation energy calculated in various theoretical schemes are compared in Table 3.6. The RPA column lists the values obtained through the density response (3.164) with the RPA assumption $G(k) = 0$. The HD column lists the values in the high-density expansion calculation where Eqs. (3.162a) and (3.162b) are summed. The discrepancy between these two sets of evaluation in the high-density limit, $r_s \to 0$, stems from RPA's neglect of the $E_c^{2x}(r_s)$ term (3.162b). The SI column lists the values obtained through the density response (3.164) with the second-order exchange local field correction (3.170a)

Table 3.6 Values of $-E_c(r_s; 0)$ in mRy calculated with various schemes

r_s	RPA	HD	SI	STLS	MC
0.01	588.1	368.4	379.1		357
0.05	573.3	315.7	361.6		356
0.1	555.9	258.7	346.4		354
0.5	460.6	107.5	284.7	250	303
1	396.4		246.8	211	220
2	324.2		205.1	155	166

[Sato & Ichimaru 1989]; the local field correction significantly improves the RPA in the high-density regime. The STLS column lists the values obtained through a 2-D version of the STLS self-consistent scheme studied in Section 3.2B [Jonson 1976]. The MC column contains the values computed by the GFMC method as parametrized in Eq. (3.161) [Tanatar & Ceperley 1989]. As with the 3-D cases, good agreement is observed between the latter two sets of calculations in a low-density regime.

The collective excitation of the spin-density fluctuations corresponding to a pole of $\chi^s(k, \omega)$ on the complex ω plane,

$$1 + J(k)\chi_0(k, \omega) = 0 \,, \tag{3.171}$$

is called a spin wave or a *magnon*. If such a relation is satisfied with a finite value of k under static circumstances, that is, if

$$1 + J(k)\chi_0(k, 0) = 0 \,, \tag{3.172}$$

then it gives the condition for an onset of the SDW instability. Explicit calculation of the spin-density local field correction such as Eq. (3.170b) makes it possible to solve Eq. (3.172). It has been concluded through such an analysis that an SDW instability with wave number $k = 1.966k_F$ may take place as r_s is increased in a 2-D layer of paramagnetic electrons to a critical value $r_s = 1.887$. Associated with the SDW instability, attractive interaction emerges effectively between the nearest-neighbor electrons with antiparallel spins [Sato & Ichimaru 1989]. An inhomogeneous, antiferromagnetically spin-ordered ground state may thus be indicated for such layered electrons. In the sequel we investigate the phase diagram in terms of antiferromagnetic order parameters associated with microscopic magnetic structures for 2-D electrons.

C. The Inhomogeneous Spin-Ordered Ground State

The possibility of a spin-inhomogeneous ground state for 2-D electrons with energy lower than a homogeneous paramagnetic state may be approached microscopically through nonlinear spin-density-functional theory (Section 1.2E), in which the presence of an effective interaction between nearest-neighbor electrons with parallel or antiparallel spins may be accounted for through first-principle calculations of second-order exchange-correlation potentials. Phase diagrams obtained for the magnetic structures predict that the onset of an SDW instability at $r_s = 1.89$ is followed by a set of distinct structural transitions into *conductive* SDW states and leads eventually to the antiferromagnetic Mott-insulator states [Mott 1990] for $r_s \geq 2.16$ [Iyetomi & Ichimaru 1993].

The energy functional of the distributions $n_\sigma(\mathbf{r})$ for the electrons with spin σ

in the absence of an external potential is expressed as a sum of the kinetic and interaction contributions,

$$E[n_\sigma(\mathbf{r})] = E_K[n_\sigma(\mathbf{r})] + \frac{1}{2} \sum_{\sigma,\tau} \int d\mathbf{r} d\mathbf{r}' \frac{e^2}{|\mathbf{r} - \mathbf{r}'|} \delta n_\sigma(\mathbf{r}) \delta n_\tau(\mathbf{r}') + E_{xc}[n_\sigma(\mathbf{r})],$$

$$(3.173)$$

where $\delta n_\sigma(\mathbf{r}) \equiv n_\sigma(\mathbf{r}) - n_\sigma$ denote the spin-density variations. The exchange-correlation energy functional is expanded up to the quadratic terms with respect to the density inhomogeneities as

$$E_{xc}[n_\sigma(\mathbf{r})] \cong E_{xc}(n_\sigma) + \frac{1}{2} \sum_{\sigma,\tau} \int d\mathbf{r} d\mathbf{r}' K_{\sigma\tau}^{xc}(|\mathbf{r} - \mathbf{r}'|) \delta n_\sigma(\mathbf{r}) \delta n_\tau(\mathbf{r}').$$

$$(3.174)$$

Here the exchange-correlation potential $K_{\sigma\tau}^{xc}(r)$ is the second functional derivative of $E_{xc}[n_\sigma(\mathbf{r})]$ with respect to $\delta n_\sigma(\mathbf{r})$, and its Fourier transform $\tilde{K}_{\sigma\tau}^{xc}(k)$ is related to the spin-dependent local field corrections $G_{\sigma\tau}(k)$ via Eq. (1.89) with $v(k) = 2\pi e^2/k$. Substitution of Eq. (3.174) in Eq. (3.173) leads to the Hamiltonian,

$$H = \sum_{\mathbf{p},\sigma} \left(\frac{p^2}{2m}\right) c_{\mathbf{p}\sigma}^\dagger c_{\mathbf{p}\sigma}$$
$$+ \frac{1}{2} \sum_{\substack{\mathbf{pp'k} \\ \sigma,\tau}}' v(k)[1 - G_{\sigma\tau}(k)] c_{\mathbf{p}+\hbar\mathbf{k},\sigma}^\dagger c_{\mathbf{p'}-\hbar\mathbf{k},\tau}^\dagger c_{\mathbf{p'}\tau} c_{\mathbf{p}\sigma} + E_{xc}(n_\sigma), \qquad (3.175)$$

where the prime on the summation means omission of terms with $\mathbf{k} = 0$. The local field corrections obtained by accounting for the second-order exchange processes take the form

$$G_{\sigma\tau}(k) = G(k)\delta_{\sigma\tau}, \qquad (3.176)$$

with $G(k)$ given by Eq. (3.170a).

The physical origin of the *net attraction between nearest-neighbor electrons with antiparallel spins* arising from the local field corrections is clear in Eqs. (3.170a), (3.175), and (3.176): We remark that the local field corrections resulting from the second-order exchange processes are positive definite $(G_{\sigma\tau}(k) \geq 0)$ and affect those interactions between parallel-spin electrons only by virtue of Eq. (3.176). Equation (3.175) then implies that the electrons with antiparallel spins are more *strongly Coulomb-coupled* than those with parallel spins. The resultant imbalance in the peak heights between the antiparallel and parallel spin-correlations induces inhomogeneity of the SDW type and thereby net attraction between antiparallel spins at a nearest-neighbor separation. Emergence of such an effective attraction has been explicitly demonstrated near the onset of a SDW instability [Sato & Ichimaru 1989]. A nonlinear analysis thus becomes necessary in the treatment of layered electrons beyond such an instability.

The inhomogeneous ground state may be approached in the variational principle through the equation $\delta E[n_\sigma(\mathbf{r})]/\delta n_\sigma(\mathbf{r}) = 0$. For a solution to the variational problem, we begin with a single-particle Schrödinger equation with a given potential $\varphi(\mathbf{r})$,

$$\left[-\frac{\hbar^2}{2m}\nabla^2 + \varphi(\mathbf{r})\right]\psi(\mathbf{r}) = \varepsilon\psi(\mathbf{r}) , \qquad (3.177)$$

where $\varphi_\uparrow(\mathbf{r}) = -\varphi_\downarrow(\mathbf{r}) = -2\Phi[\cos(\kappa k_F x) + \cos(\kappa k_F y)]$, reflecting square symmetry with spatial periodicity $2\pi/\kappa k_F$. We then separate the variables by setting $\psi(\mathbf{r}) = u_1(x)u_2(y)$ and $\varepsilon = \rho_1 + \rho_2$, and find that Eq. (3.177) turns into an eigenvalue equation of the Mathieu type,

$$\frac{d^2u(w)}{dw^2} + \left(\alpha \pm \frac{s}{2}\cos 2w\right)u(w) = 0 , \qquad (3.178)$$

where $w = \kappa k_F x/2$, $s = 16\Phi/\kappa^2 E_F$, $\alpha = (8ms/\phi)^{1/2}\rho/\kappa k_F\hbar$, and $E_F = (\hbar k_F)^2/2m$. The parameters s and n play the part of *antiferromagnetic order parameters*.

The variational calculations are carried out numerically in the following steps:

1. For a given r_s value, choose (s, κ) as the variational parameters.
2. For each spin component, solve Eq. (3.177) or (3.178) for the wave functions $\psi_{\mathbf{k}\sigma}(\mathbf{r})$ and the eigenvalues $\varepsilon_{\mathbf{k}\sigma}$.
3. Determine the chemical potential, $\mu = \mu_\uparrow(s, \kappa) = \mu_\downarrow(s, \kappa)$, of the noninteracting system in the usual way by normalization to the areal number densities, $n = n_\uparrow + n_\downarrow$, over the states $\varepsilon_{\mathbf{k}\sigma} \leq \mu$.
4. From the wave functions in accordance with Eqs. (3.173) and (3.174), calculate spin-density distributions $n_\sigma(\mathbf{r})$, the areal kinetic-energy density $E_K(s, \kappa)$, and the areal interaction-energy density $E_{int}(s, \kappa; r_s)$.
5. The kinetic-energy increment, $\Delta E_K(s, \kappa) = E_K[n_\sigma(\mathbf{r})] - E_K(n_\sigma)$, generally increases with s, while the interaction-energy increment is

given by

$$\Delta E_{\text{int}}(s, \kappa; r_s) = \frac{1}{2} \sum_{k}' v(k) \left\{ [1 - G(k)] |\delta \tilde{n}_+(k)|^2 - G(k) |\delta \tilde{n}_-(k)|^2 \right\} ,$$

(3.179)

where $\delta \tilde{n}_+(k)$ and $\delta \tilde{n}_-(k)$ are the Fourier components of the charge- and spin-density distributions, $n_+ = n_\uparrow + n_\downarrow$ and $n_- = n_\uparrow - n_\downarrow$; since $G(k) \geq 0$, $\Delta E_{\text{int}}(s, \kappa; r_s)$ in Eq. (3.179) *decreases* with the spin-density inhomogeneity $|\delta \tilde{n}_-(k)|^2$.

6. Finally, to find a variational solution, minimize the total energy increment, $\Delta E(s, \kappa; r_s) = \Delta E_K(s, \kappa) + \Delta E_{\text{int}}(s, \kappa; r_s)$, with respect to the variational parameters, s and κ.

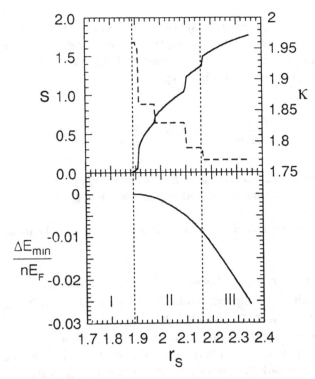

Figure 3.10 The phase diagrams represented by the variational solutions for s (upper solid curve), κ (upper dashed curve), and ΔE_{min} (lower solid curve): I ($r_s \leq 1.89$) corresponds to a homogeneous, paramagnetic, and metallic state; II ($1.89 < r_s \leq 2.16$), an inhomogeneous, antiferromagnetically spin-ordered, conductive state; and III ($2.16 \leq r_s$), an inhomogeneous, antiferromagnetic, Mott-insulator state.

As the value of r_s increases from a state with a high-density homogeneous paramagnetic electron fluid to a state with a low-density antiferromagnetic Mott insulator, the microscopic spin correlations obtained through these variational analyses change. The phase diagrams describing such changes in the resultant spin-ordered structures are shown in Fig. 3.10. Homogeneous paramagnetic electrons continuously enter an inhomogeneous state of SDW with $\kappa = 1.96$ at $r_s = 1.89$. The first Brillouin zones in the resultant SDW states are not totally occupied; the electrons in these states are therefore conductive. The system of electrons undergoes a series of successive structural transitions at $\kappa = 1.86, 1.83, 1.79$, and 1.77, with discrete steps in s. The transition with $\kappa = 1.77 \ldots (= \sqrt{\pi})$ at $r_s = 2.16$ leads the system into the antiferromagnetic Mott-insulator state. The antiferromagnetic states for $r_s \geq 2.16$ correspond to the Mott insulators, in which the adjacent electrons in the square lattices have antiparallel spin orientations; the first Brillouin zones are totally occupied.

The critical temperature T_c for an SDW state may be estimated through the aforementioned steps of variational calculations at finite temperatures. It may also be calculated approximately from the condition that an additional kinetic energy contribution, $\Delta E_K(T) = (\pi^2/3)N(0)(k_B T)^2$, is sufficient to break up the coherent state with the cohesive energy $\Delta E_{\min}$, where $N(0)$ is the state density at the Fermi level. The values of T_c so estimated range from $(1.0–4.6) \times 10^3$ K. Hence the conductive SDW states in the parametric window, $1.89 < r_s < 2.16$, are remarkably stable against thermal fluctuations.

It has not been shown as yet, however, if these inhomogeneous SDW ground states are superconductive. The Hamiltonian (3.175) implies that in these SDW states pairs of electrons $(\mathbf{p}_1 \uparrow, \mathbf{p}_2 \downarrow)$ with $|\mathbf{p}_1 - \mathbf{p}_2| \cong 2\hbar k_F$ are coupled strongly by the net attraction to produce the coherent spin-inhomogeneous structures. They are conduction electrons with state densities available near the Fermi surfaces. We may then qualitatively account for those aspects of superconductivity associated with infinite conductivity and persistent current in terms of coherent superposition of the current-carrying pair states $(\mathbf{p}_1 + \mathbf{p} \uparrow, \mathbf{p}_2 + \mathbf{p} \downarrow)$ [Bardeen, Cooper, & Schrieffer 1957] with $|\mathbf{p}_1 - \mathbf{p}_2| \cong 2\hbar k_F$, if it can be shown that these pair states cannot be destroyed by scattering of individual electrons.

PROBLEMS

3.1 Derive Eqs. (3.3), (3.4a), and (3.4b).

3.2 Show the Hartree–Fock correlation functions (3.10) and (3.12).

3.3 Calculate the Hartree–Fock exchange energy (3.14) either from Eqs. (3.10) or from Eqs. (3.12).

3.4   Derive Eq. (3.19).

3.5   Calculate the RPA correlation energy (3.21).

3.6   In the second-order perturbation theory of quantum mechanics [e.g., Schiff 1968], the contribution to the ground-state energy (in rydbergs) of the electron gas in the second order for the exchange and Coulomb interaction is given by the formula

$$E_c^{(2)}(r_s) = -\frac{2\hbar^2}{ne^4m} \sum_{\nu(\neq 0)} \frac{\langle 0 | H_{\rm int} | \nu \rangle \langle \nu | H_{\rm int} | 0 \rangle}{E_\nu - E_0} \; ,$$

where $\langle \nu | H_{\rm int} | 0 \rangle$ refers to the matrix element of the interaction Hamiltonian,

$$H_{\rm int} = \frac{1}{2} \sum_{\substack{pp'k \\ \sigma, \sigma'}}' v(k) c_{p+\hbar k.\sigma}^\dagger c_{p'-\hbar k.\sigma'}^\dagger c_{p',\sigma'} c_{p,\sigma} \; ,$$

between the unperturbed Fermi sphere (3.28) state $|0\rangle$ with kinetic energy E_0 and an intermediate state $|\nu\rangle$ with kinetic energy E_ν where two particles are excited above the Fermi sphere. Use these formulas to derive the second-order exchange contribution, Eq. (3.27).

3.7   For a classical OCP, show that the VS scheme consisting of Eqs. (3.31), (3.34), and (3.40) satisfies the compressibility sum rule exactly when $\xi = 1/2$ is chosen.

3.8   In the convolution approximation, the triple correlation function is expressed as

$$h^{(3)}(r_{12}, r_{23}, r_{31}) = h(r_{12})h(r_{23}) + h(r_{23})h(r_{31}) + h(r_{31})h(r_{12})$$
$$+ n \int dr_4 \, h(r_{14})h(r_{24})h(r_{34}) \; .$$

Derive Eq. (3.44) from this expression.

3.9   Show the relation

$$\left. \frac{\partial g(r)}{\partial r} \right|_{r=0} = \frac{3\pi}{8k_F^3} \lim_{k \to \infty} \left\{ k^4 [1 - S(k)] \right\}$$

from a direct analysis of Eq. (1.36). Using this relation in Eq. (3.65) and calculating $S(k)$ with the static local-field correction approximation (1.60), derive the short-wavelength boundary condition (3.61) [Kimball 1973].

3.10 Prove the cusp condition (3.65).

3.11 With the definition of Hamiltonians (3.96), show the relations,

$$(\Psi, (K + 2H_{int})\Psi) = -\sum_{\mathbf{k},\sigma} \left(\Psi, c_{\hbar\mathbf{k}\sigma}^\dagger [H, c_{\hbar\mathbf{k}\sigma}] \Psi\right)$$

$$= -\frac{\hbar}{2\pi i} \sum_{\mathbf{k},\sigma} \int d\omega \ \exp(i0\omega)\omega G_\sigma(\mathbf{k}, \omega) ,$$

and then derive Eq. (3.97b).

3.12 Derive Eqs. (3.122) and (3.123).

3.13 Derive Eq. (3.135c).

3.14 Derive Eqs. (3.148) and (3.149).

3.15 Derive Eq. (3.160).

CHAPTER 4

Dense Plasma Materials

A dense plasma material may be modeled as a two-component plasma (TCP) constituted by electrons and ions. These constituents were treated individually in the preceding two chapters as representatives for the classical and quantum charged systems. In a TCP, the attractive interaction between electrons and ions, an essential ingredient for formation of atoms, brings about a novel feature in physics of dense plasmas: that the strong correlations between electrons and ions be taken into account on an equal footing to the atomic and molecular processes in such a condensed environment. Atomic levels and their existence should be influenced strongly by the statistical properties of the dense plasmas. The *strong electron-ion (e-i) coupling* thus opens up new dimensions in the condensed plasma physics, where an outstanding issue will be an interplay between the atomic and molecular physics, on the one hand, and the plasma and condensed matter physics, on the other. The former may be classified as a few-particle physics, and the latter may belong to the class of many-particle physics.

Most of the treatments of plasma problems in the foregoing chapters, except those issues on solidification and the resultant Coulomb crystals, have assumed translational invariance or homogeneity of the system. Strong electron-ion correlations and potential involvement of the atomic and molecular processes now make it necessary to consider cases related to *inhomogeneous* systems. Microscopic study of *e-i* correlations in hydrogen plasmas near *metal-insulator transition* may in fact reveal emergence of an "*incipient Rydberg state* (IRS)" for the electrons even when the plasma is in the liquid-metallic phase; the IRS acts partially to localize wave functions of the electrons. Inclusion of IRS thus leads to significant modification for the equations of state and enhancement in the rates of electron scattering. Such a strong e-i coupling likewise affects ionization processes for the impurity atoms immersed in such a liquid-metallic hydrogen, by shallowing or eliminating the atomic levels and through changes in the interaction potentials. Taking those strong coupling effects into account, we investigate the degrees of various stages of ionization, the resultant equations of state, and transport processes in dense plasma materials.

4.1   FUNDAMENTALS

Various theoretical tools relevant to consideration of electron-ion coupling are briefly revisited in this section. They are strong-coupling theories of statistical correlations without resort to perturbation-theoretic expansions. Some of the perturbative approaches to statistical correlations and atomic problems have been described in the literature [e.g., Ebeling et al. 1991].

A. Dielectric Formulation for Multicomponent Systems

For the analyses of interparticle correlations in dense multicomponent plasmas, the density-response formalism or the dielectric formulation may be employed in the framework of the linear response theory (see Section 1.2D). A weak external potential field $V_\mu^{ext}(\mathbf{r}, t)$ that couples only to the density field $n_\mu(\mathbf{r})$ of the μ-species particles is applied to the plasma. The extra Hamiltonian arising from the presence of such external fields is then written as

$$H_{ext} = \sum_\mu \int d\mathbf{r} \, n_\mu(\mathbf{r}) V_\mu^{ext}(\mathbf{r}, t) \, . \qquad (4.1)$$

For a TCP, we reserve the subscript $\mu = 1$ for the electrons and $\mu = 2$ for the ions, so that $Z_1 = -1$ and $Z_2 = Z$.

The plasma state in the absence of an external field is assumed to be translationally invariant both in space and in time. In terms of the Fourier components, the linear density response relations may be written as

$$\delta n_\mu(\mathbf{k}, \omega) =$$

$$\chi_\mu^{(0)}(\mathbf{k}, \omega) \left\{ \delta V_\mu^{ext}(\mathbf{k}, \omega) + \sum_\nu Z_\mu Z_\nu v(k) \left[1 - G_{\mu\nu}(\mathbf{k}, \omega) \right] \delta n_\nu(\mathbf{k}, \omega) \right\} \, , \qquad (4.2)$$

where we recall $v(k) = 4\pi e^2 / k^2$.

The free-particle polarizability $\chi_\mu^{(0)}(\mathbf{k}, \omega)$ thus describes a density response of the μ-species particles against a renormalized external disturbance given by the braces in Eq. (4.2). Assuming the Fermi distribution, one expresses

$$
\chi_\mu^{(0)}(k, \omega) = -\frac{m_\mu (3\pi^2 n_\mu)^{1/3}}{2\pi^2 \hbar^2 K}
$$

$$
\times \int_0^\infty dx \, \frac{x}{1 + \exp\left[(x^2 - M_\mu)/\Theta_\mu\right]} \ln\left[\frac{(2xK + K^2)^2 - \Omega^2}{(2xK - K^2)^2 - \Omega^2}\right]
$$

(4.3)

with

$$
K = \frac{k}{(3\pi^2 n_\mu)^{1/3}} ,
$$
(4.4a)

$$
\Omega = \frac{2m_\mu(\omega + i0)}{\hbar(3\pi^2 n_\mu)^{2/3}} ,
$$
(4.4b)

$$
\Theta_\mu = \frac{2m_\mu k_B T}{\hbar^2(3\pi^2 n_\mu)^{2/3}} ,
$$
(4.4c)

and the dimensionless chemical potential M_μ is determined through normalization condition,

$$
\frac{1}{3} = \int_0^\infty dx \, \frac{x^2}{1 + \exp\left[(x^2 - M_\mu)/\Theta_\mu\right]} .
$$
(4.4d)

The dynamic local field corrections $G_{\mu\nu}(\mathbf{k}, \omega)$ in Eq. (4.2) are the functions that represent all the correlation effects in the plasmas. Through these functions, the linear-response formalism coupled with the fluctuation-dissipation theorem may lead to a *rigorous* description for all the higher-order nonlinear processes in strong coupling.

The density-density response functions $\chi_{\mu\nu}(\mathbf{k}, \omega)$ defined in Eq. (1.32) are now obtained from the solutions to Eq. (4.2). For a TCP with suppression of the wave number and frequency variables, we write

$$
\chi_{11} = \chi_1^{(0)}\left[1 - Z_2^2 v \chi_2^{(0)}(1 - G_{22})\right]/D ,
$$
(4.5a)

$$
\chi_{22} = \chi_2^{(0)}\left[1 - Z_1^2 v \chi_1^{(0)}(1 - G_{11})\right]/D ,
$$
(4.5b)

$$
\chi_{12} = Z_1 Z_2 v \chi_1^{(0)} \chi_2^{(0)}(1 - G_{12})/D ,
$$
(4.5c)

$$
\chi_{21} = Z_1 Z_2 v \chi_1^{(0)} \chi_2^{(0)}(1 - G_{21})/D ,
$$
(4.5d)

where

$$D = \left[1 - Z_1^2 v \chi_1^{(0)}(1 - G_{11})\right]\left[1 - Z_2^2 v \chi_2^{(0)}(1 - G_{22})\right]$$
$$- Z_1^2 Z_2^2 v^2 \chi_1^{(0)} \chi_2^{(0)}(G_{11} + G_{22} - G_{12} - G_{21}) \qquad (4.5e)$$

[e.g., Ichimaru, Iyetomi, & Tanaka 1987]. The dielectric response function $\varepsilon(\mathbf{k}, \omega)$ is then formulated as in Eq. (1.33). Partial elements of the dynamic structure factor $S_{\mu\nu}(\mathbf{k}, \omega)$, the static structure factor $S_{\mu\nu}(\mathbf{k})$, and the radial distribution function $g_{\mu\nu}(\mathbf{k}, \omega)$ are given by Eqs. (1.34)–(1.36), respectively, through the fluctuation-dissipation theorem of Appendix A.

The interaction energy per unit volume is likewise expressed as a sum of the partial contributions

$$E_{\text{int}} = \sum_{\mu,\nu} E_{\mu\nu}^{\text{int}} , \qquad (4.6a)$$

where

$$E_{\mu\nu}^{\text{int}} = \frac{\sqrt{n_\mu n_\nu}}{2} \int \frac{d\mathbf{k}}{(2\pi)^3} \frac{4\pi Z_\mu Z_\nu e^2}{k^2} \left[S_{\mu\nu}(k) - \delta_{\mu\nu}\right]$$
$$= \frac{n_\mu n_\nu}{2} \int d\mathbf{r} \frac{Z_\mu Z_\nu e^2}{r} \left[g_{\mu\nu}(r) - 1\right] . \qquad (4.6b)$$

The free energy per unit volume is then calculated by a coupling-constant integration as

$$F = F_0 + \int_0^1 \frac{d\eta}{\eta} E_{\text{int}}(\eta) . \qquad (4.7)$$

Here F_0 is the free energy of the corresponding ideal-gas system and $E_{\text{int}}(\eta)$ refers to the interaction energy (4.6a) evaluated in a system where the strength of Coulomb coupling e^2 is replaced by ηe^2.

B. Pseudopotentials in Electron-Ion Systems

Effective interactions between particles of various species are central issues in elucidating the physical properties of dense plasma materials. These issues of *pseudopotentials* have been extensively investigated in condensed matter physics [Heine, Nozières, & Wilkins 1966; Hedin & Lundqvist 1971; Kukkonen & Smith 1973; Kukkonen & Wilkins 1979]. We here reformulate these effective interactions for TCPs in conjunction with the density-response relationship (4.2).

1. Ion-Ion Interaction

To begin we assume that the coordinates of electrons and of ions can be treated separately and that the electrons follow the motions of ions adiabatically because of large differences in the masses. The ion fields are thus regarded as those of mutually distinguishable classical charges; a wave number and frequency dependent, external test-charge field $\varphi_{ion}(\mathbf{k}, \omega)$ of ions is introduced to the OCP system of electrons. The external ion-field induces a density fluctuation $\delta n_e(\mathbf{k}, \omega)$ in the electron OCP in accordance with Eq. (4.2):

$$\delta n_e(\mathbf{k}, \omega) = \chi_e^{(0)}(\mathbf{k}, \omega)\,[\delta\varphi_{ion}(\mathbf{k}, \omega) + \delta\phi_{ind}(\mathbf{k}, \omega)] \ , \tag{4.8a}$$

with

$$\delta\phi_{ind}(\mathbf{k}, \omega) = v(k)\,[1 - G_e(\mathbf{k}, \omega)]\,\delta n_e(\mathbf{k}, \omega) \ . \tag{4.8b}$$

Here $\chi_e^{(0)}(\mathbf{k}, \omega)$ is the free-electron polarizability, $G_e(\mathbf{k}, \omega)$ refers to the dynamic local field correction in the electron OCP, and $\delta\phi_{ind}(\mathbf{k}, \omega)$ represents the effective potential of electron interaction produced by the induced density fluctuation (4.8a) of electrons. This potential generally differs from the bare Coulomb potential, $v(k)\delta n_e(\mathbf{k}, \omega)$, due to the exchange and correlation effects between electrons; the difference is here measured by the dynamic local field correction $G_e(\mathbf{k}, \omega)$.

The total potential acting on another test charge (ion)—that is, the effective potential of ion-ion interaction—is now calculated as

$$\phi_{tot}(\mathbf{k}, \omega) = \varphi_{ion}(\mathbf{k}, \omega) + v(k)\delta n_e(\mathbf{k}, \omega)$$
$$= \varphi_{ion}(\mathbf{k}, \omega)/\varepsilon_e(\mathbf{k}, \omega) \ , \tag{4.9}$$

where the dielectric function for the electron is given by

$$\varepsilon_e(\mathbf{k}, \omega) = 1 - \frac{v(k)\chi_e^{(0)}(\mathbf{k}, \omega)}{1 + v(k)G_e(\mathbf{k}, \omega)\chi_e^{(0)}(\mathbf{k}, \omega)} \ . \tag{4.10}$$

The ion-ion field is screened in Eq. (4.9) by the usual dielectric function of the electrons.

The effective interaction between two ions at rest is then calculated as

$$\Phi_{eff}(r) = \int \frac{d\mathbf{k}}{(2\pi)^3}\,\frac{4\pi(Ze)^2}{k^2\varepsilon_e(k, 0)}\,\exp(i\mathbf{k}\cdot\mathbf{r}) \ . \tag{4.11}$$

Such a formula provides a basis for the pseudopotential theory of ion-ion interaction in metallic substances [Ashcroft & Stroud 1978; Singwi & Tosi 1981; Hafner 1987].

2. Electron-Ion Interaction

In light of the formulation described in Eqs. (4.8), the potential field of the test ion acting on the electrons is simply expressed as

$$\phi_{ei}(\mathbf{k}, \omega) = \varphi_{ion}(\mathbf{k}, \omega) + \delta\phi_{ind}(\mathbf{k}, \omega)$$
$$= \varphi_{ion}(\mathbf{k}, \omega)/\tilde{\varepsilon}_e(\mathbf{k}, \omega) . \qquad (4.12)$$

The screening function $\tilde{\varepsilon}_e(\mathbf{k}, \omega)$ for the electron-ion interaction is now given by

$$\tilde{\varepsilon}_e(\mathbf{k}, \omega) = 1 - v(k)[1 - G_e(\mathbf{k}, \omega)]\chi_e^{(0)}(\mathbf{k}, \omega) . \qquad (4.13)$$

As we find here, the electronic screening function for the electron-ion interaction generally differs from the usual dielectric screening function (4.10) of the electrons that enters the ion-ion interaction; the difference again stems from the local field correction of the electron-electron interaction. The formula (4.12) provides a basis for the pseudo-potential theory of electron-ion interaction in metallic substances [Ashcroft & Stroud 1978; Singwi & Tosi 1981; Hafner 1987]. The distinction of electronic screening between electron-ion and ion-ion interactions should be correctly taken into account for the analysis of pseudopotentials in condensed plasmas.

C. A Thermodynamic Variational Principle

A variational principle based on the *Gibbs–Bogoliubov inequality* for the free energy of a quantum-statistical system [e.g., Feynman 1972; Hansen & McDonald 1986] may sometimes be useful in assessing the thermodynamic quantities of a condensed matter [Mansoori & Canfield 1969; Rasaiah & Stell 1970; Ashcroft & Stroud 1978]. For the plasma under consideration, we assume the following: (1) The ions may be regarded as classical particles so that $\Lambda \ll 1$ (see Eq. 1.12). (2) The electrons are degenerate so that $\Theta \ll 1$ (see Eq. 1.8). (3) The electrons are nonrelativistic so that $x_F \ll 1$ (see Eq. 1.6).

Since Fermi velocity of the electrons is much greater than thermal velocity of the ions, an adiabatic approximation may be adopted, permitting elimination of the electronic coordinates; the system may thus be looked upon as an electron-screened ion fluid. In light of Eq. (4.11), the Hamiltonian of such a system with the total number of ions N is written as

$$H = E_e + K + \frac{Z^2}{2V}\sum_{k \neq 0}\frac{v(k)}{\varepsilon_e(k, 0)}(\rho_k\rho_{-k} - N) . \qquad (4.14)$$

Here E_e represents the ground-state energy of the electron system, K refers to the kinetic energy of ions, and ρ_k are the Fourier components of ion-density fluctuations. The terms E_e and K, being simply additive in Eq. (4.14), will be omitted hereafter.

The Helmholtz free energy F is calculated as a functional of the density matrix P with unit trace:

$$F = F[P] = \text{Tr}(PH) + \beta^{-1}\text{Tr}(P \ln P) \,. \tag{4.15}$$

The canonical ensemble, $P_{eq} = \exp(-\beta H)/\text{Tr}\exp(-\beta H)$, minimizes this functional [Mermin 1965], so that

$$F_{eq} \equiv F[P_{eq}] \le F[P] \,. \tag{4.16}$$

Generally for an interacting many-particle system, it is a difficult task to obtain F_{eq} and P_{eq} exactly. Instead one may evoke a variational principle on the basis of the inequality (4.16), by which F_{eq} and P_{eq} may be determined approximately.

One thus chooses a Hamiltonian H_0 with a known solution and regards it as a reference system containing variational parameters. F_{eq} and P_{eq} may then be approximated by $F[P_0]$ and $P_0 = \exp(-\beta H_0)/\text{Tr}\exp(-\beta H_0)$, where P_0 is determined by minimizing $F[P_0]$ with respect to the variational parameters. Substitution of P_0 in (4.16) yields a formula known as the Gibbs–Bogoliubov inequality:

$$F_{eq} \le F_0 + \langle H - H_0 \rangle_0 \,, \tag{4.17}$$

where $\langle \ldots \rangle_0$ denotes the ensemble average over the reference system, and

$$F_0 = \langle H_0 \rangle_0 + \beta^{-1}\langle \ln P_0 \rangle_0 \,. \tag{4.18}$$

With a variational principle so established, the remaining problem is the choice of reference system. Two candidates may be noted: a hard-sphere (HS) system for which the Percus–Yevick equation has the exact Thiele–Wertheim solution (see Eq. 2.64) and an OCP, whose correlation and thermodynamic properties have been accurately elucidated in Chapter 2. The HS reference system has been used extensively for the study of thermodynamic functions of liquid metals [e.g., Jones 1971; Stroud 1973; Ross & Seale 1974; Umar et al. 1974; Stevenson 1975]. Galam and Hansen [1976] have shown numerically that the OCP reference systems give lower estimates of free energy for the electron-screened ion systems than the HS systems; OCPs make a better reference system than HS systems. One may thus adopt the OCPs as the reference system and regard its "effective charge Ze'" as the variational parameter. The effective charge physically describes the extent to which the ionic charge Ze is reduced owing to the electron screening.

The Gibbs–Bogoliubov inequality (4.17) now reads

$$\frac{\beta F}{N}(\Gamma, r_s) \le \frac{\beta \tilde{F}}{N}(\Gamma', \Gamma, r_s) , \tag{4.19}$$

with

$$\frac{\beta \tilde{F}}{N}(\Gamma', \Gamma, r_s) = \frac{\beta F^{OCP}}{N}(\Gamma') + \frac{a\Gamma}{\pi} \int_0^\infty dk \, S^{OCP}(k; \Gamma') \left[\frac{1}{\varepsilon_e(k, 0)} - 1\right]$$
$$+ \frac{\Gamma - \Gamma'}{\Gamma'} \frac{\beta U^{OCP}}{N}(\Gamma') , \tag{4.20}$$

where $F^{OCP}(\Gamma')$, $U^{OCP}(\Gamma')$, and $S^{OCP}(k; \Gamma')$ refer, respectively, to the excess free energy, the excess internal energy, and the structure factor for the OCP with an effective coupling parameter

$$\Gamma' = \frac{\beta(Ze')^2}{a} . \tag{4.21}$$

The variational parameter Γ' is then determined from the condition for minimization of the variational free energy:

$$\frac{\partial}{\partial \Gamma'} \frac{\beta \tilde{F}}{N}(\Gamma', \Gamma, r_s) \bigg|_{\Gamma, r_s} = 0 . \tag{4.22}$$

The use of Γ' both as the variational parameter and as the value of solution to Eq. (4.22) should not cause any confusion.

The variational estimates of the free energy may thus be obtained by substitution of the solution Γ' to Eq. (4.22) in Eq. (4.20). Other thermodynamic quantities such as the internal energy U and the pressure P may likewise be calculated as

$$\frac{\beta U}{N}(\Gamma, r_s) = \frac{\Gamma}{\Gamma'} \frac{\beta U^{OCP}}{N}(\Gamma') + \frac{a\Gamma}{3\pi} \int_0^\infty dk \, S^{OCP}(k; \Gamma') \left[\frac{1}{\varepsilon_e(k, 0)} - 1\right] \tag{4.23}$$

and

$$\frac{\beta P}{n}(\Gamma, r_s) = \frac{1}{3} \frac{\beta U^{OCP}}{N}(\Gamma') - \frac{a\Gamma r_s}{3\pi} \int_0^\infty dk \, S^{OCP}(k; \Gamma') \frac{\partial}{\partial r_s} \left[\frac{1}{\varepsilon_e(k, 0)}\right] . \tag{4.24}$$

The thermodynamic properties of dense electron-screened plasmas have been investigated with various versions of the static–screening function $\varepsilon_e(k, 0)$ through such a variational method [Galam & Hansen 1976; Iyetomi, Utsumi, & Ichimaru 1981].

D. Self-Consistent Integral Equations

In the static local field correction approximation of Section 3.2A, one obtains a set of self-consistent integral equations describing correlation and thermodynamic properties of multicomponent plasmas. In Section 3.2B, two of such integral-equation schemes were noted: the STLS scheme proposed by Singwi and his collaborators [Singwi et al. 1968] and the MCA scheme introduced by Tago, Utsumi, and Ichimaru [1981].

For a TCP, the STLS approach follows Eqs. (3.82). For the analysis of the electron-hole liquids in semiconductors, Vashishta, Bhattacharyya, and Singwi [1974] used such an STLS scheme, the results of which were incorporated in the review by Vashishta, Kalia, and Singwi [1983].

The MCA approach for a TCP analogously follows Eqs. (3.83) and (3.84). To develop a strong coupling theory of dense hydrogen plasmas applicable in the vicinity of the metal-insulator boundaries (see Fig. 4.1 on page 175), Tanaka, Yan, and Ichimaru [1990] employed such an MCA scheme, justified both in the classical plasmas and in the quantum electron liquids, for the description of quantum mechanical electron-electron and electron-ion correlations and adopted the hypernetted-chain (HNC) approximation for the classical ion-ion correlations. The resulting HNC-MCA equations were solved self-consistently for the structure factors and the local field corrections. As we shall elaborate in Section 4.2, the equation of state thus calculated has revealed the emergence of an incipient Rydberg state (IRS) in the metallic phase, implying physically an approach toward an insulator phase. The IRS has been found to modify the electron-ion correlations remarkably and to reduce electric and thermal conductivities by enhancing the rates of electron-ion scattering.

For a multicomponent plasma, the Ornstein–Zernike relation (2.41) may be generalized to

$$h_{\mu\nu}(\mathbf{r}, \mathbf{r}') = c_{\mu\nu}(\mathbf{r}, \mathbf{r}') + \sum_{\lambda} \int d\mathbf{r}'' \, n_\lambda(\mathbf{r}'') h_{\mu\lambda}(\mathbf{r}, \mathbf{r}'') c_{\lambda\nu}(\mathbf{r}'', \mathbf{r}') . \quad (4.25)$$

Mindful of application to an inhomogeneous situation, two-particle functions have been expressed as two-point functions of the spatial coordinates. Closure of the pair distributions by the HNC approximation (2.45) yields

$$g_{\mu\nu}(\mathbf{r}, \mathbf{r}') = 1 + h_{\mu\nu}(\mathbf{r}, \mathbf{r}') \stackrel{\text{(HNC)}}{=} \exp\left[-\beta \frac{Z_\mu Z_\nu e^2}{|\mathbf{r} - \mathbf{r}'|} + h_{\mu\nu}(\mathbf{r}, \mathbf{r}') - c_{\mu\nu}(\mathbf{r}, \mathbf{r}')\right]$$

$$(4.26)$$

Since these formulations are based totally on classical approximations, the cases of attractive interaction with $Z_\mu Z_\nu < 0$ cannot be treated by such HNC integral equations. It is essential to consider quantum effects, especially when an attractive interaction is involved.

E. Density-Functional Approaches to Inhomogeneous Systems

As Eqs. (4.25) and (4.26) have illustrated, quantities describing the exchange and correlation effects in an inhomogeneous system, such as electrons in atoms and molecules and in metal surfaces, exhibit complex nonlocal characters in their dependence on spatial coordinates of the individual particles. The principal issue in formulating a density-functional theory for such an inhomogeneous system is the evaluation of exchange-correlation potentials $v_\sigma^{xc}[\mathbf{r}; n]$ of Eq. (1.77) or (1.84) acting on an electron with spin σ. We use brackets here to denote *functionals* of the inhomogeneous density distributions. Since the electron densities may exhibit steep variations reflecting the spatial inhomogeneity, nonlocality may play an essential role in the treatment of such an exchange-correlation potential.

The exchange-correlation free-energy functionals of the inhomogeneous electron systems are generally calculated as

$$F_{\sigma\tau}^{xc}\left[n_\sigma(\mathbf{r})\right] = \frac{1}{1+\delta_{\sigma\tau}} \int_0^1 d\eta \int d\mathbf{r}\,d\mathbf{r}' \frac{e^2}{|\mathbf{r}-\mathbf{r}'|} n_\sigma(\mathbf{r}) n_\tau(\mathbf{r}') \left\{ g_{\sigma\tau}\left[\mathbf{r}, \mathbf{r}'; n\right] - 1 \right\}$$

$$\text{(4.27)}$$

(see Eq. 2.69). Here $g_{\mu\nu}[\mathbf{r}, \mathbf{r}'; n]$ are the spin-dependent pair-distribution functionals evaluated in the inhomogeneous electron systems where the strength of Coulomb coupling e^2 has been replaced by ηe^2 and $\delta_{\sigma\tau}$ represents Kronecker's delta. The exchange-correlation potentials on the electrons with spin σ are therefore given by the functional derivatives,

$$v_\sigma^{xc}\left[n_\sigma(\mathbf{r})\right] = \sum_\tau \frac{\delta F_{\sigma\tau}^{xc}\left[n_\sigma(\mathbf{r})\right]}{\delta n_\sigma(\mathbf{r})} . \tag{4.28}$$

Evaluation of the exchange-correlation potentials thus depends on the assessability of the spin-dependent pair-distribution functionals in the inhomogeneous system. For the accuracy of such an assessment, various sum rule constraints, such as those considered in Section 3.2C, need to be accounted for. Particularly important among such constraints is a precaution to avoid inclusion of *self-interaction* in the exchange-correlation potentials. Since the pair distributions are by definition the joint probability densities between two *different* particles, the use of correctly evaluated $g_{\mu\nu}[\mathbf{r}, \mathbf{r}'; n]$ in Eq. (4.27) should exclude such a self-interaction in Eq. (4.28). In practice, however, approximations are involved in an evaluation of $g_{\mu\nu}[\mathbf{r}, \mathbf{r}'; n]$. To make certain that the self-interaction may be appropriately excluded, one then evokes unitality conditions for the strengths of exchange-correlation holes [e.g., Kohn & Vashishta 1983]:

$$\int d\mathbf{r}'\, n_\tau(\mathbf{r}')\left\{g_{\sigma\tau}\left[\mathbf{r}, \mathbf{r}'; n\right] - 1\right\} = -\delta_{\sigma\tau} .\tag{4.29}$$

Gunnarsson, Jonson, and Lundqvist [1979] noted the importance of securing such conservation properties in the nonlocal exchange and correlation holes. The proposed approximation schemes, however, turned out to be fairly complex, so the self-consistent solutions to the resultant Kohn–Sham equations (1.78) for the density distributions were not obtained. Subsequently, in a treatment of strongly inhomogeneous systems, such as the electrons near the metal surface, Yamashita and Ichimaru [1984] developed a theoretical scheme by which the nonlocal character of the exchange-correlation potentials may be properly considered in the evaluation of the pair-distribution functionals; a metal-surface problem has thereby been treated.

In the latter treatment, the pair-distribution functionals of the inhomogeneous system are approximated by the radial distribution functions of an equivalent homogeneous system with an *average density* $n_{\sigma\tau}^{\mathrm{av}}$, so that

$$g_{\sigma\tau}\left[\mathbf{r}, \mathbf{r}'; n\right] \rightarrow g_{\sigma\tau}\left(\left|\mathbf{r} - \mathbf{r}'\right|, n_{\sigma\tau}^{\mathrm{av}}\right) .\tag{4.30}$$

Utility of the approximation scheme (4.30) rests on finding a proper nonlocal density dependence $n_{\sigma\tau}^{\mathrm{av}}$ sufficiently accurate to represent the inhomogeneous pair-distribution functionals. For reasons of symmetry and simplicity, three such possibilities have been examined, and it was concluded that the choice of

$$n_{\sigma\tau}^{\mathrm{av}} = \frac{1}{2}\left[n_\sigma(\mathbf{r}) + n_\tau(\mathbf{r}')\right]\tag{4.31}$$

is most satisfactory in light of the sum rules and other rigorous requirements. In Section 4.3A, such a nonlocal approach will be applied to certain atomic problems.

The *local-density approximation* (LDA), developed by Kohn and Sham [1965] and used for metal-surface [e.g., Lang & Kohn 1970, 1971; see also Lang 1973] and atomic [e.g., Perdew & Zunger 1981] problems, assumes that the exchange-correlation free energies are

$$F_{\sigma\tau}^{\mathrm{xc}}\left[n_\sigma(\mathbf{r})\right] \rightarrow \int d\mathbf{r}\, f_{\sigma\tau}^{\mathrm{xc}}(\mathbf{r}) n_\sigma(\mathbf{r}) ,\tag{4.32}$$

where $f_{\sigma\tau}^{\mathrm{xc}}(\mathbf{r})$ denotes the exchange-correlation free energy between a pair of particles σ and τ for a homogeneous electron liquid with number densities $n_\sigma(\mathbf{r})$ and

$n_\tau(\mathbf{r})$. In light of Eq. (1.89) and the compressibility sum rules (1.66)–(1.67), the LDA physically amounts to taking account of the spin-dependent compressibilities appropriate to the local spin densities $n_\sigma(\mathbf{r})$. Nonlocal characters resulting from the two-point functionals in Eq. (4.27) have been entirely ignored in the evaluation of the exchange-correlation potentials (4.28) in which Eq. (4.32) is substituted. The local spin-density approximations of Eq. (4.32) for exchange-correlation free energy involve self-interaction [Gunnarsson & Lundqvist 1976]; precautions to avoid such self-interaction need to be worked out separately [Perdew & Zunger 1981].

A number of investigators have considered improvement on the LDA by including those terms stemming from the density-gradient expansions [see, for example, Kohn & Vashishta 1983]. Basically this is a perturbation-theoretic calculation in which the density variation is assumed to be weak. Opinion about the usefulness of including those gradient corrections in the exchange-correlation functionals has been divided, however. Ma and Brueckner [1968] found that for heavy atoms the correlation energy due to the density-gradient expansion overestimates the necessary correction by about a factor of 5. Gupta and Singwi [1977] estimated that with the use of the first gradient correction the remaining error in the metal surface energy was only a few percent. Lau and Kohn [1976] and Perdew, Langreth, and Sahni [1977] demonstrated that the first density-gradient correction to the LDA gives no improvement to the surface energy and worsens the density profile. Langreth and Perdew [1977], considering the surface energy in terms of fluctuations at various wavelengths, interpolated between the LDA, accurate in the short-wavelength limit, and the RPA in the long-wavelength limit. In the process they concluded that the LDA gave reasonable accuracy (better than 10%) for the surface energy.

4.2   LIQUID-METALLIC HYDROGEN

Liquid-metallic hydrogen is a statistical ensemble consisting of electrons and protons with the same number density n. The Coulomb coupling parameter for such a plasma is $\Gamma = e^2/ak_BT$, where $a = (3/4\pi n)^{1/3}$ is the ion-sphere radius; the degeneracy parameter Θ for the electrons is given by Eq. (1.8). For a nonrelativistic plasma ($x_F \ll 1$ and $\Lambda \ll 1$), the dimensionless density parameter (1.2) of the electrons is related to these parameters via

$$r_s = \frac{1}{2}\left(\frac{9\pi}{4}\right)^{2/3}\Gamma\Theta. \tag{4.33}$$

Figure 4.1 illustrates the relative magnitude of those dimensionless parameters on the n-T plane.

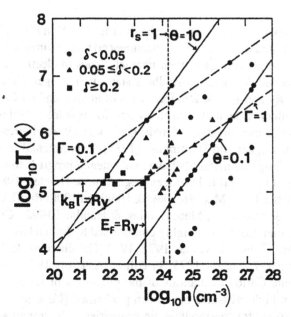

Figure 4.1 Various parameters on the number density versus temperature plane for hydrogen plasma. See the text for definition of Γ, r_s, Θ, and δ.

The condition for pressure ionization in hydrogen is approximately given by

$$E_F = Ry \, , \qquad (4.34a)$$

where $Ry = me^4/2\hbar^2 = 13.6058$ eV, the ionization energy of a hydrogen atom in the ground state. Analogously the condition for thermal ionization is taken approximately as

$$k_B T = Ry \, . \qquad (4.34b)$$

Relations (4.34) are likewise displayed in Fig. 4.1. Hydrogen is in a fully ionized, metallic state when $E_F > Ry$ or $k_B T > Ry$.

For an isolated hydrogen atom in the ground state, the value of the joint probability function between an electron and an ion at separation $r = 0$ is

$$g_{12}^0(0) = \frac{1}{n}|\psi_{1s}(0)|^2 = \frac{4}{3}r_s^3 \, , \qquad (4.35a)$$

where $\psi_{1s}(r)$ is the wave function for a 1s electron. The e-i interaction energy in such an atom is $-2Ry$, so that

$$\frac{E_{12}^0}{Nk_B T} = -\frac{1}{2}\left(\frac{9\pi}{4}\right)^{2/3}\Theta\Gamma^2 \, . \qquad (4.35b)$$

As a hydrogen plasma approaches the boundaries (4.34) in Fig. 4.1, coupling between electrons and ions becomes so pronounced that features resembling (4.35) of a bound state may emerge in their characteristics of electron-ion correlation. Such will be called the incipient Rydberg states in hydrogen plasmas.

For $E_F < Ry$ and $k_B T < Ry$, formation of atomic and molecular hydrogen becomes a primary issue; metal-insulator transitions [Mott 1990] will take place. The possibility of metallization of hydrogen under ultrahigh pressures was first pointed out by Wigner and Huntington [1935]. Experimental investigations on pressurized hydrogenic materials have made steady progress toward laboratory demonstration for a metallic hydrogen [e.g., Mao & Hemley, 1989; Lorenzana, Silvera, & Goettel 1989; Mao, Hemley, & Hanfland 1990; Ruoff & Vanderborgh 1991; Hemley & Mao 1991; Mao, Hemley, & Hanfland 1992]. Condensed matter theories of metallization in highly compressed hydrogen have been proposed [e.g., Brovman, Kagan, & Kholas 1971, 1972; Hammerberg & Ashcroft 1974; Ramaker, Kumar, & Harris 1975; Friedli & Ashcroft 1977; Ashcroft 1989]. Quantum Monte Carlo calculations of the properties of solid hydrogen in the ground state at high pressures have been performed [Ceperley & Alder 1987]. Model calculations for thermodynamic properties of hydrogen across the conditions for pressure ionization have been advanced [e.g., Ebeling et al. 1991; Saumon & Chabrier 1992]. A heuristic treatment of ionization processes for high-Z elements in dense plasmas will be considered in Section 4.3.

A. Strong Electron-Ion Coupling

The strong coupling effects in dense plasmas have been studied theoretically by a number of investigators. Strong ion-ion and electron-electron correlations were studied individually in Chapters 2 and 3. The principal new feature in dense plasma materials is an additional strong coupling effect between electrons and ions brought about by their attractive mutual interaction. In Section 4.1C, we considered application of a thermodynamic variational principle to a treatment of such an electron-ion effect, assuming that an adiabatic approximation may be applicable to coordinates of the electrons.

A density-functional theory of hydrogen plasmas was presented by Dharma-wardana and Perrot [1982] (referred to as DP). In this theory, the ion-ion correlations were analyzed in the HNC approximation. The electron-electron correlations were treated in the density-functional formalism with the LDA (see Section 4.1E); spatial dispersion (i.e., wave number dependent) effects in the exchange-correlation potentials between electrons were essentially ignored. In their LDA treatment of electron-ion correlations, the contributions of exchange-correlation potentials were neglected; no local field correction effects between electrons and ions have therefore been retained.

In an earlier series of investigations, Ichimaru et al. [1985; see also Ichimaru, Iyetomi, & Tanaka 1987] (referred to as IMTY) developed a strong-coupling theory of dense hydrogen plasmas appropriate to the interior of the main-sequence stars and to final stages of inertial-confinement fusion plasmas. In such a dense plasma, the strong exchange and Coulomb coupling between the charged particles beyond the RPA becomes essential. In the density-response formalism of Section 4.1D, on which the IMTY theory is based, such a strong coupling effect can be treated in terms of the local field corrections.

When the density and/or temperature of the plasma is lowered toward the conditions (4.34) for the onset of pressure and/or thermal ionization, Coulomb coupling between electrons and ions becomes particularly pronounced; a trend toward an incipient formation of bound pairs (i.e., neutral atoms) should be revealed in the features of the electron-ion correlations. A major shortcoming of the IMTY theory lies in its inaccuracy in treating such an effect of strong electron-ion coupling as the plasma approaches the metal-insulator boundaries. These authors expressed the joint distributions between electrons and ions approximately in terms of a linear response of the electrons against the ions; as a consequence, the local field correction $G_{12}(r)$ between electrons and ions vanished identically in their calculations. The predicted equation of state (EOS) did not show a tendency toward incipient bound pairs; the calculated values of the conductivities remained relatively high near the metal-insulator boundaries.

A strong-coupling theory of dense hydrogen plasmas applicable in the vicinity of the metal-insulator boundaries has been developed subsequently [Tanaka, Yan, & Ichimaru 1990; Ichimaru 1993b]. Strong exchange and Coulomb correlations are analyzed through an integral equation approach, which adopts the HNC approximation for the classical ion-ion correlations and the MCA for the quantum mechanical electron-electron and electron-ion correlations. The resultant HNC MCA equations are solved self-consistently for the structure factors and the local field corrections at 48 parameter points exhibited by various dots in Fig. 4.1.

To single out the difference between the HNC MCA results [where $G_{12}(k) \neq 0$] and the IMTY results [where $G_{12}(k) = 0$], we define

$$\delta \equiv \frac{E_{\text{int}}^{\text{HNC MCA}} - E_{\text{int}}^{\text{IMTY}}}{E_{\text{int}}^{\text{IMTY}}} \tag{4.36}$$

and exhibit their ranges in Fig. 4.1. For a given value of Θ, the HNC MCA results agree with those of IMTY at small values of Γ, that is, in a weak coupling regime. As Coulomb coupling increases with Γ, however, the HNC MCA predictions start to deviate significantly from the IMTY predictions; the deviation is due to IRS, which acts to modify $g_{12}(r)$. The EOS thus calculated in the HNC

MCA scheme reveals emergence of an IRS in the metallic phase; the emergence physically signals an approach toward an insulator phase. The IRS modifying the electron-ion correlations acts remarkably to reduce the electric and thermal conductivities.

B. Correlation Functions

Correlation functions in dense hydrogen plasmas have been calculated in various theoretical schemes mentioned in the preceding subsection. The partial radial distribution functions $g_{\mu\nu}(r)$ at $\Theta = 1$ and $\Gamma = 0.5$ calculated in the HNC MCA scheme [where $G_{12}(k) \neq 0$] are compared in Fig. 4.2 with those in a HNC MCA scheme where $G_{12}(k) = 0$ is set ab initio. We observe that $G_{12}(k)$ acts to increase the values of $g_{\mu\nu}(r)$ in the short ranges, implying physically an effective enhancement of attraction between particles in the plasma.

Typical values of the local field corrections calculated in the HNC MCA scheme are shown in Fig. 4.3. It is noteworthy that $G_{11}(k) \geq 0$, $G_{22}(k) \geq 0$, and $G_{12}(k) \leq 0$. These imply that differences between the effective potentials $Z_\mu Z_\nu v(k)[1 - G_{\mu\nu}(k)]$ and the bare potentials $Z_\mu Z_\nu v(k)$ are *always* negative (i.e., attractive), irrespective of the sign of $Z_\mu Z_\nu$. The physical reason for the origin of such an extra attraction may be clear as we recall that the local field cor-

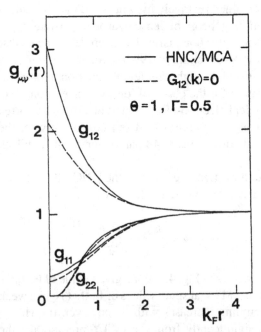

Figure 4.2 Partial radial distribution functions in a hydrogen plasma.

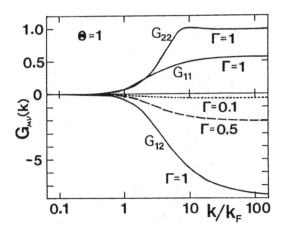

Figure 4.3 Local field corrections at $\Theta = 1$ and $\Gamma = 1$ in the HNC MCA scheme. The values of $G_{12}(k)$ at $\Gamma = 0.1$ (dotted line) and $\Gamma = 0.5$ (dashed line) are also plotted.

rections are the functions describing the correlation effects arising beyond RPA treatments and that the RPA density responses are calculated in a system where interparticle correlations are ignored. Suppose that $Z_\mu Z_\nu > 0$, so that interaction between particles is repulsive; the probabilities of finding other particles around a given particle are actually *less* than those in uncorrelated (i.e., RPA) systems. When such a correlation is taken into account, the strength of repulsive forces therefore becomes *weaker* effectively than what would be assumed in an RPA calculation. Analogous arguments apply to the cases with $Z_\mu Z_\nu < 0$, that is, attractive interaction. These arguments altogether substantiate the foregoing observation that the extra interactions produced by the local field corrections are attractive irrespective of the sign of $Z_\mu Z_\nu$. In a strong coupling regime, a remarkable increase in the magnitude of $G_{12}(k)$ takes place for large k domain, leading to enhancement of the effective electron-ion *attraction* at short distances (see Fig. 4.2); an IRS may thereby be formed.

In Fig. 4.4, values of the effective ion-ion potentials $w_{22}(r)$ in the HNC MCA scheme with $G_{12}(k) \neq 0$ are compared with those with an ab initio assumption of $G_{12}(k) = 0$ and with those of the bare Coulomb potential. Inclusion of the electron-ion local field correction $G_{12}(k)$ is found to reduce the repulsive interionic potential in short ranges, leading to enhancement of $g_{22}(r)$. The short-range enhancement of $g_{11}(r)$, shown in Fig. 4.5, can likewise be interpreted in terms of such reduction in the repulsive interelectronic potentials.

Figures 4.6 and 4.7 compare the HNC MCA results with those (solid circles) obtained by Dharma-wardana and Perrot [1982] (DP). The DP ion-ion corre-

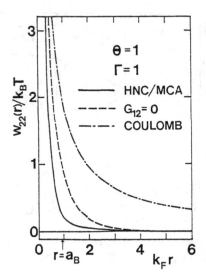

Figure 4.4 Effective ion-ion potentials.

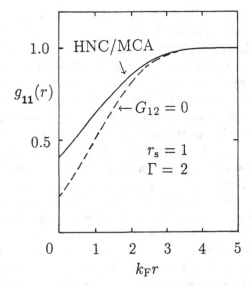

Figure 4.5 Electron-electron radial distribution functions.

lations, relying on the HNC approximation, resemble the HNC MCA results with $G_{12}(k) = 0$. It may be somewhat surprising to find, in Fig 4.7, that the DP scheme, assuming $G_{12}(k) = 0$, appears capable of predicting an IRS, as does the HNC MCA scheme with $G_{12}(k)$ fully taken into account. This feature can be

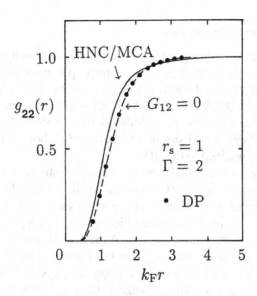

Figure 4.6 Ion-ion radial distribution functions.

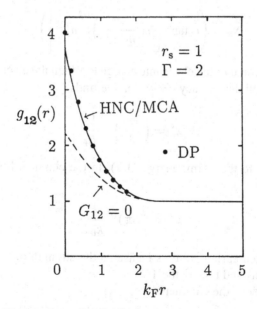

Figure 4.7 Electron-ion radial distribution functions.

attributed to the use of the Kohn–Sham equation for the electron-ion correlations in the DP scheme (albeit assuming that $F_{12}^{\text{ex}} = 0$ in their LDA).

C. Thermodynamic Properties

The ratio between the Rydberg energy Ry and a relevant kinetic energy measures the degree to which the e-i coupling affects the properties of a dense TCP. Strong Coulomb coupling beyond the RPA in a plasma may be accounted for by the local field corrections $G_{\mu\nu}(k)$. We recall that an IRS is a consequence of those strong e-i correlation effects beyond an RPA description; it stems from the effects of scattering that go beyond the Born approximation. *True* bound states of electrons, however, are *not* contained in the IRS.

We proceed to construct an IRS model for a description of the thermodynamic properties of a nonrelativistic TCP with emphasis on the effects of strong e-i coupling. The model is formulated in a general case of TCP with $Z \geq 1$, so that the electron and ion densities are distinguished by n_e and n_i. An explicit comparison of the model predictions with the results of microscopic calculations is possible for a hydrogen plasma ($Z = 1$) [Tanaka, Yan, & Ichimaru 1990]. The IRS description of a TCP with $Z > 1$ appears physically plausible, but it is presented here only as a working hypothesis [Ichimaru 1993b].

The IRS may be characterized by the parameter

$$x_b = \left(r_s \tanh \left[\hbar \left(\frac{2\pi}{mk_B T} \right)^{1/2} n_e^{1/3} \right] \right)^{1/2}, \tag{4.37}$$

which is proportional to the electronic charge e. When the electrons are in a state of complete Fermi degeneracy ($\Theta \ll 1$), one finds

$$x_b^4 = \left(\frac{9\pi}{4} \right)^{2/3} \frac{E_b}{E_F}, \tag{4.38a}$$

where E_F refers to the Fermi energy (1.7). In the classical limit ($\Theta \gg 1$), one has

$$x_b^4 = (36\pi)^{1/3} \frac{E_b}{k_B T}. \tag{4.38b}$$

In the sense stated earlier, x_b^4 thus measures the strength of Coulomb coupling between electrons and ions in the TCP.

Let us introduce the parameter

$$X = \frac{x_b}{1 + x_b}, \tag{4.39}$$

representing the fraction of electrons in the IRS. The quantity, $n_f = (1 - X)n_e$, designates that part of electron number density that may be regarded as in the ordinary plane-wavelike states, while the remainder, $n_b = X n_e$, may correspond to those IRS parts of free-electron orbitals in strong coupling with the ions.

Associated with those plane-wavelike and IRS electrons, the following screening parameters may be characteristically defined (see Section 3.2D):

$$K_f = \left[\frac{4\pi e^2}{k_B T (\partial \alpha_f / \partial n_f)_{T,V}} \right]^{1/2} , \tag{4.40}$$

$$K_b = \left(\frac{2}{a_B} \right) \frac{8 + 12 a_B K_f}{(2 + a_B K_f)^3} . \tag{4.41}$$

Here a_B is the Bohr radius (1.4) and α_f denotes the dimensionless chemical potential of the plane-wavelike electrons determined from Eq. (B.2) at the degeneracy parameter,

$$\Theta_f = \frac{2m k_B T}{(3\pi^2 n_f)^{2/3}} . \tag{4.42}$$

The screening parameter (4.40) for the plane-wavelike electrons turns into the Thomas–Fermi screening parameter k_{TF} of Eq. (1.50) when $\Theta_f \ll 1$ and into the Debye–Hückel screening parameter k_D of Eq. (1.52) when $\Theta_f \gg 1$. The screening parameter (4.41) for the IRS electrons is related to a 1s orbital and takes into account such a screening effect of the plane-wavelike electrons.

Electrons in both the free states combined then contribute a screening function to the ion fields as

$$S_c(r) = A_f \exp(-K_f r) + (A_{b0} + A_{b1} r) \exp(-K_b r) , \tag{4.43}$$

where

$$A_f = 1 - X \frac{K_b^4}{(K_b^2 - K_f^2)^2} , \tag{4.44a}$$

$$A_{b0} = X \frac{K_b^4}{(K_b^2 - K_f^2)^2} , \tag{4.44b}$$

$$A_{b1} = \frac{X}{2} \frac{K_b^3}{K_b^2 - K_f^2} . \tag{4.44c}$$

The electron-screened Coulomb-coupling parameter for the ions is given by

$$\Gamma_s = \Gamma S_c(a) \approx \frac{(Ze)^2}{a k_B T} \exp\left(-\frac{a}{D_s} \right) , \tag{4.45}$$

where

$$D_s = (A_f K_f + A_{b0} K_b - A_{b1})^{-1} . \qquad (4.46)$$

This quantity thus represents the *short-range screening distance* (see Section 3.2D) in the IRS model. Validity for such a reduced coupling parameter as Eq. (4.45) has been verified in a number of examples [Ichimaru 1993b].

The thermodynamic functions for TCP may be derived by accounting for the IRS as follows: The normalized interaction energy per unit volume of the TCP may be defined as and decomposed into

$$u_{ex}(\Gamma, \Theta; Z) \equiv \frac{E_{int}}{\sqrt{n_e n_i k_B T}} = \sqrt{Z} u_{11} + \frac{1}{\sqrt{Z}} (u_{22} + u_{12} + \delta u_{12}) . \qquad (4.47)$$

Here E_{int} is the quantity defined in Eq. (4.6a); u_{11}, u_{22}, and u_{12} refer to the partial, *e-e*, *i-i*, and *e-i*, interaction energies; and δu_{12} represents a classical, weak-coupling correction to u_{12}. The excess free energy and pressure of the TCP are then derived from Eq. (4.47) in accordance with Eq. (3.78).

The various interaction energies in Eq. (4.47) are expressed and calculated as

$$u_{11} = u_{xc}^{STLS}(\Gamma_e, \Theta) , \qquad (4.48a)$$

$$u_{22} = u_{ex}(\Gamma_s) , \qquad (4.48b)$$

$$u_{12} = (1 - X) u_f^{ei} + X u_b^{ei} , \qquad (4.48c)$$

$$\delta u_{12} = \left[\left(\sqrt{2} - 1 \right) u_{ex}^{ABE}(\Gamma_s) - (1 - X) u_f^{ei} + \frac{\sqrt{3}}{2} \Gamma^{3/2} \right] \exp \left(-\Gamma^{3/2} - \frac{30}{\sqrt{\Theta}} \right) \qquad (4.48d)$$

with

$$u_f^{ei} = -0.94 \frac{(Ze)^2 K_f}{k_B T} \left[1 + 0.16 \exp \left(-\frac{1}{\sqrt{\Theta}} \right) \right] , \qquad (4.49a)$$

$$u_b^{ei} = -\frac{1}{2} Z \Gamma_e a_B K_b x_b^2 , \qquad (4.49b)$$

Here the reader is reminded of the definition given in Eqs. (3.79), (2.75), and (2.73) for the functions, u_{ex}^{STLS}, u_{ex}, and u_{ex}^{ABE}, respectively.

Numerical accuracy of these IRS-model predictions for the thermodynamic properties of hydrogen TCP has been examined through comparison with the results of microscopic calculations on the basis of HNC MCA scheme [Tanaka, Yan, & Ichimaru 1990]. It has been confirmed that deviations in the interaction energy between the model predictions and the HNC MCA calculations remain far smaller than the relevant scales in kinetic energies for all the 48 parametric combinations plotted in Fig. 4.1 [Ichimaru 1993a, b]. Consequently, the IRS formulas provide a comprehensive physical account as well as a useful representation of thermodynamic functions for a strongly coupled TCP.

The values of various thermodynamic functions calculated in different theoretical models of hydrogen plasmas (i.e., $Z = 1$) are listed for semiclassical ($\Theta \geq 1$) cases in Table 4.1 and for degenerate ($\Theta < 1$) cases in Table 4.2. The thermodynamic functions considered are the internal energy density E, the pressure P, the specific heat at constant pressure

$$c_P \equiv \frac{1}{V}\left(\frac{\partial W}{\partial T}\right)_P , \tag{4.50}$$

and the coefficient of isobaric thermal expansion

$$\lambda_P \equiv \frac{T}{V}\left(\frac{\partial V}{\partial T}\right)_P , \tag{4.51}$$

where

$$W = (E + P)V \tag{4.52}$$

is the *enthalpy*, or the *heat function*.

Those functions are evaluated at four different plasma settings to elucidate the extents to which various physical effects, such as electron degeneracy and e-e, i-i, and e-i couplings, are involved: (1) IRS-model equation of state for hydrogen TCP; (2) IMTY equation of state [Tanaka & Ichimaru 1985] for hydrogen TCP, where $G_{12}(k) = 0$ has been set in a dielectric formulation; (3) STLS equation of state (see Section 3.2G) for equivalent electron OCP; and (4) ideal-gas fermion equation of state (see Appendix B) for equivalent electron OCP. "Equivalent electron OCP" refers to a system of hydrogen TCP where the ions are smeared out to make uniform background charges. Setting (4) can single out the effects of electron degeneracy, while (3) can account for additional exchange and Coulomb correlation effects between electrons. Setting (2) adds the classical ions to the system as distinct constituents with mutual interaction. Finally, (1) takes the IRS into account for a description of strong e-i coupling.

We observe that the exchange and Coulomb effects give rise to negative contributions in E and P substantially for those strongly coupled plasmas, in which $\Gamma > 1$ for semiclassical cases and $r_s > 1$ for degenerate cases. It is noteworthy in particular that even in strongly degenerate cases where E and P are dominated by the contributions from the electrons, the presence of strongly coupled

Table 4.1   Internal energy density, pressure, specific heat at constant pressure, and coefficient of thermal expansion for semiclassical ($\Theta \geq 1$) hydrogen plasmas

Γ Θ	$\log_{10} \rho_m (\text{g/cm}^3)$ $\log_{10} T$ (K)	$\dfrac{E}{nk_B T}$	$\dfrac{P}{nk_B T}$	$\dfrac{c_P}{nk_B}$	λ_P
0.05	0.535	**2.99**	**2.00**	**5.01**	**1.00**
		2.98	1.99	5.02	1.00
10.0	6.84	**1.50**	**1.00**	**2.50**	**0.998**
		1.51	1.00	2.49	0.99
0.1	2.63	**3.14**	**2.11**	**4.60**	**0.868**
		3.12	2.10	4.62	0.873
1.0	7.23	**1.66**	**1.12**	**2.13**	**0.759**
		1.70	1.13	2.10	0.741
0.2	−0.368	**2.85**	**1.97**	**5.12**	**1.03**
		2.82	1.95	5.18	1.05
5.0	5.93	**1.44**	**0.985**	**2.55**	**1.02**
		1.52	1.01	2.46	0.972
0.3	1.20	**2.97**	**2.06**	**4.72**	**0.901**
		2.88	2.03	4.72	0.913
1.0	6.28	**1.56**	**1.08**	**2.17**	**0.790**
		1.70	1.13	2.10	0.741
0.35	−2.0	**2.44**	**1.86**	**5.56**	**1.13**
		2.58	1.86	5.47	1.13
10.0	5.15	**1.36**	**0.953**	**2.66**	**1.09**
		1.51	1.00	2.49	0.99
0.5	−1.56	**2.20**	**1.81**	**5.68**	**1.16**
		2.34	1.79	5.68	1.21
5.0	5.14	**1.27**	**0.928**	**2.72**	**1.13**
		1.52	1.01	2.46	0.972
0.8	−0.0775	**2.17**	**1.84**	**5.72**	**1.15**
		2.12	1.79	4.93	1.03
1.0	5.43	**1.27**	**0.985**	**2.25**	**0.873**
		1.70	1.13	2.10	0.741
1.1	−0.492	**1.53**	**1.69**	**6.44**	**1.37**
		1.60	1.64	5.06	1.13
1.0	5.15	**1.08**	**0.925**	**2.30**	**0.932**
		1.70	1.13	2.10	0.741

Note: Each entry refers to one of the four settings described in the text: the bold figure to setting (1); the light figure to setting (2); the small bold figure to setting (3); and the small light figure to setting (4). The mass density ρ_m and temperature T are assumed values for the hydrogen plasma.

Table 4.2   Internal energy density, pressure, specific heat at constant pressure, and coefficient of thermal expansion for degenerate ($\Theta < 1$) hydrogen plasmas

Γ Θ	$\log_{10} \rho_m (\text{g/cm}^3)$ $\log_{10} T$ (K)	$\dfrac{E}{nk_B T}$	$\dfrac{P}{nk_B T}$	$\dfrac{c_P}{nk_B}$	λ_P
5.43	3.43	**54.8**	**38.9**	**1.40**	**0.0156**
		54.6	38.8	1.99	0.0185
0.01	5.76	**57.6**	**39.3**	**0.134**	**0.00135**
		60.1	40.1	0.135	0.00135
1.0	2.63	**6.59**	**4.78**	**2.31**	**0.185**
		6.59	4.78	2.36	0.188
0.1	6.23	**5.74**	**3.99**	**0.458**	**0.0449**
		6.22	4.15	0.472	0.0455
0.6	2.00	**3.69**	**2.67**	**3.86**	**0.540**
		3.65	2.66	3.48	0.509
0.272	6.24	**2.49**	**1.75**	**1.20**	**0.268**
		2.79	1.86	1.19	0.256
3.0	1.20	**3.84**	**3.97**	**2.06**	**0.199**
		3.75	3.89	2.63	0.238
0.1	5.28	**4.71**	**3.66**	**0.444**	**0.0465**
		6.22	4.15	0.472	0.0455
43.4	0.718	**−1.28**	**22.5**	**1.95**	**0.0299**
		−1.62	21.3	1.78	0.0288
0.01	3.96	**38.0**	**33.2**	**0.131**	**0.00152**
		60.1	40.1	0.135	0.00135
5.43	0.427	**−0.129**	**2.96**	**2.89**	**0.312**
		−0.0011	2.75	2.89	0.330
0.1	4.76	**3.40**	**3.27**	**0.437**	**0.0507**
		6.22	4.15	0.472	0.0455
2.40	0.190	**0.758**	**1.89**	**3.75**	**0.690**
		1.06	1.91	4.24	0.788
0.272	5.04	**1.50**	**1.45**	**1.20**	**0.311**
		2.79	1.86	1.19	0.256
2.50	0.137	**0.570**	**1.84**	**3.78**	**0.705**
		0.903	1.86	4.30	0.811
0.272	5.00	**1.45**	**1.43**	**1.20**	**0.314**
		2.79	1.86	1.19	0.256

Note: Each entry refers to one of the four settings described in the text: the bold figure to setting (1); the light figure to setting (2); the small bold figure to setting (3); and the small light figure to setting (4). The mass density ρ_m and temperature T are assumed values for the hydrogen plasma.

ions exerts decisive influences on the thermal and mechanical properties of the TCP as exemplified in the values of c_P and λ_P. As an example, the role that c_P should play in the evaluation of thermal resistivity will be considered in the next subsection.

For recent progress in the study of thermodynamic properties for metallic hydrogren, see Ichimaru [2001a] and references therein.

D. Electric and Thermal Resistivities

Electric and thermal resistivities arise as a consequence of scattering between electrons and ions in a plasma; a proper account of the *e-i* interaction, which would diverge in a classical treatment at short distances, is therefore essential. By taking quantum diffraction into account, an effective potential may be obtained that would converge at short distances [e. g., Deutsch, Gombert, & Minoo 1981]. A number of authors investigated the resistivities with the aid of such effective interparticle potentials. Hansen and McDonald [1981] thus computed electric conductivities through MD simulation; Sjögren, Hansen, and Pollock [1981] analyzed results of the simulation on the basis of a kinetic theory. MD calculation of thermal conductivity was carried out by Bernu and Hansen [1982; Bernu 1983]; theoretical accounts of these results were offered by Rozmus and Offenberger [1985] and by Zehnlé, Bernu, and Wallenborn [1986]. It may be remarked, however, that such a semiclassical approach may lose its validity once the effects of Fermi degeneracy start to play a role.

Hubbard and Lampe [1969] investigated thermal conduction by electrons in dense stellar matter through a Chapman–Enskog solution to the quantum-mechanical transport equation in weak Coulomb coupling. On the basis of a quantum kinetic theory for the current-current correlation functions, Boercker, Rogers, and DeWitt [1982] obtained an expression for electric resistivity; in their numerical calculations, however, the Maxwellian distribution for the electrons was assumed and the correlation functions were determined through a solution to HNC equations for the semiclassical system. Ebeling and Röpke [1979], Meister and Röpke [1982], and Lee and More [1984] adopted the Fermi–Dirac distribution in their calculations of the electronic transport coefficients; the effects of the interparticle correlations were, however, taken into account in a qualitative manner.

In a remarkable experiment, Ivanov et al. [1976] measured the Coulomb conductivity of non-ideal plasmas that were produced by a dynamic method based on compression and irreversible heating of gases in the front of high-power ionizing shock waves [Fortov 1982]. The gases used—argon, xenon, neon, and air—were regarded as forming singly ionized ($Z = 1$) plasmas. Transport properties were measured also in dense plasmas produced by metal vaporization [e.g., Mostovych et al. 1991; Benage & Shanahan 1993; DeSilva & Kunze 1993; Mostovych & Kearney 1993].

In the strong coupling regime with the quantum effects of electrons fully taken into account, the IMTY theory cited in Section 4.2A was applied to a calculation of electric and thermal resistivities [Ichimaru & Tanaka 1985]. Since the e-i local field correction $G_{12}(k)$ was neglected in the IMTY theory, the calculation basically amounted to application of Ziman [1961, 1972] formulas. Perrot and Dharma-wardana [1987] applied correlation calculations based on their LDA density-functional theory [Dharma-wardana & Perrot 1982] to evaluation of electric resistivities in hot, dense plasmas, where electron scattering against ions beyond the Born approximation was treated through the phase shift analyses.

The electric and thermal resistivities, ρ_E and ρ_T, due to electronic transport in a TCP with the ionic charge number Z may be expressed as

$$\rho_E = 4 \left(\frac{2\pi}{3} \right)^{1/2} \frac{\Gamma_{ei}^{3/2}}{\omega_p} L_E , \tag{4.53a}$$

$$\rho_T = \frac{52(6\pi)^{1/2}}{75} \frac{c_P{}^{(0)}}{c_P} \frac{e^2}{k_B^2 T} \frac{\Gamma_{ei}^{3/2}}{\omega_p} L_T , \tag{4.53b}$$

with

$$\Gamma_{ei} = \frac{Ze^2}{a_i} . \tag{4.54}$$

The generalized Coulomb logarithms, L_E and L_T, for the electric and thermal resistivities have been calculated through solution to quantum mechanical collision equations for the electrons (see Section 7.3F of Volume I) with inclusion of the dielectric screening (4.12) of the e-i interaction and the strong e-i coupling effects described by the local field correction $G_{12}(k)$ with the results [e.g., Ichimaru, Iyetomi, & Tanaka 1987],

$$L_E(\Gamma, \Theta) = \frac{3\sqrt{\pi}\,\Theta^{3/2}}{4} \int_0^\infty \frac{dk}{k} f_0 \left(\frac{k}{2} \right) \frac{1 - G_{12}(k)}{|\tilde{\varepsilon}_e(k, 0)|^2} S_{22}(k) , \tag{4.55a}$$

$$L_T(\Gamma, \Theta) = \frac{75\sqrt{\pi}\,\Theta^{9/2}}{104\Sigma^2} \int_0^\infty \frac{dk}{k} \frac{1 - G_{12}(k)}{|\tilde{\varepsilon}_e(k, 0)|^2} S_{22}(k)$$

$$\times \int_{k/2k_F}^\infty dx\, x(x^2 - \lambda)^2 \frac{\partial f_0(k_F x)}{\partial \alpha} . \tag{4.55b}$$

Here

$$f_0(k) = \frac{1}{\exp \left(\dfrac{\hbar^2 k^2}{2mk_B T} - \alpha \right) + 1} \tag{4.56}$$

is the Fermi distribution, and

$$\Sigma = \frac{7}{4}\frac{\Theta^{9/2}}{I_{1/2}(\alpha)}\left\{I_{5/2}(\alpha)I_{1/2}(\alpha) - \frac{25}{21}[I_{3/2}(\alpha)]^2\right\} , \tag{4.57a}$$

$$\lambda = \frac{5}{3}\Theta\frac{I_{3/2}(\alpha)}{I_{1/2}(\alpha)} \tag{4.57b}$$

are combinations of the Fermi integrals (B.1).

Formula (4.53b) differs from the corresponding formula (7.174) in Volume I on an important account. In the derivation of the latter formula, it was assumed that electrons carried the enthalpy (per unit volume) by the amount $n_e c_P{}^{(0)}T$ on the average, where $c_P{}^{(0)}$ referred to the specific heat at constant pressure for the ideal-gas electrons, that is, the formula (4.50) evaluated in setting (4) of Tables 4.1 and 4.2. Associated with the electronic transport in a TCP, however, the actual amount of heat energy carried by the electrons is $n_e c_P T$, where c_P now designates the specific heat at constant pressure for the TCP, that is, the formula (4.50) evaluated in the setting (1) of Tables 4.1 and 4.2; the factor $c_P{}^{(0)}/c_P$ in the formula (4.53b) stems from such a consideration. As Tables 4.1 and 4.2 illustrate, values of this factor can be significantly smaller than unity in strongly coupled TCPs; thermal transport may be enhanced by such an effect of heat capacity.

The generalized Coulomb logarithms (4.55) have been evaluated explicitly for the 48 cases (Fig. 4.1) of hydrogen plasmas ($Z = 1$) in the HNC MCA theory [Tanaka, Yan, & Ichimaru 1990]. Those values are accurately represented in terms of the IRS parameters by the following formulas:

$$L_{\rm E}^{\rm FIT}(\Gamma, \Theta) = \frac{1}{2}\ln\left[1 + \alpha_{\rm E}\left(\frac{1}{\zeta_{\rm DH}} + \frac{1}{\zeta_{\rm BORN}}\right)\right]\times(1 + 0.35x_b^2 + 0.044x_b^{10}) , \tag{4.58}$$

and

$$L_{\rm T}^{\rm FIT}(\Gamma, \Theta) = \frac{1}{2}\ln\left[1 + \alpha_{\rm T}\left(\frac{1}{\zeta_{\rm DH}} + \frac{1}{\zeta_{\rm BORN}}\right)\right]\times(1 + 0.25x_b^2 + 0.036x_b^{10}) . \tag{4.59}$$

Here

$$\zeta_{\rm DH} = \frac{C}{(12\pi^2)^{1/3}}\frac{(Z + 1)\Gamma_e}{\Theta} , \tag{4.60}$$

$$\zeta_{\mathrm{BORN}} = \frac{\exp\left(1.47\,Z^{1/3}\right)}{2.7\,\Theta^{3/2}\,Z^{4/3}}\,, \tag{4.61}$$

$$C = (Z+1)^{1/Z}\exp\gamma\,, \tag{4.62}$$

(γ is Euler's constant), $a_{\mathrm{E}} = 1$, $\alpha_{\mathrm{T}} = 75/13\pi^2 = 0.5845\ldots$ accounting for the Wiedemann–Frantz relation, Eq. (7.172) in Volume I.

In the classical ($\Theta \gg 1$) and weak coupling ($\Gamma \ll 1$) limit, the analytic formulas (4.58) and (4.59) reproduce the first two terms of the Coulomb logarithm obtained by Kivelson and DuBois [1964]. In the limit of complete Fermi degeneracy ($\Theta \ll 1$), those formulas behave proportionally to $\Theta^{3/2}$.

In Table 4.3, the values of HNC MCA generalized Coulomb logarithms are compared with those of the fitting formulas (4.58) and (4.59). The comparison implies accuracy of the formulas (4.58) and (4.59). The steep increase found in the resistivities as a function of the IRS fractional number x_{b} for the strongly degenerate ($\Theta \ll 1$) cases is remarkable. A transition toward an insulator phase is clearly indicated.

Table 4.3 HNC MCA evaluation (L_{E} and L_{T}) for the Coulomb logarithms for hydrogen plasmas compared with those ($L_{\mathrm{E}}^{\mathrm{FIT}}$ and $L_{\mathrm{T}}^{\mathrm{FIT}}$) of the fitting formulas (4.58) and (4.59) at selected combinations of Γ and θ.

Γ	Θ	x_{b}^2	L_{E}	$L_{\mathrm{E}}^{\mathrm{FIT}}$	L_{T}	$L_{\mathrm{T}}^{\mathrm{FIT}}$
0.05	10	0.312	2.732	2.819	2.363	2.451
0.2	10	1.279	3.295	3.215	2.557	2.524
0.35	10	2.238	8.041	7.906	5.573	5.724
0.4	5	1.738	3.249	3.254	2.294	2.333
0.1	1	0.150	0.956	1.129	0.751	0.875
0.6	1	0.902	0.709	0.689	0.476	0.447
1.1	1	1.654	0.914	0.871	0.583	0.516
0.6	0.2715	0.293	0.166	0.187	0.111	0.114
2.5	0.2715	1.220	0.147	0.119	9.06(−2)	6.49(−2)
0.5	0.1	9.20(−2)	4.62(−2)	7.58(−2)	2.88(−2)	4.52(−2)
5.0	0.1	0.920	2.61(−2)	2.27(−2)	1.55(−2)	1.24(−2)
10.0	0.01	0.184	5.78(−4)	7.09(−4)	3.38(−4)	4.04(−4)
43.4	0.01	0.800	4.48(−4)	5.19(−4)	2.62(−4)	2.84(−4)

Note: The numbers in the parentheses are decimal exponents.

Though the original HNC MCA calculations for the resistivities have treated only the hydrogen plasmas (i.e., $Z = 1$), the formulas such as Eqs. (4.53) include dependence on the ionic charge number Z; the dependence may be qualitatively accounted for as follows: The probability of collision for an electron against ions is proportional to $n_i \sigma_{ei}$, where σ_{ei} is the cross section of e-i scattering. For a tenuous weak-coupling plasma, the cross section is proportional to $(Ze^2/k_B T)^2$ and hence to Z^2, so that the collision frequency may behave as proportional to Z^1. For a dense strong-coupling plasma, since the ion-sphere model is applicable, the cross section becomes proportional to $Z^{2/3}$, so that the collision frequency should now behave as proportional to a_i^2 and hence to $Z^{-1/3}$. In a dense high Z plasma, on the other hand, the Born approximation provides a good description of e-i scattering, which has been accounted for by the parameter (4.61). This charge-number dependence has been properly taken into account in the resistivity formulas (4.53).

For recent progress in the study of electric and thermal resistivities in dense plasma materials, see Kitamura and Ichimaru [1995] and references therein.

4.3   ATOMS AND IONS

Electronic states in atoms and ions as well as ionization processes associated with them are essential issues in the physics of dense plasmas. In this section we consider such problems as atomic states, degrees of ionization, impurity states, and optical processes in dense plasmas.

A. Two-Electron Atoms

A two-electron atom such as neutral He and H$^-$ represents the simplest of nontrivial atomic systems involving electron-electron interaction. If such an atom is situated in a dense plasma, the electron-electron interaction will be affected by screening action of the plasma; the atomic properties are thereby modified. Since electron density distributions are inhomogeneous, two-electron atoms in plasmas offer a crucial test ground for an approximate theory of electron-electron interaction such as the LDA of the density-functional theory (see Sections 1.2E and 4.1E). For experimental data on various elementary processes in hydrogen-helium plasmas, the reader is referred to a compilation by Janev et al. [1987].

1.   He and H$^-$ in Vacuum

In the density-functional theory, the Kohn–Sham equation for an electron in the ground state of a two-electron atom ($Z = 2$ for He and $Z = 1$ for H$^-$) reads

$$\left[-\frac{\hbar^2}{2m}\nabla^2 - \frac{Ze^2}{r} + \frac{\delta}{\delta n_\sigma}\left(Y\left[n_\uparrow, n_\downarrow\right] + E_{xc}\left[n_\uparrow, n_\downarrow\right]\right)\right]\psi_\sigma^{1s}(\mathbf{r}) = \varepsilon_\sigma^{1s}\psi_\sigma^{1s}(\mathbf{r})$$

$$(4.63)$$

with

$$n_\sigma(\mathbf{r}) = \left|\psi_\sigma^{1s}(\mathbf{r})\right|^2 .\tag{4.64}$$

The quantities, $Y[n_\uparrow, n_\downarrow]$ and $E_{xc}[n_\uparrow, n_\downarrow]$, are density functionals representing Hartree energy and exchange-correlation energy, respectively. We consider three approximation schemes for the calculation of these quantities: LDA, LSD, and NSD.

Local-density approximation (LDA). In the usual LDA, one assumes

$$Y\left[n_\uparrow, n_\downarrow\right] = \frac{e^2}{2}\sum_{\sigma,\tau}\int d\mathbf{r}\,d\mathbf{r}'\,\frac{n_\sigma(\mathbf{r})n_\tau(\mathbf{r}')}{|\mathbf{r} - \mathbf{r}'|} ,\tag{4.65a}$$

$$E_{xc}\left[n_\uparrow, n_\downarrow\right] = \sum_\sigma\int d\mathbf{r}\,\varepsilon_{xc}^{PM}(n_\uparrow(\mathbf{r}), n_\downarrow(r))n_\sigma(\mathbf{r}) ,\tag{4.65b}$$

where $\varepsilon_{xc}^{PM}(n_\uparrow(\mathbf{r}), n_\downarrow(r))$ refers to the exchange-correlation free-energy per particle for a uniform paramagnetic electron system (see Section 3.1C) with local densities, $n_\uparrow(\mathbf{r}) = n_\downarrow(\mathbf{r})$. In such an approximation self-interaction of electrons cannot be avoided.

Local spin-density approximation without self-interaction (LSD). To avoid electron self-interaction in the calculation of ground-state properties for two-electron atoms, one may set

$$Y\left[n_\uparrow, n_\downarrow\right] = e^2\int d\mathbf{r}\,d\mathbf{r}'\,\frac{n_\uparrow(\mathbf{r})n_\downarrow(\mathbf{r}')}{|\mathbf{r} - \mathbf{r}'|} ,\tag{4.66a}$$

$$E_{xc}^{PM}\left[n_\uparrow, n_\downarrow\right] = \int d\mathbf{r}\,\left[\varepsilon_{xc}^{PM}(n_\uparrow(\mathbf{r}), n_\downarrow(\mathbf{r})) - \varepsilon_{xc}^{FM}(n_\uparrow(\mathbf{r}), 0)\right]n_\sigma(\mathbf{r}) ,\tag{4.66b}$$

where $\varepsilon_{xc}^{FM}(n_\uparrow(\mathbf{r}), 0)$ refers to the exchange-correlation free-energy per particle for a uniform ferromagnetic electron system (see Section 3.1C) with local densities, $n_\uparrow(\mathbf{r})$ and $n_\downarrow(\mathbf{r}) = 0$.

Nonlocal spin-density functional approximation (NSD). In this approximation, one adopts Eq. (4.66a) for the Hartree contribution and

$$E_{xc}\left[n_\uparrow, n_\downarrow\right] = \int_0^1 d\eta\int d\mathbf{r}\,d\mathbf{r}'\,\frac{e^2}{|\mathbf{r} - \mathbf{r}'|}n_\uparrow(\mathbf{r})n_\downarrow(\mathbf{r}')\left\{g_{\uparrow\downarrow}\left[\mathbf{r}, \mathbf{r}'; n\right] - 1\right\}\tag{4.67}$$

where the nonlocal radial-distribution function, $g_{\uparrow\downarrow}[\mathbf{r}, \mathbf{r}'; n]$, between antiparallel spins may be evaluated in the Yamashita–Ichimaru [1984] scheme with Eqs. (4.30) and (4.31). Specifically, we set

$$g_{\uparrow\downarrow}[\mathbf{r}, \mathbf{r}'; n] \rightarrow g_{\uparrow\downarrow}(|\mathbf{r} - \mathbf{r}'|)$$
$$= 1 + \left\{A(\eta\bar{r}_s) + B(\eta\bar{r}_s)|\mathbf{r} - \mathbf{r}'|\right\} \exp\left[-\left\{C(\eta\bar{r}_s)|\mathbf{r} - \mathbf{r}'|\right\}^2\right] ,$$

(4.68a)

where

$$\bar{r}_s = \left(\frac{3}{4\pi\bar{n}}\right)^{1/3} \frac{me^2}{\hbar^2} ,$$

(4.68b)

$$\bar{n} = \frac{1}{2}\left\{n_\uparrow(\mathbf{r}) + n_\downarrow(\mathbf{r}) + n_\uparrow(\mathbf{r}') + n_\downarrow(\mathbf{r}')\right\} .$$

(4.68c)

The coefficients, $A(r_s)$, $B(r_s)$, and $C(r_s)$, are determined from the following three conditions: (i) Short-range correlation [Yasuhara 1972]

$$g_{\uparrow\downarrow}(0) = \frac{1}{4}\left[\frac{z}{I_1(z)}\right]^2 ,$$

(4.69a)

where $z = 4\sqrt{\lambda r_s/\pi}$, $\lambda = (4/9\pi)^{1/3}$, and $I_1(z)$ is the modified Bessel function of the first kind and of the first order. (ii) Cusp condition [Kimball 1973]

$$\left[\frac{dg_{\uparrow\downarrow}(r)}{dr}\right]_{r=0} = \frac{g_{\uparrow\downarrow}(0)}{a_B} .$$

(4.69b)

Table 4.4 Ground-state energies of a helium atom in the atomic units (= 2Ry = 27.21 eV) obtained in various schemes (Errors are relative to the experimental value.)

	H-F	LDA	LSD	NSD	Experiment
E	−2.861	−2.834	−2.920	−2.897	−2.904
(error %)	(1.5)	(2.4)	(0.55)	(0.24)	
T_s	2.863	2.770	2.885	2.876	
V	−6.749	−6.625	−6.777	−6.755	
Y	2.501	1.995	1.031	1.027	
E_{xc}	−1.025	−0.9733	−0.05867	−0.02584	

Table 4.5   Ground-state energies of an H⁻ ion in the atomic units (= 2Ry = 27.21 eV) obtained in various schemes (Errors are relative to the experimental value.)

	H-F	LSD	NSD	Experiment
E	−0.4879	−0.5262	−0.5211	−0.5277
(error %)	(7.5)	(0.28)	(1.2)	
T_s	0.4876	0.5068	0.4844	
V	−1.371	−1.402	−1.373	
Y	0.7910	0.4077	0.3967	
E_{xc}	−0.3955	−0.03867	−0.02921	

(iii) Condition for self-consistency with the interaction energy between antiparallel spins

$$\frac{1}{a_B^2} \int_0^\infty dr\, r\left[g_{\uparrow\downarrow}(r) - 1\right] = \frac{4}{3} r_s^2 \frac{d}{dr_s} \left\{ r_s^2 \left[\varepsilon_{xc}^{PM}(r_s) - \varepsilon_{xc}^{FM}(r_s)\right]\right\}\,, \qquad (4.69c)$$

Table 4.4 compares the computed results for the ground-state energies of He in the foregoing three approximation schemes [Fushimi, Iyetomi, & Ichimaru 1993], those in the Hartree–Fock approximation [Fischer 1977], and the experimental value [Hotop & Lineberger 1975]. In the table, T_s and V refer to the contributions from the first two terms, kinetic energy and ionic terms, in Eq. (4.63). The values of the ground state energy predicted in the LSD and NSD schemes agree with the experimental value with an accuracy better than 1%. We thus observe superiority of LSD and NSD, which have a provision of avoiding the self-interaction, over two other approximation schemes (H-F and LDA). The H-F and LDA schemes seem to contain overestimation in the values of the Hartree and exchange-correlation energies.

Table 4.5 analogously compares the computed results for the ground-state energies of a H⁻ ion in the foregoing (LSD and NSD) approximation schemes [Fushimi, Iyetomi, & Ichimaru 1993], those in the Hartree–Fock approximation [Fischer 1977], and the experimental value [Hotop & Lineberger 1975]. The cases with LDA are not listed because a H⁻ ion would not be formed stably in this approximation. Superiority of LSD and NSD over the other two may again be visible from comparison with the experimental value.

2.   He and H⁻ in a Plasma

Having thus established the accuracy of the LSD scheme in the analyses of atomic states, we now investigate the ground-state properties of a neutral He atom

and a negative H⁻ ion immersed in a plasma as impurities. The results are
compared with those calculated in the same scheme for their single-electron
counterparts, a positive He⁺ ion and a neutral H atom, immersed in a plasma.
Stability of He and H⁻ against ionization induced by screening action of the
plasma is thereby investigated. Since no reliable information on $g_{\uparrow\downarrow}[\mathbf{r}, \mathbf{r}'; n]$ or
$g_{\uparrow\downarrow}(|\mathbf{r} - \mathbf{r}'|)$ is available under such an electron-screened circumstance and since
much mathematical complication is involved in its implementation, an equally
accurate NSD scheme will not be employed in this investigation, however.

We take into account an effect of the plasma through its screening action
by electrons. The screening parameter, $\alpha_s(r_s) \equiv a_e/D_s$, may be calculated in
conjunction with the short-range screening distance as given by Eq. (3.56). The
Kohn-Sham equation (4.63) is now modified in three ways:

$$\frac{Ze^2}{r} \rightarrow \frac{Ze^2}{r} \exp\left(-\alpha_s \frac{r}{a_e}\right) , \tag{4.70}$$

$$Y[n_\uparrow, n_\downarrow] \rightarrow \tilde{Y}[n_\uparrow, n_\downarrow] = e^2 \int d\mathbf{r}\, d\mathbf{r}'\, \frac{n_\uparrow(\mathbf{r})n_\downarrow(\mathbf{r}')}{|\mathbf{r} - \mathbf{r}'|} \exp\left(-\alpha_s \frac{|\mathbf{r} - \mathbf{r}'|}{a_e}\right) , \tag{4.71}$$

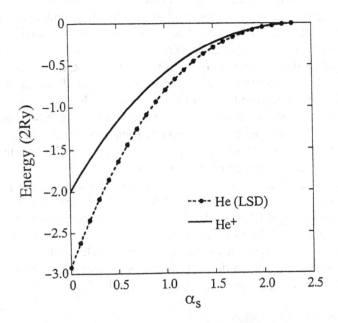

Figure 4.8 Ground-state energies of He and He⁺ as functions of screening
parameter α_s of the plasma.

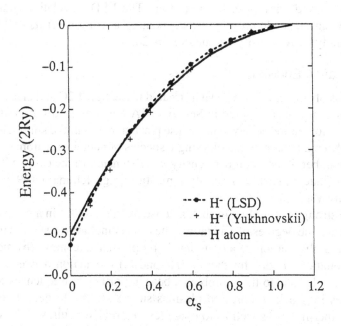

Figure 4.9 Ground-state energies of H⁻ and H as functions of screening parameter α_s of the plasma.

and the exchange-correlation energies in Eq. (4.66b) are evaluated by replacing

$$e^2 \rightarrow e^2 \exp\left(-\alpha_s r_s \frac{a_B}{a_e}\right) . \tag{4.72}$$

This last procedure is the same in principle as the use of the electron-screened Coulomb-coupling parameter (4.45) in Eq. (4.48b). Without such precaution, the plasma screening effects could not correctly be taken into account in the atomic problems [Fushimi, Iyetomi, & Ichimaru 1993].

We thus calculate the values of the ground-state energy for a He atom in a plasma as a function of the screening parameter α_s, and compare them with analogous values for a He⁺ ion in Fig. 4.8. For $\alpha_s > 2.3$, a He⁺ ion has a ground-state energy lower than a He atom, so that the latter automatically ionizes by the screening action of the plasma electrons in such a density regime. With the IU local field correction (3.71) used in the dielectric function, this would mean a dense plasma with $r_s < 0.34$.

Similarly, Fig. 4.9 plots the values of the ground-state energy calculated for a H⁻ ion in LSD as a function of the screening parameter α_s and compares them with analogous values for a H⁻ ion in Yukhnovskii's [1988] variational calculations and with those for a H atom. The LSD calculation indicates that a H^- ion can be stable only in weak screening conditions of $\alpha_s < 0.23$, which would imply a low-density regime of $r_s > 20$.

B. Ionization Equilibria

Equations of state for dense plasmas treated in Section 4.2C and electronic states in atoms and ions considered in Section 4.3A are combined for analyses of the degrees of ionization achieved in a dense plasma material. For simplicity, we here deal with a matter consisting of a single species of nuclei with atomic number A, charge number Z, and number density n_A. Extension to the cases with multiple species of nuclei, such as impurity ionization [e.g., Ichimaru 1993b], is rather straightforward.

The strong e-i coupling in such a dense matter may act in a number of ways to influence the degrees of ionization; the effects include (a) shallowing (energy shifts) and elimination of atomic levels by plasma screening, (b) modification in the equation of state for the plasma, and (c) interaction between atoms and plasmas. Solutions to these problems bear important consequences to physical processes in shock-compressed plasmas such as those in inertial-confinement fusion experiments as well as to opacities, internal structures, and evolution in various stellar objects, including giant planets, brown dwarfs, and the Sun.

1. Stages of Ionization

An atom in the jth stage of ionization has $(Z - j)$ bound electrons with an effective core diameter d_j [e.g., Allen 1973]; n_j denotes the number density of such atoms such that

$$n_A = \sum_{j=0}^{Z} n_j .$$

The number density of electrons is then given by

$$\bar{n}_e = \frac{n_e}{1 - \eta} = \frac{1}{1 - \eta} \sum_{j=0}^{Z} j n_j , \tag{4.73}$$

where

$$\eta = \sum_{j=0}^{Z} \eta_j = \frac{\pi}{6} \sum_{j=0}^{Z} n_j d_j^3 . \tag{4.74}$$

refers to the total packing fraction due to the atomic cores. The electron density $\bar{n}_e$ in Eq. (4.73) therefore represents a value renormalized by the excluded volume effect of the core electrons. In the evaluation of thermodynamic functions, analogous renormalization may be required in the consideration of ion densities as well.

We account for the shallowing and disappearance of atomic levels due to plasma screening in the following way: Let the ionization potentials of the isolated atoms [e.g., Allen 1973] be defined as the balance between the binding energies of the jth and $(j + 1)$st stages of ionization:

$$\delta E_{B0}^{(j+1)} \equiv E_{B0}^{(j)} - E_{B0}^{(j+1)} . \tag{4.75}$$

Effective Bohr radii may then be defined as

$$R_{B0}^{j,j+1} \equiv \frac{(j + 1)e^2}{2\delta E_{B0}^{(j+1)}} . \tag{4.76}$$

Due to the screening effect of electrons represented by D_s in Eq. (3.56), the ionization potentials are reduced from the vacuum values in Eq. (4.75) by a factor $f(R_{B0}^{(j+1)}/D_s)$ and are given by

$$\delta E_B^{(j+1)} = f\left(\frac{R_{B0}^{j,j+1}}{D_s}\right) \delta E_{B0}^{(j+1)} . \tag{4.77}$$

The level-shallowing factor $f(x)$ is determined from an exact numerical solution to the Schrödinger equation for a 1s electron in a screened potential, $-(Ze^2/r)\exp(-r/D_s)$, as has been done in Section 4.3A, and takes the form [Fushimi, Iyetomi, & Ichimaru 1993]

$$f(x) = 1 - 1.9585x + 1.2172x^2 - 0.24900x^3 + 0.012973x^4 . \tag{4.78}$$

For $x > 1.191$, this factor takes on a negative value, implying disappearance of a level.

2. The Process of Ionization

For those stages of atomic states that may remain against the screening action of the plasma electrons, we set the relation

$$A^{j+} \Leftrightarrow A^{(j+1)+} + e^- \tag{4.79}$$

and calculate the fractional population x_j of the jth stage of ionization such that

$$\sum_j x_j = 1 . \tag{4.80}$$

The chemical equilibria for the matter with atomic density n_A at temperature T may be determined from the condition that the free energy be minimized with respect to the distribution $\{x_j\}$ under the constraint (4.80). The free energy may be evaluated as a sum of the contributions from electrons in ionized states, ideal-gas and interaction parts from atoms in jth stage, and the bound core parts from atoms in jth stage:

$$\frac{F\left(n_A, T; \{x_j\}\right)}{n_A k_B T} = \left(\sum_j j x_j\right)\left[f_e^{\mathrm{id}}(\bar{n}_e, T) + f_e^{\mathrm{xc}}(\bar{n}_e, T)\right]$$
$$+ f_A^{\mathrm{id}}(n_A, T) + \sum_j x_j f_j^{\mathrm{int}}(\bar{n}_e, T) + \sum_j x_j f_j^{\mathrm{atom}}(\bar{n}_e, T) .$$

(4.81)

For $f_e^{\mathrm{xc}}(\bar{n}_e, T)$, we may use the STLS evaluation (3.81). The interaction free energy $f_j^{\mathrm{int}}(\bar{n}_e, T)$ of ions consists of two parts: the Coulombic part calculated from Eqs. (4.50b)–(4.50d) and the hard sphere part calculated with Carnahan–Stirling formula [e.g., Hansen & McDonald 1986],

$$f_j^{\mathrm{HS}}(\eta_j) = \frac{\eta_j(4 - 3\eta_j)}{(1 - \eta_j)^2} .$$

(4.82)

Finally the core contribution $f_j^{\mathrm{atom}}(\bar{n}_e, T)$ may be calculated with the information on atomic levels E_B^{jk} and level multiplicities g_{jk} [e.g., Moore 1971] as

$$f_j^{\mathrm{atom}}(\bar{n}_e, T) = -\ln\left[\sum_k g_{jk} \exp(-\beta E_B^{jk})\right] ,$$

(4.83)

where the level-shallowing effect described in Eq. (4.77) may be taken into account.

3. Degrees of Ionization and Equation of State

Once the set of values $\{x_j\}$ is determined by the variational method as already described, the thermodynamic functions such as the pressure may be calculated through a standard procedure. To follow the equation of state with varied degrees of ionization, we start with a matter at low density and temperature, where it may not be in an ionized state. As the density is raised along an isotherm, the pressure of such a lowly ionized matter generally increases. When the matter starts to ionize efficiently, the plasma part of its density increases. As the plasma density increases into a strong coupling regime such that the Coulomb coupling parameter takes on a value on the order of or greater than unity, the Coulombic pressure, which is negative, may take on a magnitude comparable to or greater than the kinetic pressure, which is positive. The total pressure may thus *decrease* with the density along the isotherm, indicating a thermodynamic instability. Such an instability will result in a discrete change in the density n_A, from n_{min} to n_{max}, determined from the usual condition [e.g., Landau & Lifshitz 1969],

$$\int_{n_{min}}^{n_{max}} \frac{dn}{n} \left[\frac{\partial P\left(n, T; \{x_j\}\right)}{\partial n} \right]_T = 0 \, , \qquad (4.84)$$

at the pressure

$$P = P(n_{min}, T; \{x_j\}) = P(n_{max}, T; \{x_j\}) \, . \qquad (4.85)$$

Associated with such a density discontinuity, various thermodynamic functions and the degrees of ionization change discontinuously. It is a *first-order phase transition* related closely to the *metal-insulator transition* [e.g., Mott 1990].

Figure 4.10 exhibits the equation of state and ionization diagrams for a dense helium material on a mass density versus pressure plane. Discontinuities in the density and in the degree of ionization represented by an average ionic charge,

$$\langle Z \rangle \equiv \sum_j j x_j \, , \qquad (4.86)$$

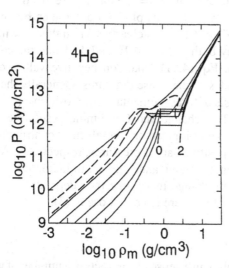

Figure 4.10 Isotherms (solid) and constant (dashed and labeled by the values of $\langle Z \rangle$) lines for dense helium material. The isotherms are at $\log_{10} T(K) = 3.2$, 3.6, 4.0, 4.4, 4.8, 5.2 from the bottom. [Ogata & Ichimaru 1993]

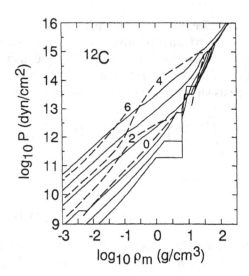

Figure 4.11 Isotherms (solid) and constant (dashed and labeled by the values of $\langle Z \rangle$) lines for dense carbon material. The isotherms are at $\log_{10} T(K) = 4.0$, 4.4, 4.8, 5.2, 5.6, 6.0 from the bottom. [Ogata & Ichimaru 1993]

are clearly manifested at lower temperatures. Due to the plasma screening effects, the atomic level at $j = 1$ is shallowed by a factor of approximately $1/5$ and consequently the states with $\langle Z \rangle = 1$ cannot be realized.

Figure 4.11 shows analogous plots for a dense carbon material. The special role that the line $\langle Z \rangle = 4$ plays is clearly seen in this drawing.

It is noteworthy both in Figs. 4.10 and 4.11 that the coefficient of isobaric thermal expansion, Eq. (4.51), takes on negative values on the high-density, low-temperature sides of the phase diagrams. Generally the pressure for such a dense matter consists of three elementary compositions: (1) degeneracy pressure of the electrons, (2) exchange and Coulombic excess pressure of the charged components, and (3) non-ionized or core-electron contributions. These are positive, negative, and positive partial pressures, respectively. As the temperature is raised, stages of ionization advance and act to increase the partial pressure (1). Sign of the resultant change in volume then depends on competition between changes in the partial pressure (2) and those in (3).

C. Impurities

The strong e-i coupling in a dense plasma acts in a number of ways to influence the degrees of ionization for the impurity atoms immersed in it. Such a problem can be treated in a way analogous to the calculations of ionization equilibria described in the preceding subsection. These theoretical considerations have been applied to the study on the levels of ionization for Fe impurities in the Sun and in the brown dwarf, and for He impurities in the giant planets [Ichimaru 1993b].

In conjunction with the issue of Fe impurity in the Sun, Pollock, and Alder [1978] suggested originally that the solar neutrino dilemma—a large discrepancy between the observed solar neutrino counts [Davis 1984] and the calculated capture rates in the standard solar model [Bahcall et al. 1982]—might be resolved if iron had a limited solubility in a hydrogen plasma at a pressure and temperature appropriate to the solar interior. Highly charged (i.e., high Z) elements such as iron are efficient scatterers of the electrons in a plasma and thus have substantial effects on its transport properties. The solar abundance (2.5×10^{-5} ionic molar fraction) of iron near the thermonuclear burn region, though extremely small in quantity, can significantly influence the rate of fusion of α particles that generates solar neutrinos monitored in the ^{37}Cl experiment [Bahcall & Ulrich 1971; Davis 1984]. Pollock and Alder [1978] actually evaluated the solubility of iron in hydrogen plasmas using the classical Debye–Hückel theory for both electrons and ions, and concluded that the plasma could possibly undergo phase separations at concentrations of iron well below the cosmic abundance.

To construct a reliable phase diagram for solar interior plasmas, however, one must take an appropriate account of both the strong coupling effects between the ions involved in the iron-rich phase and the screening effects due to the semiclassical electrons in the solar interior. Such an electron gas acts to screen the ion-ion interaction quite efficiently and hence modifies the thermodynamic properties of the plasma substantially. The miscibility problem of Fe in the solar interior has been revisited subsequently by Alder, Pollock, and Hansen [1980] and by Iyetomi and Ichimaru [1986]. The phase diagram obtained for the H^+-Fe^{24+} mixtures at $P = 0.5 \times 10^5$ Mbar exhibited the critical point for demixing at $T_c \approx 5.5 \times 10^6$K and $x_c \approx 2.4 \times 10^{-2}$ [Iyetomi & Ichimaru 1986]. These values, however, are not sufficient enough to resolve the solar-neutrino dilemma through the idea of a limited solubility of the iron atoms in the solar-interior plasma.

D. Optical Properties

States of atoms and ions in dense matter are crucial elements in controlling the optical properties such as opacities and spectral line shapes.

1. Opacities

Elastic and inelastic scattering of photon by electrons, ions, and atoms is the physical source of the radiative opacities. For astrophysical applications, the radiative opacities calculated at Los Alamos National Laboratory [see Cox & Stewart 1965; Cox & Tabor 1976, and references therein] have been the only ones available for realistic cases for quite some time. As calculations that use these opacities have improved, and as comparisons between theory and observations have become sharper, however, there have been increasing indications that those opacity calculations have needed improvements in some areas. For this reason, several groups have attacked the calculations of astrophysical opacities independently in recent years.

One of the new efforts, the OPAL project, led by Rogers and Iglesias [1992, 1993], is based at Lawrence Livermore National Laboratory. The second is the international Opacity Project, led by M.J. Seaton at University College, London [Seaton et al. 1992]. Both projects have now begun to produce results. Interested readers are referred to published materials for detail.

2. Microscopic Electric Fields

The spectral line shapes of ions radiating in plasmas provide an important diagnostic tool to probe the microscopic structures in plasmas. In the Stark broadening theory [e.g., Griem 1974], the line shapes are determined by the probability density of the electric microfields perceived by radiator atoms. Since the pioneering work of Holtsmark [1919; see also Chandrasekhar 1943], many investigators advanced statistical theories of microscopic field fluctuations with improved accuracy [e.g., Margenau 1932; Baranger & Mozer 1959; Hooper 1966, 1968; Iglesias & Hooper 1982].

To account for some of the strong coupling effects, which had not been included in the foregoing theories, Iglesias, Lebowitz, and MacGowan [1983] developed an adjustable parameter exponential approximation (APEX) method. This method exactly satisfies the sum rule requirement arising from the second moments of the electric microfield distributions, without introducing adjustable parameters. This approximation scheme was extended subsequently to the cases of TCP with classical ions and partially degenerate electrons; polarization shifts of spectral lines were thereby treated [Yan & Ichimaru 1986].

PROBLEMS

4.1   Derive the free-particle polarizability (4.3).

4.2   Derive the density-density response functions (4.5) for a TCP.

4.3   Compute the effective ion-ion interaction potential, Eq. (4.11), in the electron liquid at a metallic density $r_s = 4$ for the following two evaluations of the static dielectric function $\varepsilon_e(k, 0)$: (a) the RPA evaluation (1.47) with Eq. (1.48); (b) the static local field approximation evaluation (1.60) with Eq. (3.71).

4.4   Derive the variational free energy (4.20).

4.5   Derive Eqs. (4.35).

4.6   Show that the screening parameter (4.40) for the "free" electrons turns into the Thomas–Fermi screening parameter k_{TF} in Eq. (1.50) when $\Theta_f \ll 1$ and into the Debye–Hückel screening parameter k_D in Eq. (1.52) when $\Theta_f \gg 1$.

CHAPTER 5

Nuclear Fusion Reactions

The rate of fusion reactions between nuclear species, i and j, is proportional to $g_{ij}(r_N)$, the value of a joint probability density $g_{ij}(r)$ for the reacting pairs at a nuclear reaction radius r_N. Either in vacuum or in a dense plasma, such a contact probability is given generally as the square of the wave functions that describe scattering between the reacting pairs. Being a *correlation function*, albeit at short distances, the contact probability may depend quite sensitively on the changes in the microscopic, macroscopic, and thermodynamic states of the environment in case it consists of a *condensed matter*. Thus the rates of nuclear reactions in dense plasmas can differ drastically from those expected in vacuum, due to many-body correlation effects or statistical-mechanical effects, inherent in such a condensed matter system.

In this chapter we study current stages of understanding with regard to those effects of enhancement in the rates of nuclear reactions expected from the correlation and thermodynamic effects, for various realizations of the condensed plasmas both in the astrophysical and laboratory settings.[*] Consequently we shall be concerned with the aspects of statistical condensed matter physics rather than with those of nuclear reaction physics per se. The subject of *nuclear fusion in dense plasmas* is here viewed as a forum in which the interplay between nuclear physics and statistical physics may be studied usefully through the concept of such correlation functions.

---

[*] For further progress in the recent study of pycnonuclear reactions in dense astrophysical and fusion plasmas, see Ichimaru and Kitamura [1999] and references therein.

5.1   NUCLEAR FUSION IN DENSE PLASMAS: AN OVERVIEW

Nuclear reactions may be grouped into two elementary classifications: the usual *binary processes* and *few-particle processes*. Binary processes are those that are expected without the effects of the environment and include the celebrated Gamow rates [Gamow 1928; Gurney & Condon 1929] of *thermonuclear reactions*. In few-particle processes, one includes possible effects of screening, or modification of internuclear forces by light particles such as electrons; a possibility of *electron-screened cold fusion* emerges.

The nuclear reaction rates are related to the short-range behavior of the pair correlation functions, where the quantum mechanical effects are essential. The special features of nuclear fusion in dense plasmas rest in *enhancement of the reaction rates over those fundamental processes due to internuclear many-particle processes* [e.g., Ichimaru 1993a]. Many-particle processes arise from modification of the short-range correlations between reacting nuclei and are the effects related closely to differences between Coulombic chemical potentials before and after the nuclear reactions. In these connections, we recall the Monte Carlo simulation study as well as the analytic theories on the short-range correlations in various realizations of dense plasmas, considered in Chapter 2.

In his pioneering work, Schatzman [1948] pointed out that potential barriers between reacting nuclei may be significantly lowered in dense stellar matter, so that the probabilities of wave functions tunneling through the barrier would be greatly enhanced. Cameron [1959] argued subsequently that at very high densities, electron shielding (see Section 3.2D) cuts off nuclear Coulomb potential barriers quite close to the nuclear surface. Under these circumstances the classical turning points of low-energy ions are very insensitive to the relative energy of collision. It is in such a context that Cameron [1959] coined the term "pycnonuclear reactions" from the Greek *pyknos*, "compact, dense," to describe nuclear reactions under the strong electron screening conditions in that the rates become very insensitive to temperature but very sensitive to density. Salpeter [1954] presented an analytic treatment of the weak screening effect in a low-density, high-temperature plasma such that $\Gamma < 1$ and introduced the ion sphere model to describe the effects of interionic correlations in the strong coupling regime. Salpeter and Van Horn [1969] then derived general expressions for nuclear reaction rates appropriate to various stellar interior conditions. "Correction factors" or the enhancement factors due to the weak or strong screening over the rates of ordinary thermonuclear reactions were thereby evaluated.

From the point of view of a general statistical-mechanical theory, a significant development took place when Widom [1963] showed how certain thermodynamic functions, and also radial distribution functions or joint probability densities, can be expressed in terms of the potential energy distributions in a fluid. These statistical mechanical theories for calculations of the reaction rates were subsequently refined by Jancovici [1977], by Alastuey and Jancovici [1978], and by Ogata, Iyetomi, and Ichimaru [1991], through careful examination of the short-range behavior of the internuclear correlation functions.

Physics of nuclear fusion in dense plasmas is therefore intimately related to physics of strongly coupled plasmas [Ichimaru 1982] with the Coulomb coupling parameters $\Gamma_{ij} > 1$. The study of correlations and thermodynamic properties in such a plasma has progressed significantly, through advancement in the analytic theories coupled with accumulation of Monte Carlo simulation data for the OCP [Brush, Sahlin, & Teller 1966; Hansen 1973; Slattery, Doolen, & DeWitt 1980, 1982; Ogata, Iyetomi, & Ichimaru 1991], for the BIM [Ogata et al. 1993], for the electron-screened OCP [Ichimaru & Ogata 1991], and for deuterons in metals [Ichimaru, Ogata, & Nakano 1990]. In these connections it is important to recognize that the nuclear reaction rates in dense plasmas are intimately related to the thermodynamic functions through the screening properties and the Coulombic chemical potentials. Some of those subjects have been considered in detail in the foregoing chapters.

A condensed plasma solidifies at low temperatures (see Section 2.5D). In the ground state the nuclei form a quantum solid and perform zero-point vibrations about their equilibrium sites. The reaction rates, proportional to the square of the ground-state wave functions $|\psi_{ij}(r_N)|^2$ for nearest-neighbor pairs, thus depend very sensitively on the density via the nearest-neighbor separation but are independent of the temperature. They are therefore the pycnonuclear reactions in solids [Ichimaru, Ogata, & Van Horn 1992]. At an elevated temperature, the nuclei may occupy excited states of lattice vibrations, and the reaction rates may take on significantly enhanced values near the melting conditions. For astrophysical applications, enhancement of the pycnonuclear rates due to these thermal excitations needs also to be considered [Ichimaru, Kitamura, & Ogata 1993].

Those analyses are then applied to estimation of the nuclear reaction rates and enhancement factors in specific examples of the dense astrophysical and laboratory plasmas: Astrophysical condensed plasmas to be considered include the solar interior (SI), interior of a brown dwarf (BD), interior of a giant planet (GP), a white dwarf (WD) progenitor of supernova, and surfaces of accreting white dwarfs and neutron stars. Examples of the condensed plasmas in laboratories are those found in the inertial-confinement fusion (ICF) experiments, in metal hydride (MH) such as PdD and TiD_2, in cluster-impact fusion experiments, and in ultrahigh-pressure liquid metals. The essential similarity between the nuclear reactions in supernovae and those projected in a pressurized metal (PM) will be particularly remarked.

5.2   RATES OF ELEMENTARY PROCESSES

This section begins with the consideration of fundamental nuclear cross section with Coulomb scattering and the resultant rates of thermonuclear reactions. The essential modification of such a reaction rate arising from electron screening in a condensed matter is then studied, leading to the concept of electron-screened cold fusion. Pycnonuclear rates in crystalline solids are finally calculated as an elementary process of nuclear reactions.

A. The Gamow Rates of Thermonuclear Reactions

Consider scattering between nuclei, i and j, with relative velocity v and reduced mass

$$\mu_{ij} = \frac{A_i A_j}{A_i + A_j} m_N \tag{5.1}$$

via the Coulomb potential

$$W_{ij}^{(0)}(r) \equiv Z_i Z_j \phi(r) = Z_i Z_j e^2 / r , \tag{5.2}$$

where $m_N \approx 1.6605 \times 10^{-24}$ g is an average mass per nucleon [Allen 1973], A and Z denote mass and charge numbers. Events of scattering are described by the wave functions $\psi_{ij}(\mathbf{r})$ for the colliding pairs at internuclear separation $\mathbf{r}$. An important parameter characterizing such a Coulomb scattering is the *nuclear Bohr radius*

$$r_{ij}^* = \frac{\hbar^2}{2\mu_{ij} Z_i Z_j e^2} . \tag{5.3}$$

Another characteristic length entering the scattering problems in condensed plasmas is the internuclear spacings that scale with the ion sphere radius defined in Eqs. (1.11).

The controlling factor in the analysis of such a scattering event is the effective potential between the nuclei in the short-range domain, where the potential may be regarded as isotropic. A calculation of reaction rates may then be facilitated by the observation that the major contributions to the contact probabilities arise from the s-wave scattering acts between the reacting nuclei. Such an observation stems from the fact that the wave function of scattering in a spherically symmetric potential with the azimuthal quantum number l is proportional to r^l in short ranges [e.g., Schiff 1968]. Since one can generally assume that

$$r_N < r_{ij}^* \ll a_{ij} , \tag{5.4}$$

the s-wave scattering gives the major contribution to the reaction rate; hence $r_N \approx 0$ may be taken for the calculations of contact probabilities.

The central quantity in the theory of nuclear fusion in dense plasmas is therefore $g_{ij}(0) \propto |\Psi_{ij}(0)^2|$, the joint probability density at zero separation. It is an *equal-time*, two-particle distribution function evaluated in an equilibrium ensemble; the reaction rates depend on the static correlations. In the usual statistical treatment (see Section 1.2B), such a correlation function is expressed in terms of the static and dynamic structure factors, $S_{ij}(\mathbf{k})$ and $S_{ij}(\mathbf{k}, \omega)$, via the sum rule integrations as

$$g_{ij}(0) = 1 + \frac{1}{\sqrt{n_i n_j}} \int \frac{d\mathbf{k}}{(2\pi)^3} \left[S_{ij}(\mathbf{k}) - \delta_{ij} \right]$$

$$= 1 + \frac{1}{\sqrt{n_i n_j}} \int \frac{d\mathbf{k}}{(2\pi)^3} \left[\frac{1}{\sqrt{n_i n_j}} \int_{-\infty}^{\infty} d\omega \, S_{ij}(\mathbf{k}, \omega) - \delta_{ij} \right] .$$

Since the dynamic structure factor represents the power spectrum of the density-fluctuation excitations in the frequency and wave-vector space, all the dynamic processes are duly taken into account through the sum rule integrations in the calculation of such a static correlation. In most cases it is therefore incorrect to evoke an additional account of so-called "dynamic" or "nonequilibrium" processes in a calculation of the reaction rates. Nuclear reactions are extremely rare events compared with other scattering and relaxation processes, so ample opportunity is available for "statistical averages" in a time scale of nuclear reactions; in principle, all the dynamic effects may be incorporated in a correct evaluation of the joint probability density.

1. Contact Probabilities

The usual boundary condition in the treatment of scattering problems assumes an incident plane wave in the z direction. The asymptotic ($r \to \infty$) form of the Coulomb wave function is then calculated as

$$\Psi_{ij}(\mathbf{r}) \to \exp\left[i\kappa z + i\eta \ln \kappa(r - z) \right] \left[1 + \frac{\eta^2}{i\kappa(r - z)} \right]$$
$$+ \frac{f_c(\theta)}{r} \exp\left[i(\kappa z - \eta \ln 2\kappa r) \right] , \tag{5.5}$$

where

$$f_c(\theta) = \frac{\Gamma(1 + i\eta)}{i\Gamma(-i\eta)} \frac{\exp\left[-i\eta \ln\left(\sin^2 \frac{1}{2}\theta\right) \right]}{2\kappa \sin^2 \frac{1}{2}\theta} \tag{5.6a}$$

is the angular function of the scattered wave,

$$\Gamma(z) = \int_0^\infty dt\, t^{z-1} \exp(-t)\,, \qquad \mathrm{Re}\, z > 0\,, \tag{5.6b}$$

is the gamma function,

$$\kappa \equiv \frac{\mu_{ij} v}{\hbar}\,, \qquad \eta \equiv \frac{Z_i Z_j e^2}{\hbar v}\,, \tag{5.6c}$$

and r represents the radial coordinate with respect to the scattering center.
With a normalization such that

$$\int_\Omega d\mathbf{r}\, |\Psi_{ij}(\mathbf{r})|^2 = \Omega \tag{5.7a}$$

over a spherical volume Ω with a radius greater than $2a_{ij}$, the incident flux
$(z \to -\infty)$ is

$$\frac{\hbar}{2i\mu_{ij}} \left[\Psi_{ij}^*(\nabla\Psi_{ij}) - \Psi_{ij}(\nabla\Psi_{ij})^*\right] = v\,. \tag{5.7b}$$

The *contact probability*, or the square of the wave function at the origin, then
takes on the value

$$\begin{aligned}
|\Psi_{ij}(0)|^2 &= \frac{1}{v}|\Gamma(1+i\eta)|^2 \exp(-\pi\eta) \\
&= \frac{2\pi\eta}{\exp(2\pi\eta) - 1} \\
&= \frac{\pi\sqrt{E_G/E}}{\exp\left(\pi\sqrt{E_G/E}\right) - 1}\,.
\end{aligned} \tag{5.8}$$

Here $E = (\mu_{ij}/2)v^2$ is the relative kinetic energy, and

$$E_G = \frac{Z_i Z_j e^2}{r_{ij}^*} \approx 49.5(\mathrm{keV})(Z_i Z_j)^2 \frac{2\mu_{ij}}{m_N} \tag{5.9}$$

is called the *Gamow energy*.

2. Nuclear Cross-Section Factor

For small collision speed ($E \ll E_G$), Eq. (5.8) is approximated by

$$|\Psi_{ij}(0)|^2 = \pi\sqrt{E_G/E} \exp\left(-\pi\sqrt{E_G/E}\right)\,. \tag{5.10}$$

Because of this energy dependence, one expresses the cross section for the nuclear reactions as

$$\sigma_{ij}(E) = \frac{S_{ij}(E)}{E} \exp\left(-\pi\sqrt{E_G/E}\right) ,$$ (5.11)

which defines the nuclear cross-section factor $S_{ij}(E)$. The reader is referred to Fowler, Caughlan, and Zimmerman [1967, 1975] for compilation of cross-section parameters for nuclear reactions of astrophysical interest.

Reaction rate (in units of reactions/cm^3/s) between nuclei of i and j species at number densities n_i and n_j with a relative kinetic energy E is thus calculated as [e.g., Ichimaru 1993a]

$$R_{ij}(E) = \frac{2S_{ij}(E)r_{ij}^* n_i n_j}{\pi(1+\delta_{ij})\hbar}\left|\Psi_{ij}(0)\right|^2 ,$$ (5.12)

where the cases with i $=$ j are accounted for by Kronecker's delta δ_{ij}.

3.   The Gamow Rates

The rates of thermonuclear reactions stemming from those binary processes may be calculated by substituting Eq. (5.10) in Eq. (5.12) and by carrying out the average in Eq. (2.8) with the Boltzmann distribution of E at temperature T:

$$f_B(E) = \frac{2}{k_B T}\left(\frac{E}{\pi k_B T}\right)^{1/2}\exp\left(-\frac{E}{k_B T}\right) .$$ (5.13)

The result yields the Gamow rate [Gamow & Teller 1938; Thompson 1957],

$$R_G(T) = \frac{16S_{ij}(T)_G r_{ij}^* \tau_{ij}^2}{3^{5/2}\pi(1+\delta_{ij})\hbar} n_i n_j \exp(-\tau_{ij}) ,$$ (5.14)

where

$$\tau_{ij} = 3\left(\frac{\pi}{2}\right)^{2/3}\left(\frac{E_G}{k_B T}\right)^{1/3}$$

$$\approx 33.70(Z_i Z_j)^{2/3}\left(\frac{2\mu_{ij}}{m_N}\right)^{1/3}\left(\frac{T}{10^6 \text{ K}}\right)^{-1/3} ,$$ (5.15)

and $S_{ij}(T)_G$ is a thermal average of $S_{ij}(E)$.

In the derivation of Eq. (5.14), it has been assumed that $E_G \gg k_B T$, consistent with the use of Eq. (5.10), which in turn implies that $\tau_{ij} \gg 1$. In these circumstances, the integration leading to the Gamow rate (5.14) contains in its integrand a product between a steeply rising term $\exp[-\pi\sqrt{E_G/E}]$ in Eq. (5.10) and a steeply decreasing Boltzmann factor $\exp(-E/k_BT)$ in Eq. (5.13). The product thus exhibits a *Gamow peak* at the energy

$$E_{GP} = \frac{1}{3}\tau_{ij}k_BT .$$ (5.16)

The thermal average of $S_{ij}(E)$ should therefore be performed with such a distribution taken into account. The radius $r_{TP}^{(0)}$ of the classical turning point for a colliding pair with the Gamow peak energy is thus given by

$$\frac{r_{TP}^{(0)}}{a_{ij}} = \frac{3\Gamma_{ij}}{\tau_{ij}} .$$ (5.17)

The Gamow reaction rate (5.14) contains a factor $\exp(-\tau_{ij})$ that decreases very steeply at lower temperatures. The magnitude of τ_{ij} increases with the charge numbers and with the reduced masses. They are typical features in the Gamow rates of thermonuclear reactions.

B. Electron-Screened Cold Fusion

The presence of electrons or other light particles such as muons may act to modify the internuclear Coulombic potential from $W_{ij}^{(0)}(r)$ in Eq. (5.2) to

$$W_{ij}^{(s)}(r) = W_{ij}^{(0)}(r)S_c(r) .$$ (5.18)

Here the function $S_c(r)$ describes the screening action of the light particles, which can follow the motion of the nuclei adiabatically. Such a screening function for the Coulomb potential between nuclei may exist, irrespective of whether the reacting nuclei are in itinerant (i.e., fluid), molecular, or cluster states.

For concreteness, in the balance of this section we confine ourselves to the cases of screening effects due to electrons, free or bound, in condensed materials. As Eq. (5.18) implies, such a screening action stems from the *density variation out of uniformity*, or *polarization*, of the electrons induced by the presence of distinctive nuclear charges. These screening effects should therefore be clearly distinguished from the screening potentials, considered in Section 2.1B, due to the *internuclear many-particle correlations*, that can exist even when the electrons may be regarded as uniform background charges—for example, as in the OCP or BIM models of dense plasmas. Salpeter's [1954] ion sphere model belonging to the latter, it would be a misnomer physically to call such a model a case of strong *electron* screening. Interplay between the internuclear Coulomb correlations and the electron screening will be considered in Section 5.3C.

Since $S_c(r)$ should take on unity at $r = 0$, one expands

$$S_c(r) = 1 - \frac{r}{D_s} + \cdots \tag{5.19}$$

so that

$$W_{ij}^{(s)}(r) = W_{ij}^{(0)}(r) - E_s + \tag{5.20}$$

with

$$E_s = \frac{Z_i Z_j e^2}{D_s} \approx 144.0 \, (\text{eV}) \, Z_i Z_j \left(\frac{D_s}{10^{-9} \text{cm}} \right)^{-1} . \tag{5.21}$$

The expansion (5.19) defines the short-range screening length D_s. These lengths for the electron liquids and for the relativistic electron gases were treated in Sections 3.2D and 3.2E.

Solution for the contact probabilities to the Schrödinger equation with the potential (5.20) may be obtained by replacing E by $E + E_s$ in Eq. (5.8) or (5.10) [Salpeter 1954]. One thus finds

$$
|\Psi_{ij}(0)|^2 = \frac{\pi \sqrt{\dfrac{E_G}{E + E_s}}}{\exp\left(\pi \sqrt{\dfrac{E_G}{E + E_s}} \right) - 1}
$$

$$
= \pi \left(\frac{E_G}{E + E_s} \right)^{1/2} \exp\left[-\pi \left(\frac{E_G}{E + E_s} \right)^{1/2} \right], \qquad E_G \gg E + E_s . \tag{5.22}
$$

In performing a thermal average of the contact probability with respect to the Boltzmann distribution (5.13) for the reaction rates, it is useful to introduce a critical temperature for electron screening determined from $E_{GP} = E_s$:

$$
T_{cs} = \frac{2}{\pi} \left(\frac{r_{ij}^*}{D_s} \right)^{1/2} \frac{Z_i Z_j e^2}{k_B D_s}
$$

$$
\approx 5.7 \times 10^4 \, (\text{K}) \, \sqrt{Z_i Z_j} \left(\frac{2\mu_{ij}}{m_N} \right)^{-1/2} \left(\frac{D_s}{10^{-9} \text{ cm}} \right)^{-3/2} . \tag{5.23}
$$

Relative to this temperature, the strength of the electron screening effect on nuclear reactions may be classified.

In a high-temperature regime (i.e., $E_{GP} > E_s$) such that

$$T > T_{cs} , \tag{5.24}$$

the effects of the electron screening are weak; they can be treated perturbation theoretically. Reaction rates under such weak electron screening conditions are thus calculated as

$$R_{ws}(T) = \frac{16 S_{ij}(T)_{ws} r_{ij}^* \tau_{ij}^2}{3^{5/2} \pi (1 + \delta_{ij}) \hbar} n_i n_j A_{ws}^{(e)} \exp(-\tau_{ij}) , \tag{5.25a}$$

with

$$A_{ws}^{(e)} = \left(1 - \frac{3 E_s}{\tau_{ij} k_B T}\right) \exp\left(\frac{E_s}{k_B T}\right) . \tag{5.25b}$$

Here $S_{ij}(T)_{ws}$ is another thermal average of $S_{ij}(E)$ appropriate to the weak screening conditions; the classical turning point takes place at the value given by Eq. (5.17).

If, on the other hand, the condition for a strong electron screening, that is,

$$T < T_{cs} , \tag{5.26}$$

is satisfied, the reaction rate (5.25) is replaced by

$$R_{ss}(T) = \frac{2 S_{ij}(T)_{ss} r_{ij}^*}{(1 + \delta_{ij}) \hbar} n_i n_j \left(\frac{E_G}{E_s}\right)^{1/2} \exp\left[-\pi \left(\frac{E_G}{E_s}\right)^{1/2}\right] , \tag{5.27}$$

where $S_{ij}(T)_{ss}$ is a thermal average of $S_{ij}(E)$ under the strong electron screening conditions, different generally from either $S_{ij}(T)_G$ or $S_{ij}(T)_{ws}$. The classical turning point now takes place at

$$r_{TP}^{(s)} \approx D_s . \tag{5.28}$$

The turning radius is here given basically by the short-range screening length D_s; hence it is almost independent of the relative energy E.

Contrary to the Gamow rates (5.14) or their weakly electron-screened counterparts (5.25), which change sharply with the temperature via τ_{ij}, the reaction rates of Eq. (5.27) depend weakly on temperature only through $S_{ij}(T)_{ss}$ and possibly through E_s; it may increase rather steeply with the electron density via D_s. It is in these contexts that Cameron [1959] coined the term "pycnonuclear reactions" to describe nuclear reactions under the strong electron screening conditions

Table 5.1 lists some of the parameters pertinent to electron-screened nuclear fusion reactions in examples of dense astrophysical and laboratory plasmas. The screening distances are estimated here through the schemes of Sections 3.2D and 3.2E.

We remark that the first four astrophysical examples (WD1, WD2, BD, and GP) have turned out to be the cases of weak electron screening, while the last two terrestrial examples (MH and PM) describe strong electron screening. This is somewhat ironic, since Cameron's original idea of pycnonuclear processes was for the interiors of degenerate stars such as WD. In these stars, however, the actual temperatures T are usually higher than the critical temperatures T_{cs} of electron screening; hence the effect of electron screening on reaction rates is relatively weak. Huge enhancement in the nuclear reaction rates expected in those degenerate stars stems principally from the screening potentials produced by the internuclear many-particle processes without electron screening, which we shall treat in detail in the subsequent sections.

In an ultradense stellar matter, a different sort of density-sensitive, but temperature-insensitive, nuclear reactions is expected when the plasma freezes

Table 5.1   Electron screening effects in nuclear reactions

Case	WD1	WD2	BD	GP	MH	PM
Reaction	^{12}C-^{12}C	^{4}He-^{4}He	p-p	d-p	d-d	^{7}Li-p
$\rho_m(g/cm^3)$	2×10^9	1×10^8	1×10^3	5	0.23	30
Z_1, Z_2	6, 6	2, 2	1, 1	1, 1	1, 1	3, 1
A_1, A_2	12, 12	4, 4	1, 1	2, 1	2, 2	7, 1
r_s	0.0014	0.0038	0.14	0.93	2.9	0.71
$D_s(10^{-9}cm)$	0.042	0.11	1.3	3.2	2.0	2.1
$E_s(eV)$	1.2×10^5	5.1×10^3	114	45	72	203
$E_G(keV)$	7.7×10^5	3.2×10^3	49.5	66.1	99	781
τ_{12}	228	62.6	23.3	137	634	845
$\pi\sqrt{E_G/E_s}$	250	78.5	65.7	121	117	195
T_{cs} (K)	1.1×10^7	1.5×10^6	4.0×10^4	8.6×10^3	1.4×10^4	2.4×10^4
T (K)	5×10^7	1×10^7	3×10^6	2×10^4	3×10^2	1×10^3
$\log_{10}(R_s/R_G)$	12.1	2.4	0.19	11.0	222	280

Note: Mass densities ρ_m and temperatures T are those assumed for the reacting nuclei. R_s in $\log_{10}(R_s/R_G)$ is either R_{ws} or R_{ss}, depending on relative magnitude between T and T_{cs}.

into a solid state. Here one usually finds $Y \gg 1$, so that the zero-point oscillations prevail over the thermal motion of the nuclei (see Section 2.3D). We shall consider pycnonuclear reactions for such a quantum solid in the next subsection.

The effects of electron screening in sustaining nuclear reactions at low temperatures are definitely at work for dense laboratory plasmas such as MH and PM; here a term such as "electron-screened cold fusion" seems more appropriate. Huge numbers predicted on $R_{ss}(T)/R_G(T)$ are misleading, however, since the basic thermonuclear reaction rates $R_G(T)$ are minuscule for those cases of MH and PM.

The foregoing elementary fusion reactions, with or without electron screening, are still binary as far as the internuclear processes are concerned. Enhancement of the reaction rates due to internuclear correlation processes is in fact the principal feature in dense plasmas. The extent of such many-particle effects depends in turn on the nature of binary interactions with or without electron screening. We shall elucidate these many-particle aspects in Section 5.3.

C. Pycnonuclear Rates in Solids

The body-centered cubic (bcc) crystalline structures are known to have the lowest values in the Madelung energies for the Coulombic crystals (see Section 2.3C); hence a bcc structure is usually assumed for a dense OCP solid. The principal problem then is the evaluation of contact probabilities $|\Psi(0)|^2$ arising from s-wave scattering between nearest-neighbor particles. The factor that crucially controls such a contact probability is the effective potential between the nuclei in the short-range domain. Such a potential is called the lattice potential; it may be approximated as isotropic for the s-wave scattering. Pycnonuclear rates are obtained from a solution to the relevant Schrödinger equation; both OCP and BIM cases will be treated.

1. Lattice Potentials

The study of nuclear reactions in solids begins with bcc crystalline lattice in the ground state. The lattice potential $W^L(r)$ of such a bcc-crystalline plasma is defined as the effective potential between two nearest-neighbor particles at an interparticle separation r. The screening potential $H^L(r)$ for the Coulomb solid is then calculated in accordance with

$$W^L(r) = \frac{(Ze)^2}{r} - H^L(r) .\tag{5.29}$$

The lattice potential takes on its minimum value at the nearest-neighbor distance

$$r_m = d = (3\pi^2)^{1/6}a \approx 1.7589a ,\tag{5.30}$$

where a is the ion sphere radius (1.11a) of an OCP.

Salpeter and Van Horn [1969] originally evaluated the lattice potential, by choosing a pair of nearest-neighbor particles in a bcc crystal (with the lattice constant b) and then calculating the electrostatic energy as a function of the interparticle separation r with the center of mass fixed. In their fully relaxed approximation, these authors adjusted the resultant screening potential near $r = 0$ in accordance with the ion sphere model of Salpeter [1954]. The screening potential for an OCP solid obtained in such a relaxed lattice model has been expressed as

$$\frac{H^{\text{SVH}}(r)}{(Ze)^2/b} = 1.1547 + 1.1602(1 - y) - 1.0394(1 - y)^2$$
$$+ 2.5690(1 - y)^3 - 1.6971(1 - y)^4 , \qquad (5.31)$$

where $y = r/d$.

The screening potentials can be analyzed through MC sampling methods in a way analogous to the fluid cases (see Section 2.1). Here one deals with the joint probability densities between those pairs of particles located around

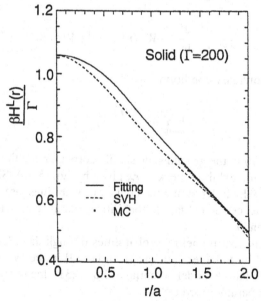

Figure 5.1 Screening potential between pairs of the nearest-neighbor particles for an OCP solid at $\Gamma = 200$. The maximum extent of uncertainties in the MC sampled points is 10^{-4} unless explicitly shown by vertical bars. SVH refers to Eq. (5.31); Fitting, Eq. (5.32) [Ogata, Iyetomi, & Ichimaru 1991].

the nearest-neighbor sites of the bcc lattice [Ogata, Iyetomi, & Ichimaru 1991], since only these can in fact constitute the reacting pairs. Extrapolation into the short-range domain can be executed with the aid of Eq. (2.9); one thus finds

$$\frac{\beta H^L(r)}{\Gamma} =$$
$$\begin{cases} 1.061 - 0.250x^2, & \text{for } x \le 0.70 , \\ 1.183 - 0.350x + \dfrac{1}{x}\exp\left(13.2\sqrt{x} - 22.1\right), & \text{for } 0.70 < x < 2 , \end{cases} \quad (5.32)$$

where $x = r/a$. Figure 5.1 exhibits MC data of such a screening potential for an OCP solid at $\Gamma = 200$ as well as evaluations (5.31) and (5.32); a reasonable similarity is observed between these two evaluations.

2. Pycnonuclear Rates in OCP Solids

Pycnonuclear rates may be obtained from Eq. (5.12) with the contact probability evaluated through the ground-state wave function in the lattice potential. One thus solves the Schrödinger equation for the s-wave scattering,

$$\left[-\frac{\hbar^2}{2\mu}\frac{d^2}{dr^2} + W^L(r) - E \right] r\Psi(r) = 0 , \qquad (5.33)$$

with the cusp boundary condition

$$\lim_{r \to 0} \frac{d \ln \Psi(r)}{dr} = \frac{1}{2r^*} . \qquad (5.34)$$

The solution $\psi(r)$ in the ground state should correspond to the zero-point vibrations of nuclei and exhibit a peak at r_m given by Eq. (5.30). Short-range values of the solution, which determine the contact probabilities, are governed by condition (5.34). Hence the contact probabilities should scale with the two length parameters r_m and r^*.

The pycnonuclear tunneling probabilities through the OCP lattice potential were calculated by Salpeter and Van Horn [1969] in the WKB approximation. With the relaxed lattice model potential of Eq. (5.31), the results for the reaction rates have been parametrized as

$$R_{\text{SVH}}(\text{reactions/cm}^3/\text{s}) = 3.52 \times 10^{35}\left(\frac{\rho_m}{A}\right)^2 \frac{S}{AZ^2}\lambda^{-5/4}\exp(-2.516\lambda^{-1/2}) . \qquad (5.35)$$

Here the quantity λ is defined by

$$\lambda \equiv \left(\frac{3}{4}\right)^{1/2} \frac{r^*}{r_m} = \frac{1}{AZ^2}\left(\frac{1}{A}\frac{\rho_m}{1.3574 \times 10^{11}\text{g/cm}^3}\right)^{1/3}, \qquad (5.36)$$

S is the cross-section factor expressed in MeV · barns, and ρ_m is the mass density.

Instead of using the WKB approximation, Ogata, Iyetomi, and Ichimaru [1991] calculated the contact probabilities by computing the exact numerical solution to Eq. (5.33) in the MC lattice potential (5.32) with the boundary condition (5.34). The parametrized expression for the reaction rate so obtained is

$$R_{Ol^2}(\text{reactions/cm}^3/\text{s}) = 1.34 \times 10^{32}\left(\frac{\rho_m}{A}\right)^2 \frac{S}{AZ^2}\lambda^{-1.809}\exp(-2.460\lambda^{-1/2})$$

$$(5.37)$$

The reaction rates, Eqs. (5.35) and (5.37), are independent of the temperature but increase steeply as the nearest-neighbor distance r_m decreases. As has been shown [Ogata, Iyetomi, & Ichimaru 1991], the numerical agreement between Eqs. (5.35) and (5.37) appears remarkably good, despite differences in the lattice potentials and in the ways the Schrödinger equation was solved. In particular, the density dependences of the important exponential factors are the same and proportional to $\rho_m^{-1/6}$.

3. Compositional Scaling of Pycnonuclear Rates in BIM Solids

We first consider the generalization of the OCP pycnonuclear rates to the cases of BIM solids consisting of nuclei with charge and mass numbers Z_1 and $A_i (i = 1, 2)$. Let x_i and X_i denote the molar and mass fractions, respectively, with $x_1 + x_2 = X_1 + X_2 = 1$. For simplicity we shall generally write $x \equiv x_2$ and $R_Z \equiv Z_2/Z_1$, assuming $Z_2 \geq Z_1$.

In a BIM, the electron background is taken to be completely uniform, with electron density n_e, as given by Eq. (1.1). Under these conditions, the distance between two neighboring nuclei—which are the only ones capable of participating in pycnonuclear reactions—is approximately equal to the sum of the ion sphere radii of the two nuclei. Since the radius a_i of the ion sphere is defined by Eq. (1.11a), r_m is approximately given by $a_i + a_j$. In the following, we actually use the mean ion sphere radius a_{ij} in Eq. (1.11b).

To examine the accuracy of such an ion sphere scaling, Ogata et al. [1993] have performed a series of MC samplings for the joint probabilities $g_{ij}(r)$ between i and j nuclei in ground-state BIM solids for 20 different cases, with combinations of the parameter values $x = 5/432$, 1/4, 1/2, 3/4 and $R_Z = 4/3$, 5/3, 2, 3, 4. Generally in BIM solids, the ground-state configurations deviate significantly from the simple periodic bcc lattice structure due mainly to $a_i \neq a_j$ for $Z_i \neq Z_j$. In fact for $R_Z \geq 5/3$, the ground states observed are characterized more appropriately in terms of *aperiodic glassy solids*. Under these circumstances, the exact MC nearest-neighbor separation $r_{m,ij}$ may be determined from the observed peak position of $g_{ij}(r)$; the results are then expressed as the sum of the ion sphere scaling contribution and a deviation therefrom:

$$r_{m,ij} = 1.76 a_{ij} + \Delta r_{m,ij} . \tag{5.38}$$

The deviations $\Delta r_{m,ij}$, characterizing extra distortions in the particle configurations due to the charge disparities in the BIM solids, have been measured in the MC data; the results can be summarized in the following parametrized forms for $0 \leq x < 1$ and $1 \leq R_Z < 4.5$:

$$\frac{\Delta r_{m,11}}{a_{11}} = 0.44 \frac{(R_Z - 1)(2.3 - R_Z)}{R_Z^2} x^{1.3} , \tag{5.39a}$$

$$\frac{\Delta r_{m,12}}{a_{12}} = -0.043 \frac{\sqrt{R_Z - 1}}{1 + x^{1.3}} , \tag{5.39b}$$

$$\frac{\Delta r_{m,22}}{a_{22}} = -0.17 \frac{(R_Z - 1)(2.3 - R_Z)}{R_Z^2} (1 - x)^{1.3} . \tag{5.39c}$$

A number of observations are in order concerning these results:

1. The corrections to the ion sphere scaling are small but nonnegligible.
2. For the relative magnitudes of the deviations, we find
   $|\Delta r_{m,11}| > |\Delta r_{m,22}| \gg |\Delta r_{m,12}|$.
3. The quantities $\Delta r_{m,11}$ and $\Delta r_{m,22}$ are opposite in sign.
4. The quantity $\Delta r_{m,11}$ is *positive* for $R_Z < 2.3$, corresponding to the "blocking effect" of nuclear reactions discovered for C-O solids [Ogata, Iyetomi, & Ichimaru 1991]. When R_Z is not excessively large, the phase-space reduction effect of the heavier (i.e., higher-Z) nuclei is likewise small; they simply act as obstacles for reactions between the lighter (i.e., lower-Z) nuclei.
5. For $R_Z > 2.3$, on the other hand, $\Delta r_{m,11}$ takes on *negative* values, implying enhancement in the pycnonuclear rates for the lighter species. This enhancement stems from a "catalyzing action" of the heavier elements, which reduces the effective volume available to the lower-Z nuclei, thus reducing the internuclear separations. This catalyzing action is a new feature discovered in these MC simulation studies of BIM solids.

We may thus generalize Eqs. (5.35) and (5.37) to BIMs simply by replacing

$r_m \rightarrow r_{m,ij}$ and $r^* \rightarrow r_{ij}^*$. The validity of this replacement can be justified by comparison with the explicit BIM calculations of pycnonuclear rates for C-O systems. From the more accurate reaction rates (5.37), we finally obtain the pycnonuclear rates per unit volume in BIM solids as

$$R_{ij}(\text{reactions/cm}^3/\text{s})$$
$$= \frac{1.34 \times 10^{32}}{1 + \delta_{ij}} \frac{X_i X_j (A_i + A_j)}{Z_i Z_j (A_i A_j)^2} \rho_m^2 S_{ij} \lambda_{ij}^{-1.809} \exp(-2.460 \lambda_{ij}^{-1/2}) \quad (5.40)$$

[Ichimaru, Ogata, & Van Horn 1992].

5.3   ENHANCEMENT FACTORS

The central theme of this chapter is enhancement of the reaction rates due to many-body correlations. This section begins with laying a quantum-statistical foundation for such an enhancement effect and proceeds to discuss its relation to the thermodynamic properties of the condensed plasmas. As specific examples, statistical-mechanical enhancement in one-component and binary-ionic fluids as well as thermal enhancement in solids are considered.

A. Enhancement by Many-Body Correlation Processes

Rates of nuclear reactions in dense plasmas depend on quantum-statistical correlations between reacting nuclei at extremely short distances. Multiparticle correlations act to enhance the reaction rates through the screening potentials, which are closely related to the thermodynamic functions of dense plasmas. This relationship is elucidated in this subsection.

1. Quantum Mechanical Correlation Functions

We consider an OCP of N particles in a volume V with Hamiltonian H. Nuclei 1 and 2 are designated as the *reacting (R) pair*; the rest, the *spectator (S) nuclei*. The relative coordinates and those of the center of the "R" pair are denoted by $\mathbf{r}$ and $\mathbf{R}$. The joint probability density of the "R" pair at zero separation is then expressed as [Alastuey & Jancovici 1978]

$$g(0) = V^2 \frac{\langle 00 | \langle "S" | \exp(-\beta H) | "S" \rangle | 00 \rangle}{\langle \mathbf{rR} | \langle "S" | \exp(-\beta H) | "S" \rangle | \mathbf{rR} \rangle}. \quad (5.41)$$

Here $\beta = 1/k_B T$, and integrations over the respective coordinates are implied when $\mathbf{r}$, $\mathbf{R}$, or "S" are left in the bracket notation. Assuming a translational invariance of the system (i.e., fluid), we have set $\mathbf{R} = 0$ in the numerator of Eq. (5.41).

In the range of parameters under present study, the thermal de Broglie wavelengths Λa are small enough for the thermodynamics of the system to be described by classical statistical mechanics (see Section 1.1C). This means that those configurations that contribute a nonnegligible weight to the denominator of Eq. (5.41) are classical, and this denominator can be replaced by $Q/(\Lambda a)^{3N}$, where Q is the classical configuration integral

$$Q = \int d\mathbf{r}_1 \cdots d\mathbf{r}_N \exp\left[-\beta H_{\text{int}}(\mathbf{r}; \text{``}S\text{''})\right] . \tag{5.42}$$

We express the total potential energy as

$$H_{\text{int}}(\mathbf{r}; \text{``}S\text{''}) = \frac{(Ze)^2}{r} + W(\mathbf{r}; \text{``}S\text{''}) , \tag{5.43}$$

where $W(\mathbf{r}; \text{``}S\text{''})$ is the sum of all interactions except the one between the "R" pair.

The numerator of Eq. (5.41), however, is dominated by configurations in the neighborhood of $\mathbf{r} = 0$, where the potential is very steep in light of the cusp boundary condition (5.34) applicable generally for the Coulomb scattering; hence quantum effects are essential for the relative motion between the "R" pair. Most of the motion of its center of mass, and of "S" particles, on the other hand, occurs in regions of configuration space where the potential is smooth, and these motions can be considered classical. It is therefore sufficient to keep the kinetic energy K associated with $\mathbf{r}$ and the total potential energy explicitly in the Hamiltonian H; one thus finds

$$g(0; W) =$$
$$\frac{2^{3/2} V^2 (\Lambda a)^3}{Q} \left\langle 0 \left| \left\langle \text{``}S\text{''} \left| \exp\left\{ -\beta \left[ K + \frac{(Ze)^2}{r} + W(\mathbf{r}; \text{``}S\text{''}) \right] \right\} \right| \text{``}S\text{''} \right\rangle \right| 0 \right\rangle . \tag{5.44}$$

In the framework of these approximations, the quantum many-body problem has been reduced to a quantum one-body problem, which, however, involves a complicated potential $W(\mathbf{r}, \text{``}S\text{''})$ depending on the coordinates of $N - 2$ "S" particles. In principle, one must first compute the matrix element in Eq. (5.44) for every value of the set of parameters "S" and then perform the integration upon "S." Alastuey and Jancovici [1978] have shown a way to accomplish this through a systematic method of successive approximations, in which the matrix elements may be evaluated through a path integral approach [Feynman & Hibbs 1965; Feynman 1972].

Let us note that in the case of an infinitely dilute plasma, the case treated in

Section 5.2A, $g(0; W)$ becomes $g(0; 0)$ and is obtained by using Eq. (5.44) with $Q = V^N$ and $W = 0$. An enhancement factor of the reaction rate due to many nuclear processes is therefore defined and calculated as

$$A \equiv \frac{g(0; W)}{g(0; 0)} = \frac{V^2}{Q} \left\langle \text{``}S\text{''} \left| \exp\left[-\frac{S - S_0}{\hbar}\right] \right| \text{``}S\text{''} \right\rangle . \qquad (5.45)$$

Here the action S for a particle with mass $M/2$ is

$$S = \int_0^{\hbar\beta} dt \left[\frac{M}{4}\left(\frac{d\mathbf{r}}{dt}\right)^2 + \frac{(Ze)^2}{r} + W(\mathbf{r}; \text{``}S\text{''}) \right] \qquad (5.46)$$

along the trajectory $\mathbf{r}(t)$ that minimizes S; the trajectories are to be taken from the origin back to the origin, in a time $\hbar\beta$, in the potential with the reversed sign. The action S_0 is an analogous quantity with $W = 0$ in Eq. (5.46). If only the classical trajectories are kept, the calculation will correspond to using the WKB method [e.g., Salpeter & Van Horn 1969; Alastuey & Jancovici 1978] to a solution for a Schrödinger equation at the energy of the Gamow peak (Eq. 5.16). Ogata, Iyetomi, and Ichimaru [1991] have advanced a method by which the path-integral average, appropriate to the contact probability, can be performed through an exact numerical solution to the Schrödinger equation. We shall consider this method in Section 5.3B.

2. Enhancement Factors

From the complicated many-body potential $W(\mathbf{r}; \text{``}S\text{''})$, one constructs a two-body potential $H(r)$ defined by

$$\exp[\beta H(r)] = \frac{V^2}{Q} \left\langle \text{``}S\text{''} \left| \exp[-\beta W(\mathbf{r}; \text{``}S\text{''})] \right| \text{``}S\text{''} \right\rangle . \qquad (5.47)$$

This function is the screening potential of Section 2.1B, and is related to the classical radial distribution function via Eq. (2.8).

In terms of the screening potentials, the enhancement factors due to many-particle processes are now expressed from Eq. (5.45) in compact form as

$$A = \exp\left[\beta \langle H(r) \rangle_R\right] , \qquad (5.48)$$

where $\langle \cdots \rangle_R$ means a path-integral average with respect to the penetrating wave functions $\Psi(r)$ from $r = 0$ to the classical turning point Eq. (5.17) and back. The wave functions are calculated from a solution to the Schrödinger equation without accounting for the screening potentials. This is justifiable when the major contributions to $\langle H(r)\rangle_R$ stem from the vicinity of $r = 0$, since the steep nuclear potentials dominate over the fluctuating many-body potentials near $r = 0$. Decoupling between the "S" and "R" averages is thus completed.

3. Chemical Potentials

If the classical turning point is far shorter than the internuclear spacings, that is,

$$r_{TP} \ll a \,, \tag{5.49}$$

then the enhancement factors are further simplified as

$$A = \exp\left[\beta H(0)\right] \,. \tag{5.50}$$

It is through these formulas that one establishes an intimate connection between the nuclear reaction rates and the thermodynamic functions in dense plasmas.

The screening potentials have the short-range expansion (Eq. 2.9) in power series of r^2 due to Widom [1963]. It has been proven in Eq. (2.12) that $H(0)$ corresponds to the increment in the excess chemical potentials for the "R" pair before and after the reactions. In the notation of Eq. (2.89), one finds

$$\beta H(0) = 2f_{ex}^{OCP}(\Gamma) - f_{ex}^{OCP}(2^{5/2}\Gamma) - \frac{\partial}{\partial x}\Delta f_{ex}^{BIM}(2, x, \Gamma)\Big|_{x=0} \,. \tag{5.51}$$

In Salpeter's [1954] ion sphere model, one sets

$$f_{ex}^{OCP}(\Gamma) = -0.9\Gamma \,, \tag{5.52a}$$

$$\Delta f_{ex}^{BIM}(R_Z, x, \Gamma) = 0 \,, \tag{5.52b}$$

to find

$$\beta H(0)_S = 1.057\Gamma \,. \tag{5.53}$$

The MC equation of state for BIMs such as Eqs. (2.89)–(2.92) may be used in Eq. (5.51) for a thermodynamic evaluation of $\beta H(0)$. The result of such a calculation closely corroborates the extrapolated values (2.19) in the MC simulation. This implies importance of the departure from the linear mixing law, the last term in Eq. (2.89), for the estimation of enhancement factors.

B. Multi-Ionic Fluids Without Electron Screening

The path-integral average of the screening potential that goes into the evaluation of the enhancement factor (5.48) can be performed through a method of quantum

MC sampling. Such a method usually evokes a prescription, called *importance sampling*, by which those parts of the wave functions that carry significant weights are sampled most accurately. Such a prescription is thus essential for the evaluation of the equation of state for a quantum many-body system.

In the calculations of reaction rates that depend on the contact probabilities $|\Psi_{ij}(0)|^2$, however, such a prescription of importance sampling does not generally assure the accuracy of evaluation, since $|\Psi_{ij}(0)|^2$ actually represent the parts where overlap of the wave functions is infinitesimally small. An illustrative case may be offered in the treatment of pycnonuclear rates in solids, where one knows that the bulk of the nuclear wave functions in such a solid is quite accurately written in terms of the Gaussian representing the harmonic oscillators. A calculation based on the overlap of such wave functions between two neighboring nuclei does not, however, lead to a correct evaluation of the pycnonuclear rates. Superposition of such harmonic oscillator wave functions for a quantum solid fails to account for the steep Coulomb repulsion that arises when two nuclei approach near a reaction distance; consequently, the resultant two-particle wave functions violate the cusp condition (5.34).

The contact probabilities $|\Psi_{ij}(0)|^2$ in multi-ionic fluids may be calculated from the solution to the Schrödinger equation

$$\left[-\frac{\hbar^2}{2\mu} \frac{d^2}{dr^2} + \frac{Z_i Z_j e^2}{r} - H_{ij}(r) - E \right] r \Psi_{ij}(r) = 0 , \qquad (5.54a)$$

with the cusp boundary condition

$$\lim_{r \to 0} \frac{d \ln \Psi_{ij}(r)}{dr} = \frac{1}{2r_{ij}^*} , \qquad (5.54b)$$

where $H_{ij}(r)$ is the screening potential given by Eq. (2.20). A rigorous calculation of the associated enhancement factors A_{ij} may be performed through a numerical solution to this equation in the following way [Ogata, Iyetomi, & Ichimaru 1991].

We first note that with the bare Coulomb potentials—that is, when $H_{ij}(r) = 0$—the solution to Eq. (5.54) leads to Eq. (5.8) and thus is expressed as

$$U^{(0)}(0) = -\frac{1}{2} \ln \left[\exp \left(\frac{t}{\sqrt{\varepsilon}} \right) - 1 \right] + \frac{1}{2} \ln \left(\frac{t}{\sqrt{\varepsilon}} \right) , \qquad (5.55)$$

where

$$t = \pi \sqrt{\frac{a_{ij}}{r_{ij}^*}} , \qquad (5.56a)$$

$$\varepsilon = \frac{E}{Z_i Z_j e^2 / a_{ij}} , \tag{5.56b}$$

and for conciseness we have adopted a notation $U(r) \equiv \ln \Psi_{ij}(r)$. We next consider the solution with a shifted Coulomb potential for which $H_{ij}(r)$ is replaced by $H_{ij}(0)$ in Eq. (5.54); the result then is

$$U^{(1)}(0) = -\frac{1}{2} \ln \left[\exp \left(\frac{t}{\sqrt{\varepsilon + h}} \right) - 1 \right] + \frac{1}{2} \ln \left(\frac{t}{\sqrt{\varepsilon + h}} \right) , \tag{5.57}$$

with

$$h = \frac{H_{ij}(0)}{Z_i Z_j e^2 / a_{ij}} . \tag{5.58}$$

Through the thermal averages of the contact probabilities, the enhancement factors are evaluated as

$$A_{ij} = \frac{P(T)}{P^{(0)}(T)} , \tag{5.59}$$

where

$$P(T) = \int_0^\infty dE \sqrt{E} \exp\left[-\beta E + 2U^{(1)}(0) \right] \exp(2\Delta U) , \tag{5.60a}$$

$$P^{(0)}(T) = \int_0^\infty dE \sqrt{E} \exp\left[-\beta E + 2U^{(0)}(0) \right] , \tag{5.60b}$$

and

$$\Delta U \equiv U(0) - U^{(1)}(0) . \tag{5.60c}$$

These increments are the quantities accessible accurately through numerical solutions to Eq. (5.54a) with the boundary condition (5.54b) as follows: Equation (5.54a) is integrated with the boundary conditions $U(0) = 0$ and Eq. (5.54b) starting from $r = 0$; the integrated result of $U(r)$ takes on a maximum value X near the classical turning radius (see Eq. 5.17). Next Eq. (5.54a) is solved with the shifted Coulomb potential $H_{ij}(r) \to H_{ij}(0)$ under the same boundary conditions, and the corresponding maximum value X_1 is obtained. We now note that both solutions should approach similar wave functions representing the behavior of a free particle asymptotically for large r, with differences only in their normalization. The differences between X and X_1 may thus be used for the assessment of ΔU:

$$\Delta U = \ln(X/X_1) . \qquad (5.61)$$

These processes of calculation have been followed for 120 cases of parameter combinations in the ranges $t/\pi = 20$–200 and $\varepsilon = 10^{-4}$–10^2 [Ogata, Iyetomi, & Ichimaru 1991]; the results have been fitted in an analytic formula,

$$\Delta U = -\frac{t}{\pi} \exp\left(-\sqrt{\varepsilon + 0.001}\right) \left[u_1 + u_2 \ln \Gamma_{ij} + \frac{u_3 + u_4 \ln \Gamma_{ij}}{\varepsilon + 0.8073 - 0.1335 \ln \Gamma_{ij}} \right.$$
$$\left. + \frac{u_5 + u_6 \ln \Gamma_{ij}}{(\varepsilon + 0.8073 - 0.1335 \ln \Gamma_{ij})^2} \right] , \qquad (5.62)$$

with $u_1 = 0.0011685$, $u_2 = -0.0023531$, $u_3 = 0.0020104$, $u_4 = 0.0087516$, $u_5 = 0.066453$, and $u_6 = -0.013581$. Fitting errors of Eq. (5.62) are confined to 0.3 in absolute magnitude.

The enhancement factors are evaluated in accordance with Eq. (5.59). Writing

$$A_{ij} = \exp(Q_{ij}) , \qquad (5.63)$$

we find

$$Q_{ij} = \beta H_{ij}(0) - \frac{5}{32} \Gamma_{ij} \left(\frac{3\Gamma_{ij}}{\tau_{ij}}\right)^2 \left[1 + (C_1 + C_2 \ln \Gamma_{ij}) \left(\frac{3\Gamma_{ij}}{\tau_{ij}}\right) + C_3 \left(\frac{3\Gamma_{ij}}{\tau_{ij}}\right)^2 \right] ,$$
$$(5.64)$$

with $C_1 = 1.1858$, $C_2 = -0.2472$, and $C_3 = -0.07009$; we recall Eqs. (1.15), (2.22), and (5.15) for Γ_{ij}, $H_{ij}(0)$, and τ_{ij}, respectively.

The first term on the right-hand side of Eq. (5.64) is the classical contribution (5.50). The quantum corrections are expressed in a polynomial of the classical turning radius $(3\Gamma_{ij}/\tau_{ij})$. The first term,

$$\Delta Q_{ij}\big|_{Q1} = -\frac{5}{32} \Gamma_{ij} \left(\frac{3\Gamma_{ij}}{\tau_{ij}}\right)^2 , \qquad (5.65)$$

was derived originally by Alastuey and Jancovici [1978].

C. Electron-Screened Ionic Fluids

The screening potentials in electron-screened multi-ionic plasmas are defined in terms of the radial distribution functions $g_{ij}^{(e)}(r)$ and the screened binary potentials (5.18) as

$$H_{ij}^{(e)}(r) = W_{ij}^{(s)}(r) + \frac{1}{\beta} \ln g_{ij}^{(e)}(r) \, . \tag{5.66}$$

They are the crucial quantities in the theoretical estimation for enhancement factors of the nuclear reaction rates in dense plasmas.

The short-range screening effects of relativistic degenerate electrons on Coulomb repulsion between the reacting nuclei were treated in Section 3.2E. To elucidate salient features in such an electron-screened system, an MC simulation study of the screening potential for an ultradense carbon matter at $\rho_m = 2 \times 10^9 \mathrm{g/cm^3}$ and $T = 10^8$ K was performed [Ichimaru & Ogata 1991], where it was assumed that the interparticle potential was given by

$$W_{ij}^{(s)}(r) = \frac{(Ze)^2}{r} \exp\left[-\frac{r}{D_s}\right] \, , \tag{5.67}$$

with $Z = 6$ and $D_s/a = 3.2$ as computed in Eq. (3.58). The screening potential $H^{(e)}(r)$ defined in accordance with Eq. (5.66) was sampled over 5×10^6 MC configurations generated with 500 particles in the periodic boundary conditions; the result is plotted in Fig. 5.2.

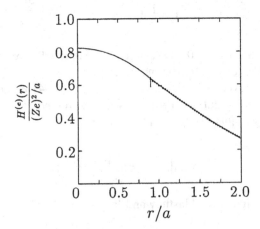

Figure 5.2 Screening potential in an ultradense carbon matter with the electron-screened internuclear potential (5.67) with $Z = 6$ and $D_s/a = 3.2$. The maximum extent of uncertainties in the MC sampled points is 10^{-5} unless explicitly shown by vertical bars. [Ichimaru & Ogata 1991]

The MC data are then fitted with an analytic formula as

$$\frac{H^{(e)}(r)}{(Ze)^2/a} = \begin{cases} 0.8252 - 0.2312(r/a)^2 \,, & \text{for } r/a \leq 0.8427 \,, \\ -1.048 + 2.071 \exp\left[-0.228(r/a)\right] \,, & \text{for } 0.8427 < r/a < 2 \,. \end{cases}$$
$$(5.68)$$

The expression for the intermediate range—the second line of Eq. (5.68)—stems from an actual fit to the MC data as shown in Fig. 5.2.

The first line, on the other hand, derives from the short-range expansion

$$H^{(e)}(r) = H^{(e)}(0) - \frac{(Ze)^2}{12a}\left(\frac{a}{D_s}\right)^2 \left\langle \sum_{p \neq 1} \frac{a_{1p}}{r_{1p}} \exp\left(-\frac{r_{1p}}{D_s}\right) \right\rangle \left(\frac{r}{a}\right)^2 \,, \quad (5.69)$$

where $r_{ij} = |\mathbf{r}_i - \mathbf{r}_j|$; it is a screened version of the expansion (2.9) with Eqs. (2.14). For the statistical average involved in the second term, the technique of MC sampling in a test-charge system as employed for the evaluation of $H^{(2)}$ in Eq. (2.15) may be used; one thus performs the sampling in a system where the test-particle 1 interacts with other 498 p-particles via a potential $2W^{(s)}(r)$, while the 498 p-particles interact with one another by the potential $W^{(s)}(r)$. The coefficient (0.2312) in the second term of Eq. (5.68) was obtained through evaluation of the statistical average in Eq. (5.69) over 3×10^6 MC configurations generated in such a system. With this coefficient determined, the first term of Eq. (5.68), 0.8252, may be derived from a smooth extrapolation toward the short-range domain.

In the absence of electron screening, Eq. (2.22) yields $H(0) = 1.110(Ze)^2/a$ for the carbon OCP under consideration. With the MC estimate (5.68), one thus calculates

$$H(0) - H^{(e)}(0) = \frac{0.285(Ze)^2}{a} \,. \tag{5.70a}$$

On the other hand, an application of Salpeter's formula (5.53) based on the ion sphere model and the concept of the electron-screened Coulomb coupling parameter (see Section 4.2C) yields

$$H(0) - H^{(e)}(0) = 1.057\left[1 - \exp\left(-\frac{a}{D_s}\right)\right]\frac{(Ze)^2}{a} = 0.284\frac{(Ze)^2}{a} \,, \quad (5.70b)$$

a value in good agreement with the MC value (5.70a). This result illustrates the validity of the concept for the electron-screened Coulomb coupling parameters.

The screening by electrons acts in two ways to influence the rates of nuclear reactions [Ichimaru 1993a]. First, as we have considered in Section 5.2B, the binary repulsive potentials between reacting nuclei are reduced by electrons, thus enhancing reaction rates; these are the short-range screening effects. The reduction in the strength of particle interactions by the screening, in turn, affects the many-body correlation processes and generally acts to lower values of the resultant enhancement factors, when compared with the enhancement through many-body processes without electron screening, as Eqs. (5.70) illustrate. These long-range effects of electron screening therefore counteract the gain obtained in the short-range processes. Thus the net gain due to electron screening may not be as large as the calculation with direct screening alone in Section 5.2B might imply.

1. Weak Electron Screening

When the temperature is high enough so that condition (5.24) is satisfied, the effects of electron screening on nuclear reactions are small and may be treated as a weak perturbation. The basic reaction rates are the Gamow rates $R_G(T)$ in Eq. (5.14). They are enhanced through the many-body correlation processes without electron screening by factors A_{ij} given by Eq. (5.63). The total reaction rates are expressed as a product,

$$R(T) = R_G(T) \cdot A_{ij} \cdot A_{ij}^{(e)} \,, \tag{5.71}$$

that defines the extra enhancement factor $A_{ij}^{(e)}$ due to electron screening.

Since the electron screening is weak, a WKB approximation may be applicable to an evaluation of the extra enhancement factor. The WKB penetration probabilities $P_{ij}(E)$ between nuclei with relative kinetic energy E are calculated as [e.g., Salpeter & Van Horn 1969]

$$P_{ij}(E) = \exp\left[-\frac{2\sqrt{2\mu_{ij}}}{\hbar} \int_0^{r_{TP}} dr \, \sqrt{V_{ij}(r) - E}\right] \,, \tag{5.72}$$

where r_{TP} is the classical turning radius satisfying $V_{ij}(r_{TP}) = E$. The factor $A_{ij}^{(e)}$ is then expressed and calculated as

$$A_{ij}^{(e)} \equiv \exp(Q_{ij}^{(e)}) = \frac{P_{ij}^{(e)}(T)}{P_{ij}^{(0)}(T)} \,, \tag{5.73}$$

where $P_{ij}^{(e)}(T)$ and $P_{ij}^{(0)}(E)$, respectively, are the thermal averages of Eq. (5.72) with the potentials,

$$V_{ij}^{(e)}(r) = \frac{Z_i Z_j e^2}{r} \exp\left[-\frac{r}{D_s}\right] - H_{ij}^{(e)}(r) \tag{5.74a}$$

and

$$V_{ij}^{(0)}(r) = \frac{Z_i Z_j e^2}{r} - H_{ij}(r) \,. \tag{5.74b}$$

To derive a parametrized expression for the enhancement factor (5.73), Ichimaru and Ogata [1991] carried out the relevant WKB integrations and thermal averages for 12 cases of the combination between $a_e \langle Z \rangle^{1/3} / D_s = 0.2, 0.4, 0.6,$ and $3\Gamma_{ij}/\tau_{ij} = 0.5, 1.0, 1.5, 2.0$, where the average charge number

$$\langle Z \rangle = \sum_i x_i Z_i \,, \tag{5.75}$$

with x_i denoting the molar fraction of the i nuclei. The result finally takes the form

$$
\begin{aligned}
Q_{ij}^{(e)} = {} & \langle Z \rangle^{1/3} \frac{a_e}{D_s} \left\{ 1 - 1.057 \frac{D_s}{a_{ij}} \left[1 - \exp\left(-\frac{a_{ij}}{D_s} \right) \right] \right\} \Gamma_{ij} \\
& + \left[0.342 - 0.354 \exp\left(-\frac{3\Gamma_{ij}}{\tau_{ij}} \right) \right] \Gamma_{ij} \frac{3\Gamma_{ij}}{\tau_{ij}} \\
& - \frac{3}{8} \left(\langle Z \rangle^{1/3} \frac{a_e}{D_s} \right)^2 \Gamma_{ij} \frac{3\Gamma_{ij}}{\tau_{ij}} + 0.091 \left(\langle Z \rangle^{1/3} \frac{a_e}{D_s} \right)^{2.923} \Gamma_{ij} \left(\frac{3\Gamma_{ij}}{\tau_{ij}} \right)^{1.897} .
\end{aligned}
\tag{5.76}
$$

Except for the last term, which has been determined by a fitting process, all terms in this expression can be obtained by perturbation-theoretic calculations.

The enhancement factor (5.76) increases rather steeply with $\langle Z \rangle$ and the mass densities. Enhancement of nuclear reactions due to the electron screening becomes significant in high-Z materials such as carbon and oxygen at high densities near ignition conditions for supernova explosion.

2. Strong Electron Screening

When the temperature is low such that the condition (5.26) for the electron-screened cold fusion is satisfied, electron screening has a dominant effect on nuclear reactions; the basic reaction rates are now given by $R_{ss}(T)$ in Eq. (5.27). These represent enormously enhanced rates over $R_G(T)$, as the MH and PM cases in Table 5.1 numerically illustrate. The total reaction rates may be expressed as the product

$$R(T) = R_{ss}(T) \cdot A_{ij}^{(s)} \,, \tag{5.77}$$

which defines the enhancement factor $A_{ij}^{(s)}$ arising from the electron-screened internuclear correlation processes over the fundamental electron-screened cold fusion rates.

Recalling that the radius of the classical turning point has been given by $r_{TP}^{(s)}$ in Eq. (5.28) and that the electron-screened Coulomb coupling parameters may be estimated as

$$\Gamma_{ij}^{(s)} = \Gamma_{ij} \exp\left(-\frac{a_{ij}}{D_{ij}}\right) , \tag{5.78}$$

we find an explicit expression for the enhancement factor derived from Eqs. (5.63) and (5.64) as

$$A_{ij}^{(s)} = \exp(Q_{ij}^{(s)}) , \tag{5.79a}$$

where

$$Q_{ij}^{(s)} = \beta H_{ij}^{(s)}(0) - \frac{5}{32}\Gamma_{ij}^{(s)}\left(\frac{r_{TP}^{(s)}}{a_{ij}}\right)^2\left[1 + (C_1 + C_2 \ln \Gamma_{ij}^{(s)})\frac{r_{TP}^{(s)}}{a_{ij}} + C_3\left(\frac{r_{TP}^{(s)}}{a_{ij}}\right)^2\right] , \tag{5.79b}$$

with $C_1 = 1.1858$, $C_2 = -0.2472$, and $C_3 = -0.07009$. The first term on the right-hand side of Eq. (5.79b) may be calculated as

$$\beta H_{ij}^{(s)}(0)/\Gamma_{ij}^{(s)} = 1.148 - 0.00944 \ln \Gamma_{ij}^{(s)} - 0.000168(\ln \Gamma_{ij}^{(s)})^2 , \tag{5.80}$$

in light of Eq. (2.22), with an assumption for the validity of the electron-screened Coulomb coupling parameters (5.78) [see Ichimaru, Ogata, & Nakano 1990].

D. Thermal Enhancement of Pycnonuclear Rates

The nuclei forming a quantum solid perform zero-point vibrations about their equilibrium sites in the ground state. Their reaction rates proportional to the contact probabilities between nearest-neighbor pairs depend only on the density but are independent of the temperature. These pycnonuclear rates have been treated in Section 5.2C.

At elevated temperatures, the nuclei may occupy excited states of lattice vibrations and the reaction rates may take on significantly enhanced values over those in the ground state, especially near the melting conditions. Enhancement factors arising from such thermal excitations may likewise be evaluated through the solutions to s-wave Schrödinger equation

$$\left[-\frac{\hbar^2}{2\mu}\frac{d^2}{dr^2} + W^L(r) - E_\nu\right] r\Psi_\nu(r) = 0 \tag{5.81}$$

for the excited states with the cusp boundary condition

$$\lim_{r \to 0} \frac{d \ln \Psi_\nu(r)}{dr} = \frac{1}{2r^*} . \tag{5.82}$$

Here $\Psi_\nu(r)$ refers to the wave function of the s-wave scattering for the νth excited state of nuclei with energy eigenvalue E_ν.

The lattice potential $W^L(r)$ between nearest-neighbor particles is basically the same as that used in Eq. (5.33); it is an angular-averaged quantity obtained by the MC sampling methods. Figure 5.3 shows the MC data on the lattice potential sampled at $\Gamma = 400$ and 200. For a treatment of higher levels in the excited states, it is necessary to extend the potential into the domain $x(= r/a) > 2.3$, where probabilities of the MC sampling are vanishingly small. For $\Gamma = 400$, we find a fitting formula,

$$\frac{\beta W^L(r)}{\Gamma} =$$

$$\begin{cases} \dfrac{1}{x} - 1.0605 + 0.250x^2 , & \text{for } x \le 0.70 , \\[2mm] \dfrac{1}{x} - 1.183 + 0.350x - \dfrac{1}{x} \exp\left(13.2\sqrt{x} - 22.1\right) , & \text{for } 0.70 < x \le x_m \\[2mm] 0.154(x - x_m)^2 - 0.080(x - x_m)^3 - 4.6026 \times 10^{-3} , & \text{for } x_m < x \le 2.22 \\[2mm] \dfrac{0.00705}{2.50 - x} - 4.6026 \times 10^{-3} , & \text{for } 2.22 < x , \end{cases} \tag{5.83a}$$

where $x_m = 1.7536$. For $\Gamma = 200$, we likewise find

$$\frac{\beta W^L(r)}{\Gamma} =$$

$$\begin{cases} \dfrac{1}{x} - 1.0605 + 0.250x^2 , & \text{for } x \le 0.70 , \\[2mm] \dfrac{1}{x} - 1.183 + 0.350x - \dfrac{1}{x} \exp\left(13.2\sqrt{x} - 22.1\right) , & \text{for } 0.70 < x \le x_m , \\[2mm] 0.153(x - x_m)^2 - 0.104(x - x_m)^3 + 0.015(x - x_m)^4 \\ \quad + 0.017(x - x_m)^5 - 4.6026 \times 10^{-3} , & \text{for } x_m < x \le 2.6 \\[2mm] \dfrac{0.02816}{3.094 - x} , & \text{for } 2.6 < x . \end{cases} \tag{5.83b}$$

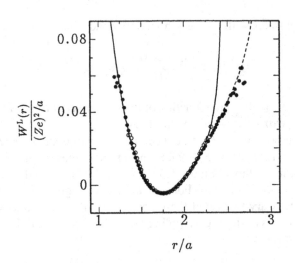

Figure 5.3 Lattice potentials for a bcc configuration averaged over angular directions. The filled circles represent the MC data at $\Gamma = 200$; the open circles, $\Gamma = 400$. The solid and dashed lines correspond to the fitting formulas Eqs. (5.83a) and (5.83b).

These formulas are also plotted in Fig. 5.3 for comparison. It has been confirmed that choice of potential form for $x > 2.3$ between these two formulas is rather immaterial for the final results.

The temperature-dependent enhancement factor $A(T)$ of the pycnonuclear rates is thus calculated in terms of the solutions to Eq. (5.81) as

$$A(T) = \frac{\sum_{\nu=0}^{\infty} \exp\left[-\beta(E_\nu - E_0)\right] |\Psi_\nu(0)|^2}{\sum_{\nu=0}^{\infty} \exp\left[-\beta(E_\nu - E_0)\right] |\Psi_0(0)|^2}. \tag{5.84}$$

Salpeter and Van Horn [1969] estimated such an enhancement factor approximately by calculating the WKB tunneling probabilities for $|\psi_\nu(0)|^2$ and by performing the ν-summation through a saddle-point integration, as one does in calculations such as Eqs. (5.14) and (5.73). Their final results for $A(T)$ turned out to be fairly accurate except for the vicinity of the melting conditions (see Fig. 5.4).

The states of a quantum OCP solid may be characterized by two dimensionless parameters, R_s and Y, defined in Eq. (2.105). Numerically they take on

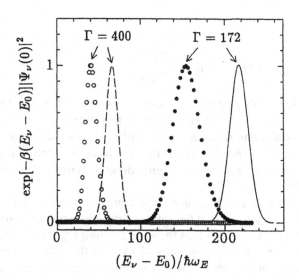

Figure 5.4 Distributions of elementary strengths, $\exp[-\beta(E_\nu - E_0)]|\Psi_\nu(0)|^2$, for a density $R_s = 5 \times 10^4$. The filled circles represent the cases with $\Gamma = 172$; the open circles, $\Gamma = 400$. The solid and dashed lines are the corresponding values in the Salpeter–Van Horn [1969] saddle-point integration; ω_E denotes the Einstein frequency (2.94a). The scale on the vertical axis is linear and arbitrary.

values

$$R_s = \frac{0.4924}{\lambda} = 0.4924 A Z^2 \left(\frac{1}{A} \frac{\rho_m}{1.3574 \times 10^{11} \text{ g/cm}^3} \right)^{-1/3}, \quad (5.85a)$$

$$Y = 14.30 \frac{Z}{A} \left(\frac{\rho_m}{10^7 \text{ g/cm}^3} \right)^{1/2} \left(\frac{T}{10^6 \text{ K}} \right)^{-1}, \quad (5.85b)$$

where λ is the parameter related to the ratio of r^* to r_m, defined in Eq. (5.36). The parameters in Eqs. (5.85) are related to the Coulomb coupling parameter via $\Gamma = Y\sqrt{R_s}$. The Schrödinger equation (5.81) with Eqs. (5.82) and (5.83) has been solved for the excited states [Ichimaru, Kitamura, & Ogata 1994] in the ranges of density-temperature combinations:

$$10^3 \leq R_s \leq 5 \times 10^4, \qquad Y \leq 20, \qquad \text{and} \qquad \Gamma \geq 172. \quad (5.86)$$

Distributions of the elementary strengths, $\exp[-\beta(E_\nu - E_0)]|\Psi_\nu(0)|^2$, for a density $R_s = 5 \times 10^4$ at two different temperatures corresponding to $\Gamma = 172$ and 400 are plotted in Fig. 5.4.

The values of $A(T)$ so computed are then fitted in an analytic formula with

$$A(T) = \exp[F(T) R_s^{1/2}], \quad (5.87a)$$

where

$$F(Y) = 0.613 \frac{1.678 + Y^2}{0.779 + Y^2} \exp(-0.780Y) . \qquad (5.87b)$$

Errors in fitting by this formula are confined within 0.5 in absolute magnitude. In this formula, we see that enhancement factor can be as large as 10^{54} near the melting conditions when $R_s = 5 \times 10^4$. This does not mean a large value of the reaction rate itself, however, since the basic pycnonuclear rate at $T = 0$, (Eq. 5.37) takes on a minuscule magnitude at a relatively low density such as $R_s = 5 \times 10^4$.

A principal effect of the huge enhancement predicted in Eqs. (5.87) is to make the total reaction rates in the solid regime *join smoothly across the thermal melting lines* (see Fig. 2.25) into those in the fluid regime, which are given by a product between the Gamow rates (5.14) and the enhancement (5.64). The smooth connection has been achieved in terms of both physical understanding and numerical results. Since evolution lines for a considerable class of degenerate cores [e.g., Nomoto 1982] follow close to the melting lines, it is a significant issue to elucidate pycnonuclear enhancement in the vicinity of the melting conditions.

Finally, combining Eq. (5.87) with the pycnonuclear rates (5.37), we obtain the pycnonuclear rates at finite temperatures expressed as

$$R_{PN}(T)(\text{reactions/cm}^3/\text{s}) = A(T) R_{OI^2} =$$
$$4.83 \times 10^{32} \left[\frac{\rho_m(\text{g/cm}^3)}{A} \right]^2 \frac{S(\text{MeV} \cdot \text{barns})}{A Z^2} R_s^{1.809} \exp \left\{ [F(Y) - 3.506] R_s^{1/2} \right\} .$$
$$(5.88)$$

As we have noted in conjunction with the pycnonuclear rates (5.37), the density dependence of the exponential factor reflects the features germane in solids in that particle motions are confined around their lattice sites. Since $F(Y)$ and consequently the exponent of Eq. (5.88) increase rapidly with the temperature for $Y < 1$, the thermal effects in the pycnonuclear rates become very pronounced near the classical (i.e., thermal) melting conditions. The formula (5.88) can cover the pycnonuclear rates over the entire parameter regime of OCP solids.

5.4   RATES OF NUCLEAR FUSION REACTIONS

The analyses for the basic binary reactions and the many-body enhancement factors presented in the foregoing sections are applied to estimation of reaction rates for specific examples of dense plasmas in astrophysical and laboratory settings: Astrophysical condensed plasmas under consideration include the solar interior (SI), interior of a brown dwarf (BD), interior of a giant planet (GP), a white-dwarf (WD) progenitor of supernova, and surfaces of accreting white dwarfs and neutron stars. Examples of the condensed plasmas in the laboratory are those found in the inertial confinement fusion (ICF) experiments, in metal hydrides (MH) such as PdD and TiD_2, in cluster-impact fusion experiments, and in ultrahigh-pressure liquid metals (PM).

A. Solar Interior and Inertial Confinement Fusion

One of the proton-proton (p-p) chains consists in

$$^1H(p, e^+\nu_e)^2D(p, \gamma)^3He(^3He, 2p)^4He , \qquad (5.89)$$

which altogether yields

$$4p \rightarrow \alpha + 2e^+ + 2\nu_e + 26.2 \text{ (MeV)} . \qquad (5.90)$$

The cross-section parameters S_{ij} and the Q values, the nuclear energy released by the reaction, are [Bahcall & Ulrich 1988]

$$S_{pp} = 4.07 \times 10^{-22} (\text{keV} \cdot \text{barns}), \qquad Q(\text{p-p}) = 1.442(\text{MeV}) , \quad (5.91a)$$
$$S_{dp} = 2.5 \times 10^{-4} (\text{keV} \cdot \text{barns}), \qquad Q(\text{d-p}) = 5.494(\text{MeV}) , \quad (5.91b)$$
$$S_{^3He^3He} = 5.15 \times 10^3 (\text{keV} \cdot \text{barns}), \qquad Q(^3He^3He) = 12.860(\text{MeV}) . \quad (5.91c)$$

The chain (5.89), starting with $^1H(p, e^+\nu_e)^2D$, involves a β process and thus is extremely slow; the rates are controlled by these slow processes.

In Table 5.2, two cases of fusion rates and enhancement factors associated with p-p reactions in the central parts of SI are treated; the case SI1 corresponds to conditions near the center, and SI2, those at $r \approx 0.2R_S$. The mass densities and temperatures are the values appropriate to hydrogen. The enhancement factors, $A_{ij}^{(e)}$ and A_{ij}, are calculated with Eqs. (5.25b) and by replacing Γ_{ij} in Eqs. (5.63) and (5.64) for $\Gamma_{ij}^{(s)}$ of Eq. (5.78). The net reaction rates are then given by Eq. (5.71). The enhancement $A_{ij}^{(e)}$ due to weak electron screening near the center is approximately 2.2%, and that A_{ij} due to many-body correlations in electron-screened protons amounts to 4.8%.

In calculating the local densities P of fusion power generated by p-p reactions per unit mass (including those of other elements as well) in Table 5.2, the effective Q value has been taken at 13.1 MeV on account of (5.90). Those values in the central parts naturally exceed the average value 1.93×10^{-7} W/g of the solar luminosity per mass cited in Section 1.1A.

In Table 5.2, we also list examples of calculations for ICF plasmas. Here, reactions $t(d, n)^4He$ with parameters [Jackson 1957],

$$S_{td} = 1.7 \times 10^4 (\text{keV} \cdot \text{barns}) , \qquad Q(\text{t-d}) = 17.6 (\text{MeV}) , \qquad (5.92)$$

Table 5.2   Reaction rates and enhancement factors in SI and ICF plasmas

Case	SI1	SI2	ICF1	ICF2
Reaction	p-p	p-p	t-d	t-d
ρ_m (g/cm^3)	56.2	24.9	30.0	2.0
T (K)	1.55×10^7	9.3×10^6	5.0×10^7	1.0×10^8
τ_{ij}	13.7	16.1	12.3	9.72
$\log_{10} R_G$ (s^{-1})	-17.69	-18.93	7.40	7.13
Γ_{ij}	0.072	0.076	0.010	0.002
Λ_{ij}	0.023	0.023	0.0049	0.0014
Θ	2.23	3.21	31.8	387
$D_s(10^{-9}$ cm)	3.97	5.48	11.9	20.0
$\Gamma_{ij}^{(s)}$	0.044	0.047	0.0079	0.0014
$\log_{10} A_{ij}^{(e)}$	0.010	0.012	0.000	0.000
$\log_{10} A_{ij}$	0.020	0.022	0.004	0.000
$\log_{10} R$ (s^{-1})	-17.66	-18.90	7.41	7.33
$\log_{10} P$ (W/g)	-5.56	-6.80	19.23	18.96

Note: Mass densities ρ_m and temperatures T are those assumed for the reacting nuclei. Reaction rates are expressed in units of reactions per second per reacting pair.

are considered in a mixture of deuterium and tritium with equal molar concentrations. Enhancement due to electron screening and ion-ion correlations is negligible in such a relatively dilute, high-temperature plasma. As with the solar interior, the nuclear reaction rates in the ICF plasmas in such a density-temperature domain are correctly described by the Gamow reaction rates.

B. Interiors of Giant Planets and Brown Dwarfs

Nuclear reactions in GP and BD are related to the issues of accounting for the excess luminosities and of the critical masses for hydrogen burning. To study the rates of p-p reactions and those of possible deuteron burning, we consider a hydrogen plasma with an admixture of deuterons at a molar fraction of 3×10^{-5} [Anders & Grevesse 1989]. In addition to the p-p and d-p reactions, two branches—d(d, n)^{3}He and d(d, p)t—of the d-d reactions are analyzed with a total cross-section parameter and an average Q value [Krauss et al. 1987]:

$$S_{dd} = 103 \text{ (keV} \cdot \text{barns)} , \qquad Q(\text{d-d}) = 3.6 \text{ (MeV)} . \qquad (5.93)$$

Table 5.3   Reaction rates and enhancement factors in BD and GP plasmas

Case Reaction	BD1 p-p	BD2 d-p	BD3 d-d	GP1 d-p	GP2 d-d
τ_{ij}	23.4	25.8	29.5	137	157
$\log_{10} R_G$ (s^{-1})	-20.20	-3.18	-4.07	-52.36	-60.25
Γ_{ij}	0.76	0.76	0.76	19.4	19.4
Λ_{ij}	0.14	0.12	0.003	0.25	0.006
Θ	0.010	0.010	0.010	0.023	0.023
$D_s(10^{-9}\ cm)$	1.30	1.30	1.30	3.21	3.21
$\Gamma_{ij}^{(s)}$	0.43	0.43	0.00	5.07	0.00
$\log_{10} A_{ij}^{(e)}$	0.162	0.164	0.167	10.94	11.01
$\log_{10} A_{ij}$	0.197	0.197	0.000	2.32	0.000
$\log_{10} R$ (s^{-1})	-19.84	-2.82	-3.90	-39.10	-49.24
$\log_{10} P$ (W/g)	-7.75	4.38	3.11	-31.91	-42.23

Note: Mass densities ρ_m and temperatures T assumed for BD cases are 10^3 g/cm^3 and 3×10^6 K; those for GP cases, 5 g/cm^3 and 2×10^4 K.

The cases of BD and GP satisfy the conditions for weak electron screening (5.24) (see Table 5.1); calculations of enhancement factors and the resultant reaction rates are analogous to those in the cases of SI and ICF. Table 5.3 lists the results of such calculations for BD and GP plasmas.

We note that the enhancement factors for the p-p reactions arising from combined effects between electron screening and ion-ion correlations in the assumed BD conditions take on a fairly large magnitude of 2.4; such an enhancement should be properly taken into account in the estimation of critical masses for hydrogen burning in the very low mass stars and BDs. Deuteron burning is quite efficient in such stellar objects, with characteristic times of reaction on the order of a few minutes; hence deuterons should have burned away in the initial stages of stellar evolution.

Despite a substantial enhancement by 10 to 11 orders of magnitude due to electron screening, rates of deuteron burning in Jovian planets remain minuscule. The estimated power production rates are far smaller than the average Jovian luminosity per unit mass, 2.4×10^{-13} W/s [Hubbard 1980] and thus cannot take part in accounting for the excess infrared luminosity.

C. White Dwarf Progenitors of Supernovae

Nuclear reaction rates in dense BIMs of carbon (C) and oxygen (O) are essential quantities governing the evolution and ignition in white dwarf progenitors of type I supernovae [Barkat, Wheeler, & Buchler 1972; Graboske 1973; Couch & Arnett 1975]. Phase diagrams associated with freezing transitions in such BIMs have been elucidated (see Section 2.5E). The short-range correlations responsible for nuclear reactions in dense matter are influenced strongly by such phase properties as well as by the quantum and classical many-body effects.

First-principle calculations of nuclear reaction rates in dense C-O BIMs are available and have been described in the preceding sections both for the fluid phases and solid phases. Examples of such calculations for reaction rates and enhancement factors are presented in Table 5.4. The fundamental reaction rates are the Gamow rates (5.14). On top of these fundamental rates, the enhancement factors A_{ij} of 23 to 39 orders of magnitude are predicted by the use of Eqs. (5.63) and (5.64).

It should be noted that these high enhancement rates have been obtained under conditions in which the density distributions of the electrons remained uniform and constant. The enhancement stems solely from the many-particle

Table 5.4   Reaction rates and enhancement factors for dense carbon-oxygen matter in WD

Reaction	^{12}C-^{12}C	^{12}C-^{16}O	^{16}O-^{16}O
S_{ij} (MeV · barn)	8.83×10^{16}	1.15×10^{21}	2.31×10^{27}
Q (MeV)	13.931	16.754	16.541
τ_{ij}	181.827	230.294	293.691
$\log_{10} R_G$ (s^{-1})	–43.35	–59.96	–81.47
Γ_{ij}	56.6	71.9	91.5
Λ_{ij}	0.47	0.42	0.37
$\Gamma_{ij}^{(s)}$	41.4	51.7	64.7
$\log_{10} A_{ij}$	1.18	1.57	2.08
$\log_{10} A_{ij}^{(e)}$	23.49	29.95	38.34
$\log_{10} R$ (s^{-1})	–18.69	–28.44	–40.96
$\log_{10} P$ (W/g)	–8.01	–17.68	–30.31

Note: Mass density ρ_m and temperature T are assumed at 4×10^9 g/cm^3 and 1×10^8 K; the molar fraction of oxygen, 50%. The degeneracy parameter of the electrons is $\Theta = 2.1 \times 10^{-4}$; their screening distance, $D_s = 3.4 \times 10^{-11}$ cm.

correlation effects between ionic nuclei elucidated in Section 5.3B; electrons do not participate in the act of screening in the derivation of A_{ij}. It is therefore a misnomer physically to label these results as cases of strong electron screening as in some of the astrophysical literature.

The short- and intermediate-range screening effects of relativistic degenerate electrons on the enhancement of nuclear reactions have been elucidated in Section 5.3C. Electron screening is weak under such relativistic conditions. It has been pointed out nonetheless that enhancement of nuclear reactions due to the electron screening becomes significant in high-Z materials such as carbon and oxygen at high densities near ignition [Ichimaru & Ogata 1991]. The extra enhancement factors resulting from the electron screening, applicable to the WD cases, have been formulated in Eqs. (5.73) and (5.76). The electronic enhancement factors $A_{ij}^{(e)}$ computed for the WD cases of Table 5.4 amount to 1.2 to 2.1 orders of magnitude.

Carbon ignition curves are the loci of the points on the density-temperature plane for which the ^{12}C-^{12}C energy release equals the neutrino loss [e.g., Ichimaru 1993a]. Conventional treatments [Arnett & Truran 1969; Nomoto 1982] assume the rates of such an energy release to be approximately 3×10^{17} erg/g; it is then found that the conditions assumed in Table 5.4 correspond to those near the ignition. The huge enhancement factors A_{ij} stemming from the many-particle processes are the vital elements responsible for the ignition. It is in this regard that the nuclear reactions in supernova processes should be clearly distinguished from those in the main-sequence stars such as the Sun.

D. Helium Burning: Triple α Reactions

Helium burning is one of the major processes of nuclear reactions in stellar evolution. The triple α reactions [Salpeter 1952; Hoyle 1954] take place in three steps:

$$^4\text{He} + {}^4\text{He} \leftrightarrow {}^8\text{Be} - 0.092\,(\text{MeV}) , \tag{5.94a}$$

$$^4\text{He} + {}^8\text{Be} \leftrightarrow {}^{12}\text{C}^* - 0.278\,(\text{MeV}) , \tag{5.94b}$$

$$^{12}\text{C}^* \rightarrow {}^{12}\text{C} + 7.644\,(\text{MeV}) . \tag{5.94c}$$

Altogether the processes yield

$$3\alpha \rightarrow {}^{12}\text{C} + 7.274\,(\text{MeV}). \tag{5.94d}$$

Here ^{12}C* denotes a ^{12}C nucleus in its second excited state, and the energy differences are taken from Ajzenberg-Selove and Busch [1980] and Ajzenberg-Selove [1984]. Since the first two reactions in (5.94) are endoergic, it is essential that these processes take place as resonant reactions; the 3α rate becomes exponentially small for temperatures less than about 10^8 K, at densities typical of normal stellar interiors [e.g., Clayton 1968].

Helium burning is expected to be one of the major reaction processes in white dwarfs and on neutron stars accreting in close binary systems [e.g., Nomoto, Thielemann, & Miyaji 1985]. The density and temperature conditions in which helium burning would occur in compact stars may differ widely from those in normal stars. In some cases, helium burning is so explosive as to give rise to x-ray bursters in neutron stars [Lewin & Joss 1983] and type I supernovae in white dwarfs [Nomoto 1982].

Accumulation of helium thus eventually leads to ignition, the conditions of which depend mainly on the accretion rate. For slower accretion, the temperature in the accreted matter is lower because of slower compressional heating relative to radiative cooling; as a result the ignition is delayed to higher density. Cameron [1959] pointed out that the 3α reaction at temperatures as low as $T \ll 10^8$ K is no longer a resonant but rather a nonresonant reaction; the Gamow peak energy (5.16) falls far smaller than the threshold energy of resonance.

The rates of resonant 3α reactions have been given in Fowler, Caughlan, and Zimmerman [1975] and in Harris et al. [1983], who made use of latest experimental information. Nomoto [1982] and subsequently Nomoto, Thielemann, and Miyaji [1985], using updated nuclear data, presented approximate analytic formulas for the rates of $\alpha + \alpha$ and ^{8}Be $+\alpha$ reactions. Denoting

$$\langle ij \rangle \equiv \left\langle \sigma_{ij}\sqrt{2E/\mu_{ij}} \right\rangle_T \tag{5.95}$$

with $\langle \cdots \rangle_T$ signifying the thermal average for the reacting pairs, these authors thus find the rates for nonresonant $\alpha + \alpha$ and ^{8}Be $+\alpha$ reactions as

$$\langle \alpha\alpha \rangle^*(cm^3/s) =$$
$$6.914 \times 10^{-15} T_9^{-2/3} \exp\left(-13.489 T_9^{-1/3}\right)$$
$$\times \left(1 + 0.031 T_9^{1/3} + 8.009 T_9^{2/3} + 1.732 T_9 + 49.883 T_9^{4/3} + 27.426 T_9^{5/3}\right),$$
$$\tag{5.96a}$$

$$\langle \alpha^8 Be \rangle(cm^3/s) =$$
$$4.167 \times 10^{-17} T_9^{-2/3} \exp\left(-23.567 T_9^{-1/3}\right)$$
$$\times \left(1 + 0.018 T_9^{1/3} + 5.249 T_9^{2/3} + 0.650 T_9 + 19.176 T_9^{4/3} + 6.034 T_9^{5/3}\right).$$
$$\tag{5.96b}$$

Here T_9 denotes the temperature in units of 10^9 K, and the asterisk in Eq. (5.96a) means that the rate $\langle \alpha\alpha \rangle$ has been calculated by including the E-dependent part of the α width for ^{8}Be.

The enhancement factors $A(3\alpha)$ in 3α reactions have been formulated by Ogata, Ichimaru, and Van Horn [1993] through the following argument: The lifetime of ^{8}Be may be estimated as

$$\tau_{Be} \approx \frac{\hbar}{\gamma_{Be}} \approx 10^{-16}\,(s)\,, \tag{5.97a}$$

with the half-width of Be, $\gamma_{Be} = 6.8\,eV$ [Ajzenberg-Selove & Busch 1980]. Relaxation times for reacting ^{4}He and ^{8}Be may be estimated through the ion-sphere cross sections, a_{ij}^2, as

$$\tau_{rel} \approx 10^{-18}\left(\frac{\rho_m}{10^8 g/cm^3}\right)^{-1/3}\left(\frac{T}{10^7\,K}\right)^{-1/2}\,(s)\,. \tag{5.97b}$$

One thus finds $\tau_{rel} \ll \tau_{Be}$ in the density-temperature regime of interest. Correlations between ^{4}He and ^{8}Be nuclei may thus be treated as those in equilibrated ^{4}He and ^{8}Be BIMs. The 3α reaction rates are proportional to the product of the ^{4}He-^{4}He and ^{4}He-^{8}Be contact probabilities—that is, $g_{^4He^4He}(0) \cdot g_{^4He^8Be}(0)$; it is not necessary to consider the triple correlations under these circumstances. The enhancement factors are therefore calculated as

$$A(3\alpha) = \exp\left[Q_{^4He^4He} + Q_{^4He^8Be} + Q_{^4He^4He}^{(e)} + Q_{^4He^8Be}^{(e)}\right]\,, \tag{5.98}$$

where Q_{ij} and $Q_{ij}^{(e)}$ have been given by Eqs. (5.64) and (5.76).

E. Metal Hydrides: PdD and TiD$_2$

Observation of nuclear fusion reactions between itinerant hydrogen in metal hydrides (MH$_x$) claimed in some of the earlier experiments [e.g., Jones et al. 1989] has created a challenge to condensed matter physics, calling for a theoretical account of how two hydrogen nuclei can come to fuse by overcoming the Coulombic repulsive forces in such a metallic environment. Other experiments [e.g., Ziegler et al. 1989; Gai et al. 1989] performed under analogous settings, however, have not shown any sign of nuclear reactions. One may therefore speculate that the rates of nuclear reactions could depend extremely delicately on the states of reacting pairs at short distances.

1. Hydrogen in Metal Hydrides

The itinerant hydrogen in metal differs from either stellar interiors or ICF plasmas in two important aspects: Hydrogen nuclei are strongly screened by valence electrons and by nearly localized electrons in hybridized states [Alefeld & Völkl

1978]. The metal atoms situated at periodic or aperiodic (due to defects) lattice sites create inhomogeneous fields that act to trap (or to localize) the hydrogen nuclei and thereby to alter microscopic features of the short-range correlations. Owing to these influences of the screening electrons and the inhomogeneous lattice fields, the hydrogen in metal hydride bears a dual character of itinerant and trapped particles.

The lattice fields of metal atoms in which hydrogen nuclei "move" are constructed so that the following observed features may be taken into account: In densely hydrated phases, both Pd and Ti assume a face-centered cubic (fcc) structure with lattice constants, $d = 4$Å and 4.4 Å, respectively. In Pd lattice, hydrogen sits around the octahedral (O) sites, where the potential assumes local minima with curvature, $\Phi'' \approx 1.1$eV Å^{-2} [Drexel et al. 1976; Alefeld & Völkl 1978]. Barrier height between the minima, $\Delta\Phi \approx 0.23$eV, is inferred from diffusivity. The separation between nearest-neighbor O sites is 2.8 Å. In Ti lattice, the tetrahedral (T) sites are the local minima with curvature, $\Phi'' \approx 5.1$eVÅ^{-2}, barrier height, $\Delta\Phi \approx 0.51$eV, and the nearest-neighbor separation 2.2Å [Pan & Webb 1965; Korn & Zamir 1970; Alefeld & Völkl 1978].

Since heights of the potential barriers substantially exceed ambient temperatures, bulk of the hydrogen nuclei would be in trapped states around O or T sites. As finiteness of diffusivity implies, a small fraction of the hydrogen remains in itinerant states. In a nonequilibrium situation, this fraction may take on a larger value, approaching unity.

Rates of nuclear reactions are most effectively influenced by interparticle correlations between such itinerant hydrogen. The screening potential associated with such an interparticle correlation may be formulated as

$$H^{\text{MH}}(r) = V^{\text{MH}}(r) + k_{\text{B}}T \ln g^{\text{MH}}(r) . \tag{5.99}$$

Here $g^{\text{MH}}(r)$ refers to a joint probability density for the system of hydrogenic nuclei interacting via $V^{\text{MH}}(r)$ in the inhomogeneous lattice fields of the MH.

Such a correlation function can be sampled through an MC simulation method designed appropriately for the hydrogen-in-metal cases [Ichimaru, Ogata, & Nakano 1990]. In a lattice simulation (L), periodic lattice fields were determined by placing 500 metal atoms at fcc sites in a MC cell with periodic boundary conditions. In a defect simulation (D), fields with defects were produced by removing 8 metal atoms randomly in such a way that no pairs of defects occupy nearest neighbor sites. The corresponding numbers, 500 for PdH and 1000 for TiH$_2$, of hydrogen atoms were placed in the cell at random. A sequence of MC configurations were then generated through the random displacements of hydrogen positions, in the Metropolis algorithm [Metropolis et al. 1953] with the canonical distribution for the sum of interaction energies

Table 5.5   Values of $\beta H^{MH}(0)$ determined by the Monte Carlo sampling method for hydrogen atoms in metals

Metal Hydride	T (K)		
	1200	600	300
PdH [L]	5.8	12.6	27.1
PdH [D]	4.4	8.7	11.6
TiH$_2$ [L]	9.6	21.6	48.2
TiH$_2$ [D]	16.6	38.1	81.5

Note: [L] refers to a case with periodic lattice fields; [D], a case with defects.

between metal and hydrogen and between hydrogen atoms. Several runs of such simulations were performed to cover various cases of metal hydride at temperatures 300, 600, 1200 K. Each run consisted typically of $(1-3) \times 10^4$ configurations per hydrogen atom to ensure an equilibrated metastable state in the system.

The joint probability densities $g^{MH}(r)$ were sampled in the statistical ensemble of particle configurations generated by such a simulation. It has been found that the lattice fields act to develop humps in $g^{MH}(r)$ at short distances inside the major humps corresponding to the nearest-neighbor O or T sites. Those short-distance humps are the consequences of the potential "dimples" in the lattice fields, which induce effective attraction between hydrogen atoms in trapping sites. The "visible" short-range parts can be fitted quite accurately by a functional form, $H^{MH}(r) = A + B \exp(-Cr)$, so that extrapolation of such a fit to $r = 0$ would yield $H^{MH}(0) = A + B$. Table 5.5 lists the values of $\beta H^{MH}(0)$ so determined.

2. Enhancement Factors

The Coulomb fields around deuterons in PdD and TiD$_2$ are heavily screened by the metallic electrons within a short-range screening distance of Eq. (3.56) at $D_s = 0.19$Å and 0.28Å, respectively [Ichimaru, Ogata, & Nakano 1990]. The critical temperature (5.23) of electron screening for d-d reactions is far greater than ambient temperatures, so the fundamental reaction rates are given by $R_{ss}(T)$ in Eq. (5.27).

Hydrogen nuclei (protons, deuterons, or tritons) in metals are thus strongly screened by metallic electrons. Enhancement factors $A_{ij}^{(MH)}$ (i, j = p, d, t) over these fundamental rates are then calculated with the screening potentials $H^{MH}(r)$

and the screening lengths D_s in a way elucidated in Sections 5.3B and 5.3C. Writing thus

$$A_{ij}^{(MH)} = \exp[Q_{ij}^{(MH)}] \,, \tag{5.100}$$

one finds

$$Q_{ij}^{MH} = \beta H_{ij}^{MH}(0) - \frac{5}{32} \Gamma_{ij}^{MH} \left(\frac{3\Gamma_{ij}^{MH}}{\tau_{ij}^{MH}} \right)^2$$

$$\times \left[1 + (1.1858 - 0.2472 \ln \Gamma_{ij}^{MH}) \left(\frac{3\Gamma_{ij}^{MH}}{\tau_{ij}^{MH}} \right) - 0.07009 \left(\frac{3\Gamma_{ij}^{MH}}{\tau_{ij}^{MH}} \right)^2 \right] \,, \tag{5.101}$$

with

$$\Gamma_{ij}^{MH} = 0.94 \beta H_{ij}^{MH}(0) \,, \tag{5.102}$$

$$\tau_{ij}^{MH} = \pi \left(\frac{D_s}{r_{ij}^*} \right)^{1/2} - \ln \left[\pi \left(\frac{D_s}{r_{ij}^*} \right)^{1/2} \right] \,. \tag{5.103}$$

Equation (5.101) stems from Eq. (5.64), in which Eq. (5.102) is set by the use of Salpeter's ion sphere model (5.53), and Eq. (5.103) is derived from a comparison between Eqs. (5.17) and (5.28).

3. Reaction Rates

In thermodynamic equilibrium, only a fraction of deuterons are in itinerant fluid states in metal deuterides. The fraction f_{itin} of such itinerant deuterons is estimated as

$$f_{itin} \approx \exp \left[-\beta \left(\Delta\Phi - \frac{3}{2}\hbar \sqrt{\frac{\Phi''}{m_d}} \right)^{1/2} \right] \,, \tag{5.104}$$

where m_d refers to the mass of a deuteron.

Table 5.6 lists the values of the reaction rates, enhancement factors, and the fractions of itinerant deuterons, computed for the lattice cases of PdD and TiD$_2$ at two different temperatures. A number of observations are in order:

1. The electron-screened cold-fusion rates R_{ss} are independent of temperature; the values are well below the upper bounds set forth by Leggett and Baym [1989].

Table 5.6 Reaction rates end enhancement factors for deuterons in metal deuterides

Case	MH1	MH2	MH3	MH4
Material		PdD		TiD_2
ρ_m (g/cm³)		11.3		4.1
n_d (cm⁻³)		6.25×10^{22}		9.4×10^{22}
E_s (eV)		75.79		51.43
$\log_{10} R_{ss}$ (s⁻¹)		−40.79		−51.13
T (K)	300	600	300	600
Γ_{ij}	356.3	178.2	408.3	204.1
Λ_{ij}	0.45	0.32	0.52	0.37
$\Gamma_{ij}^{(s)}$	0.09	0.05	3.1	1.6
$\log_{10} A_{ij}^{(MH)}$	10.90	5.39	17.71	9.09
$\log_{10} R$ (s⁻¹)	−29.89	−35.40	−33.42	−42.04
$\log_{10} P$ (W/g)	−20.38	−25.89	−23.29	−31.92
$\log_{10} f_{itin}$	−3.76	−2.43	−7.08	−4.09

Note: Mass densities ρ_m and temperatures T are assumed parameters; n_d, the number density for deuterons; f_{itin}, the fraction of itinerant deuterons in equilibrium.

2. The electron screening is more efficient, and R_{ss} consequently takes on a magnitude larger in Pd than in Ti.

3. The efficient electronic screening, in turn, shaves off the Coulomb fields around deuterons more effectively, as reflected in the values of $\Gamma_{ij}^{(s)}$, and thereby results in a smaller value of the many-body enhancement factor $A_{ij}^{(MH)}$ in Pd than in Ti.

4. Being a statistical-correlation effect, the enhancement factor decreases steeply as the temperature increases.

5. The microscopic lattice fields act to trap deuterons more effectively in Ti than in Pd.

6. The largest reaction rate in the table is $R \approx 7.2 \times 10^{-31}$ (s⁻¹) in PdD at $T = 300$ K with an assumption that $f_{itin} \approx 1$ (meaning in a nonequilibrium state). This rate implies approximately one to two d-d reactions per year per unit volume (cm³) of PdD.

F. Cluster-Impact Fusion

In a context somewhat similar to that of the aforementioned fusion in metal hydrides, experiments on *cluster-impact fusion* were performed [Beuhler, Friedlander, & Friedman 1989]: Deuteron-deuteron fusion, detected via the 3 MeV protons produced, was shown to occur when singly charged clusters of 25 to 1300 D_2O molecules, accelerated to 200 to 325 keV, impinged on TiD targets, with high fusion rates observed at approximately $1-10\,s^{-1}$/D-D; an experimental confirmation has followed [Bae, Lorents, & Young 1991; Beuhler et al. 1991]. A theoretical account for the high fusion yields has been proposed on a basis of a "thermonuclear" model [Carraro et al. 1990] at an elevated effective density [Echenique, Manson, & Ritchie 1990] or temperature [Kim et al. 1992] of deuterons. An experiment was then reported [Vandenbosch et al. 1991] showing that the enhanced fusion rates fell off rapidly with cluster size. Finally, the possibility has been confirmed that traces of high velocity beam contaminants (artifacts) could account for the experimental results [Beuhler, Friedlander, & Friedman 1992].

The analyses described in this chapter can be applied to an approximate estimation of the reaction rates expected in the cluster-impact fusion experiments. Suppose that N-molecule clusters $(D_2O)_N$ impinge on a titanium deuteride target with a kinetic energy $E_{cluster}$ per a cluster, break up into deuterons and oxygen ions, and thermalize with atoms in the target. The resultant effective temperature of deuterons may be approximately estimated as

$$T_{eff} \approx \frac{E_{cluster}}{30 N k_B} . \tag{5.105}$$

Deuterons with such a temperature can approach each other classically against the mutual Coulomb repulsion to a distance:

$$D_{cl} \approx \frac{e^2}{k_B T_{eff}} . \tag{5.106}$$

The Coulomb fields around deuterons in Ti, however, are screened by the metallic electrons within a distance D_s, which has been estimated as 2.8×10^{-9} cm in the preceding subsection. Hence the estimated maximum density that those thermalized deuterons may attain is given by

$$n_{max} \approx \frac{3}{4\pi D_{min}^3} , \tag{5.107}$$

where D_{min} takes on the smaller value between D_{cl} and D_s.

One can substitute these density-temperature estimates for calculations of the reaction rates and possible enhancement factors, using the formulas obtained in the foregoing subsections. In the ranges of parameters used in the experiments, the weak electron-screening condition of (5.24) applies; the d-d reactions here can be treated in a way analogous to the BD3 and GP2 cases of Table 5.3. For example, at $E_{\text{cluster}} = 250$ keV, the Gamow reaction rate R_G (s^{-1}) and the total enhancement factor $A = A_{ij}^{(e)} \cdot A_{ij}$ as functions of N take on values: $R_G = 1.1 \times 10^{-2}$, $A = 3.0$ ($N = 50$); $R_G = 3.1 \times 10^{-7}$, $A = 3.2$ ($N = 100$); $R_G = 1.1 \times 10^{-20}$, $A = 63$ ($N = 500$). It does not appear that these values can account for the "cluster" effects implied in those experiments. The traces of high velocity beam contaminants (artifacts) found in Beuhler, Friedlander, and Friedman [1992] should be the cause of the apparent enhancement.

G. Ultrahigh-Pressure Liquid Metals

The nuclear reactions in pressurized metals (PM) containing such nuclear species as protons, deuterons, and lithium differ in an essential way from those in the metal-hydride cases of Section 5.4E. The reacting nuclei being in fluid metallic states, a substantial enhancement of the reaction rate is expected in PM due to the many-particle processes, as in the cases of white-dwarf progenitors of supernovae [Ichimaru 1991, 1993a].

At an ambient temperature and below, the metallic hydrogen forms a quantum solid rather than a semiclassical fluid. Protons and deuterons perform zero-point vibrations around their lattice sites; their nuclear reaction rates are determined as in Eq. (4.48) from the contact probabilities between adjacent nuclei. A calculation along these lines may reveal that the fundamental pycnonuclear rates in D-H or Li-H matter at a metallic density take on a minuscule magnitude far below a possible level of observation (Problem 5.8).

For exploitation of the enhancement factors arising from the many-particle processes of Section 5.3C, at least one of the reacting nuclear species needs to be in an itinerant or fluid state; a melting criterion such as those considered in Section 2.5D plays a role here. Temperatures assumed for PM cases in Table 5.7 are in the vicinity of such melting conditions. The resultant values of Λ_{ij} then illustrate that quantum effects now become significant in such an itinerant nuclear system.

In the metalized d-p mixtures, the electron-screened internuclear potential is expressed as

$$V_{\text{dp}}(r) = \frac{e^2}{r} S_c(r) , \qquad (5.108)$$

where $S_c(r)$ is the screening function introduced in Eq. (5.18). For the treatment of short-range screening length D_s of the electrons, we may follow the approaches elucidated in Sections 3.2D and 4.2C. The total pressure P_e of the conduction electrons may likewise be estimated from the equation of state in Section 3.2G.

The critical temperature (5.23) associated with such electron screening for the cases of d-p BIM fluids exemplified in Table 5.7 exceeds 10^4 K, so that the basic reaction rates before the enhancement by many-nuclear processes are given by R_{ss} of Eq. (5.27). Since the ratio E_G/E_s in the exponential decay factor of R_{ss} is proportional to μ_{ij}, the reduced mass between the pair of reacting nuclei, it is advantageous from the point of view of nuclear fusion to choose one of the species as proton and thereby to minimize such a decay. Thus for metallic hydrogen the combination of d and p may be preferred. The enhancement factors due to many-body correlations in the electron-screened nuclei may then be given by Eqs. (5.78) and (5.79).

PM1 exemplifies a case in which a detectable level of reactions is predicted, while a power production may be expected at substantially elevated pressures in PM2 and PM3. Significant levels of the reaction rates are obtained in those cases owing to enhancement by approximately 30 to 40 orders of magnitude arising from the many-particle correlation processes. These values of pressures

Table 5.7   Reaction rates and enhancement factors in ultrahigh pressure liquid metals

Case	PM1	PM2	PM3	PM4	PM5
Reaction	d-p	d-p	d-p	^{7}Li-p	^{7}Li-p
ρ_m (g/cm^3)	2.5	8.5	20.0	16.0	140
T (K)	650	1200	1800	700	2000
P_e (Mbar)	17.2	148	649	80.9	3375
Λ_{ij}	0.95	1.05	1.14	0.88	1.07
r_s	1.17	0.78	0.59	0.87	0.43
D_s (10^{-9}cm)	3.85	3.21	2.79	2.24	1.75
$\log_{10} R_{ss}$ (s^{-1})	−53.28	−47.75	−43.89	−77.53	−66.58
Γ_{ij}	413.8	337.1	298.9	514.5	371.1
$\Gamma_{ij}^{(s)}$	68.19	87.30	95.76	118.5	133.5
$\log_{10} A_{ij}$	29.41	36.44	38.74	52.55	56.15
$\log_{10} R$ (s^{-1})	−23.87	−11.32	−5.15	−24.97	−10.43
$\log_{10} P$ (W/g)	−12.63	−0.07	6.09	−13.66	0.89

Note: Mass densities ρ_m and temperatures T are assumed parameters; P_e, the total pressure of the conduction electrons; in each case the molar fraction of the BIM is taken to be 50%.

and temperatures may suggest a possible experiment for d-p reactions by powerful, adiabatic compression of liquid metallic hydrogen [Ichimaru 2002].

For a study of nuclear fusion, lithium hydride under ultrahigh pressure appears to offer another system of interest, though the ranges of temperatures and pressures required for achieving a significant level of reactions are substantially greater than those in the cases of liquid metallic hydrogen. Here one considers reactions

$$^7\text{Li}\,(p, \alpha)^4\,\text{He} \tag{5.109}$$

with parameters [Bahcall & Ulrich 1988]

$$S_{^7\text{Lip}} = 52 \text{ (keV} \cdot \text{barns)} , \qquad Q(^7\text{Li-p}) = 17.347 \text{ (MeV)} . \tag{5.110}$$

It is an interesting system because lithium hydrides are stable compounds under laboratory conditions and because the reaction products in (5.109) are α particles, which are easier to handle. The reduced mass between Li and H is about the same as that between D and H. The nuclear charge $Z_{\text{Li}} = 3$, however, acts to reduce the reaction rate considerably; consequently, higher mass densities, temperatures, and pressures are required.

Lithium atom has the ionization potentials of 5.39 eV (first) and 75.64 eV (second, denoted as E_{2b}) [e.g., Allen 1973]. The cases PM4 and PM5 of ultrahigh pressure LiH in Table 5.7 may be looked upon as BIMs consisting of Li^+, p, and the corresponding number of free electrons. The short-range screening lengths in the table have been obtained by taking account of the two 1s electrons bound in Li^+ as well as of the screening parameters of the free electrons.

PM4 represents a case in which a detectable level of reactions is expected, while fusion power may be generated in PM5. Here again, these significant levels of reaction rates are obtained owing to enhancement by approximately 52 to 57 orders of magnitude due to many-body correlations in the systems of electron-screened nuclei.

The values of electronic pressures and temperatures implied in Table 5.7 are high. It is hoped that extension of current high-pressure experimental techniques may make such an experiment possible in near future. A detection of such a nuclear reaction in ultrahigh-pressure liquid metal will then make the first laboratory demonstration for the nuclear processes in astrophysical condensed matter such as interiors of the degenerate stars and may lead to an examination on the validity of extrapolating cross sections, such as Eq. (5.11), into regimes of extremely low energies on the order of 0.1 eV.

For recent progress in the study of radiative proton-capture nuclear reactions in metallic hyrdrogen, see Ichimaru [2001b] and references therein.

PROBLEMS

5.1   Derive the Gamow reaction rate (5.14).

5.2   Derive Eq. (5.23) for the critical temperature on the effect of electron screening.

5.3   Derive Eq. (5.25) for the rate of reactions under weak electron-screening conditions.

5.4   Derive Eq. (5.27) for the rate of reactions under strong electron-screening conditions.

5.5   Compute the values of screening potential at zero separation (5.51) for $\Gamma = 10, 100$, and 160 with the aid of Eqs. (2.89)–(2.92).

5.6   Compute the values of the enhancement factor (5.63) for helium and carbon OCPs at $\rho_m = 5 \times 10^6$ g/cm^3 and $T = 2 \times 10^7$ K.

5.7   Derive the terms in Eq. (5.76) except for its last term by a perturbation-theoretic method.

5.8   Use Eqs. (5.88) and (5.91b) for the estimation of a pycnonuclear rate in D-H matter at $\rho_m = 5$ g/cm^3 and $T = 300$ K. In the computation, assume that D-H consists of 50%-50% in the atomic fractions, and use mass number for the reduced mass.

The Fluctuation-Dissipation Theorem

In the theory of the many-particle system in thermodynamic equilibrium, the fluctuation-dissipation theorem provides a rigorous connection between the spectral functions of fluctuations and the imaginary parts of the relevant linear response functions [Callen & Welton 1951; Kubo 1957]. The theorem relates the canonically (or grand canonically) averaged commutator [,] and anticommutator { , } of any pair of Hermitian operators, such as the number densities evaluated at two different points in space and time. The average of such a commutator is related to a response function, while the average of an anticommutator gives a correlation function, which turns into a structure factor or a spectral function of fluctuations after Fourier transformations. It may therefore be said that the theorem possesses a form unique in physics, relating the properties of the system in equilibrium (i.e., fluctuations) with the parameters that characterize the irreversible processes, that is, the imaginary parts of the response functions. Here we summarize the contents of the fluctuation-dissipation theorem.

Consider a set of external disturbance fields represented by

$$a(\mathbf{r}, t) = a \exp[i(\mathbf{k} \cdot \mathbf{r} - \omega t) + 0t] + \text{cc} , \tag{A.1}$$

applied to a many-particle system in thermodynamic equilibrium, which would be uniform without the disturbances (cc stands for the complex conjugate and 0 implies a positive infinitesimal). These disturbances produce an external Hamiltonian,

$$H_{\text{ext}}(t) = -\sum \int_V d\mathbf{r} \, A(\mathbf{r}) a(\mathbf{r}, t) , \tag{A.2}$$

where A represents the physical quantity observable in the system that is coupled to the disturbance (A.1); the sum in Eq. (A.2) goes over the set of disturbances (A.1). The total Hamiltonian written as the sum of the unperturbed and external contributions,

$$H_{tot} = H + H_{ext}(t),\qquad (A.3)$$

then drives the system out of equilibrium according to the Heisenberg equation of motion (see Eq. 1.27).

A physical quantity B, which can be the same as A, of the system is perturbed and thereby deviates from its average value by $\delta B(\mathbf{r}, t)$. Within the framework of the linear response formalism (e.g., Ichimaru 1992), the deviations may be expressed as

$$\delta B(\mathbf{r}, t) = B \exp[i(\mathbf{k} \cdot \mathbf{r} - \omega t) + 0t] + \text{cc} .\qquad (A.4)$$

In this case

$$\chi_{BA}(\mathbf{k}, \omega) = \frac{B}{a}\qquad (A.5)$$

gives a linear response function in its general form. Explicit calculations with the Hamiltonian (A.3) yield a conmmutator expression for the response function:

$$\chi_{BA}(\mathbf{k}, \omega) = \frac{i}{\hbar} \int d\mathbf{r} \int_0^\infty dt \, \langle [B(\mathbf{r}' + \mathbf{r}, t' + t), A(\mathbf{r}', t')] \rangle \exp[-i(\mathbf{k} \cdot \mathbf{r} - \omega t)] .$$
$$(A.6)$$

Here $A(\mathbf{r}, t)$ and $B(\mathbf{r}, t)$ are the Heisenberg operators evolving with the unperturbed Hamiltonian and $\langle \cdots \rangle$ refers to the expectation value in the unperturbed equilibrium state. The linear response functions therefore depend only on the system properties without purturbations.

The physical quantities A and B fluctuate in space and time even in a system under thermodynamic equilibrium. The correlation function between them is defined in terms of the statistical average of the anticommutator as

$$C_{BA}(\mathbf{r}, t) = \frac{1}{2} \langle \{B(\mathbf{r}' + \mathbf{r}, t' + t), A(\mathbf{r}', t')\} \rangle .\qquad (A.7)$$

Fourier transformation of such a correlation function yields a structure factor or a spectral function of the fluctuations:

$$S_{BA}(\mathbf{k}, \omega) = \frac{1}{2\pi} \int d\mathbf{r} \int_{-\infty}^\infty dt \, C_{BA}(\mathbf{r}, t) \exp[-i(\mathbf{k} \cdot \mathbf{r} - \omega t)] .\qquad (A.8)$$

It is then connected with the linear response function (A.6) via the *fluctuation-dissipation theorem*,

$$S_{BA}(\mathbf{k}, \omega) = -\frac{i\hbar}{4\pi} \coth\left(\frac{\hbar\omega}{2k_B T}\right) [\chi_{BA}(\mathbf{k}, \omega) - \chi_{AB}(-\mathbf{k}, -\omega)] \ . \qquad (A.9)$$

In conjunction with the considerations related to static response problems in many-particle systems, it is instructive to treat a specific case of the external disturbance, which in terms of the spatial Fourier component takes the form

$$a(\mathbf{k}, t) = \begin{cases} a(\mathbf{k}) \exp(0t), & (t \le 0) \ , \\ 0, & (t < 0) \ . \end{cases} \qquad (A.10)$$

The Fourier component $\delta B(\mathbf{k}, t = 0)$ of the induced fluctuation is then calculated in terms of the *relaxation functions*

$$\chi_{BA}''(\mathbf{k}, \omega) = \frac{1}{2\hbar} \int d\mathbf{r} \int_{-\infty}^{\infty} dt \, \langle [B(\mathbf{r}' + \mathbf{r}, t' + t), A(\mathbf{r}', t')] \rangle \exp[-i(\mathbf{k} \cdot \mathbf{r} - \omega t)]$$

$$(A.11)$$

as

$$\frac{\delta B(\mathbf{k}, t = 0)}{a(\mathbf{k})} = \frac{1}{\pi} \int_{-\infty}^{\infty} d\omega \, \frac{\chi_{BA}''(\mathbf{k}, \omega)}{\omega} = \frac{1}{2k_B T} \langle [B(\mathbf{k}, t = 0), A(-\mathbf{k}, t = 0)] \rangle \ .$$

$$(A.12)$$

Thermodynamic sum rules for the relaxation functions are finally obtained in the long-wavelength limit of Eq. (A.12) as

$$\frac{\partial B}{\partial a} \equiv \lim_{\mathbf{k} \to 0} \frac{\delta B(\mathbf{k}, t = 0)}{a(\mathbf{k})} = \lim_{\mathbf{k} \to 0} \frac{1}{\pi} \int_{-\infty}^{\infty} d\omega \, \frac{\chi_{BA}''(\mathbf{k}, \omega)}{\omega} \ . \qquad (A.13)$$

The compressibility sum rule and the spin-susceptibility sum rule in Section 1.2D are typical examples of these thermodynamic sum rules.

Fermi Integrals

In the treatment of a free-electron gas at finite temperatures, it is useful to define the Fermi integrals:

$$I_\nu(\alpha) \equiv \int_0^\infty dt \, \frac{t^\nu}{\exp(t-\alpha)+1} \, . \tag{B.1}$$

For the electrons in the paramagnetic state (1.56), the normalization condition (1.44) is then expressed as

$$I_{1/2}(\alpha) = \frac{2}{3}\Theta^{-3/2} \, , \tag{B.2}$$

where $\alpha = \beta\mu_\sigma$ and Eq. (1.8) are assumed.

The Fermi pressure P_0 of the free-electron gas is likewise expressed as

$$\beta P_0 = n\Theta^{3/2} I_{3/2}(\alpha) \, . \tag{B.3}$$

The ideal-gas contribution to the free energy per unit volume is then calculated as

$$\beta F_0 = n\alpha - \beta P_0 \, . \tag{B.4}$$

Useful fitting formulas for the chemical potential and the Fermi pressure are

$$\beta\mu_\sigma = -\frac{3}{2}\ln\Theta + \ln\frac{4}{3\sqrt{\pi}} + \frac{A\Theta^{-(b+1)} + B\Theta^{-(b+1)/2}}{1+A\Theta^{-b}} \, , \tag{B.5}$$

with $A = 0.25954$, $B = 0.072$, and $b = 0.858$; and

$$\frac{\beta P_0}{n} = 1 + \frac{2}{5} \frac{X\Theta^{-(y+1)} + Y\Theta^{-(y+1)/2}}{1 + X\Theta^{-y}} , \tag{B.6}$$

with $X = 0.27232$, $Y = 0.145$, and $y = 1.044$. Maximum deviations of Eq. (B.5) from the exact values determined from Eq. (B.2) are about 0.19% at $\theta \sim 0.05$; those of Eq. (B.6) from the exact values determined from Eq. (B.3) are about 0.26% at $\theta \sim 5$.

In the classical limit—when $\Theta \gg 1$—the Fermi integrals may be expanded as [e.g., Pathria 1972]

$$I_\nu(\alpha) = \Gamma(\nu + 1) \sum_{s=1}^{\infty} (-1)^{s+1} \exp(s\alpha) s^{-(\nu+1)} , \tag{B.7}$$

where

$$\Gamma(z) = \int_0^\infty dt\, t^{z-1} \exp(-t) \tag{B.8}$$

is the gamma function. In this limit one thus has

$$\alpha = -\frac{3}{2} \ln \Theta + \ln \frac{4}{3\sqrt{\pi}} , \tag{B.9}$$

$$\frac{\beta P_0}{n} = 1 . \tag{B.10}$$

In the quantum limit of strong degeneracy—when $\Theta \ll 1$—

$$I_\nu(\alpha) = \frac{\alpha^{\nu+1}}{\nu+1} \left[1 + \sum_{s=1}^{\infty} 2(1 - 2^{1-2s})\zeta(2s)\frac{(\nu+1)!}{(\nu+1-2s)!}\alpha^{-2s} \right] \tag{B.11}$$

[Landau & Lifshitz 1969], where

$$\zeta(x) = \sum_{\nu=1}^{\infty} \frac{1}{\nu^x} , \qquad x > 1 \tag{B.12}$$

is Riemann's zeta function. Hence

$$\mu_\sigma = E_F , \tag{B.13}$$

$$P_0 = \frac{2}{5} n E_F , \tag{B.14}$$

in this limit. For reference, we list some values of those functions:

$$\zeta(\tfrac{3}{2})=2.612, \qquad \zeta(\tfrac{5}{2})=1.341, \qquad \zeta(3)=1.202,$$
$$\zeta(5)=1.037, \qquad \Gamma(\tfrac{3}{2})=\sqrt{\pi}/2, \qquad \Gamma(\tfrac{3}{2})=\tfrac{3}{4}\sqrt{\pi}/2.$$

Functional Derivatives

The functional derivative technique is closely related to the dielectric formulation, the density-functional theory, and Green's function formalism in Chapter 1 and to the multiparticle distributions in Chapter 2. In this appendix, fundamental relations in the functional derivatives are summarized.

Let $F[f(x)]$ be a functional of a function $f(x)$ with a variable x defined over a domain, $a \leq x \leq b$. The functional derivative, $\delta F/\delta f(x)$, is given in terms of the increment δF produced by an infinitesimal variation $\delta f(x)$ of $f(x)$ as

$$\delta F = \int_a^b \mathrm{d}x \, \frac{\delta F}{\delta f(x)} \delta f(x) . \qquad \text{(C.1)}$$

It has the following properties:

1. *Identity relation*:

$$\frac{\delta f(x')}{\delta f(x)} = \delta(x - x') . \qquad \text{(C.2)}$$

2. *Product rule*: When a functional is expressed by a simple product of the functions $f(x_i)$ as

$$F = \int_a^b \mathrm{d}x_1 \cdots \mathrm{d}x_N \prod_{i=1}^N f(x_i) , \qquad \text{(C.3)}$$

then

$$\frac{\delta F}{\delta f(x)} = N \int_a^b dx_2 \cdots dx_N \prod_{i=2}^{N} f(x_i) \,. \tag{C.4}$$

3. *Chain rule*: When a functional F is a functional of $G(x')$ which in turn is a functional of $f(x)$, then

$$\frac{\delta F}{\delta f(x)} = \int_a^b dx' \, \frac{\delta F}{\delta G(x')} \frac{\delta G(x')}{\delta f(x)} \,. \tag{C.5}$$

4. When the functional is given by

$$F = \int_a^b dx \, F\left(f(x)\right) \tag{C.6}$$

where $F\left(f(x)\right)$ is a function of $f(x)$, then

$$\frac{\delta F}{\delta f(x)} = \frac{dF\left(f(x)\right)}{df(x)} \,. \tag{C.7}$$

Bibliography

Abe, R., 1959, *Progr. Theor. Phys.* **21**, 475.

Ajzenberg-Selove, F., 1984, *Nucl. Phys. A* **413**, 1.

Ajzenberg-Selove, F., & C.L. Busch, 1980, *Nucl. Phys. A* **336**, 1.

Alastuey, A., & B. Jancovici, 1978, *Astrophys. J.* **226**, 1034.

Alder, B.J., E.L. Pollock, & J.-P. Hansen, 1980, *Proc. Natl. Acad. Sci. USA* **77**, 6272.

Alder, B.J., & T.E. Wainwright, 1957, *J. Chem. Phys.* **27**, 1207.

Alder, B.J., & T.E. Wainwright, 1959, *J. Chem. Phys.* **31**, 459.

Alder, B.J., & T.E. Wainwright, 1970, *Phys. Rev. A* **1**, 18.

Alefeld, G., & J. Völkl, eds., 1978, *Hydrogen in Metals* (Springer, Berlin), Vols. I & II.

Allen, C.W., 1973, *Astrophysical Quantities*, 3rd ed. (Athlone, London).

Anders, E., & N. Grevesse, 1989, *Geochem. Cosmochem. Acta* **53**, 197.

Andersen, H.C., 1980, *J. Chem. Phys.* **72**, 2384.

Anderson, P.W., 1987, *Science* **235**, 1196.

Ando, T., 1990, in *Strongly Coupled Plasma Physics*, edited by S. Ichimaru (North-Holland/Yamada Science Foundation, Amsterdam), p. 263.

Ando, T., A.B. Fowler, & F. Stern, 1982, *Rev. Mod. Phys.* **54**, 437.

Armstrong, K.R., D.A. Harper, Jr., & F.J. Low, 1972, *Astrophys. J. Lett.* **178**, L89.

Arnett, W.D., & J.W. Truran, 1969, *Astrophys. J.* **157**, 339.

Ashcroft, N.W., 1989, *Nature* **340**, 345.

Ashcroft, N.W., & D. Stroud, 1978, in *Solid State Physics*, edited by H. Ehrenreich, F.

Seitz, and D. Turnbull (Academic, New York), Vol. 33, p. 1.

Bae, Y.K., D.C. Lorents, & S.E. Young, 1991, *Phys. Rev. A* **44**, R4051.

Bahcall, J.N., W.F. Huebner, S.H. Lubow, P.D. Parker, & R.K. Ulrich, 1982, *Rev. Mod. Phys.* **54**, 767.

Bahcall, J.N., & M.H. Pinsonneault, 1992, *Rev. Mod. Phys.* **64**, 885.

Bahcall, J.N., & R.K. Ulrich, 1971, *Astrophys. J.* **170**, 593.

Bahcall, J.N., & R.K. Ulrich, 1988, *Rev. Mod. Phys.* **60**, 297.

Baranger, M., & B. Mozer, 1959, *Phys. Rev.* **115**, 521.

Bardeen, J., L.N. Cooper, & J.R. Schrieffer, 1957, *Phys. Rev.* **108**, 1175.

Barkat, J.L, C. Wheeler, & J.-R. Buchler, 1972, *Astrophys. J.* **171**, 651.

Barnea, G., 1979, *J. Phys. C* **12**, L263.

Barrat, J.L., J.-P. Hansen, & R. Mochkovitch, 1988, *Astron. Astrophys.* **199**, L15.

Batson, P.E., C.H. Chen, & J. Silcox, 1976, *Phys. Rev. Lett.* **37**, 937.

Baus, M., & J.-P. Hansen, 1980, *Phys. Rep.* **59**, 1.

Baym, G., & C.J. Pethick, 1975, *Ann. Rev. Nucl. Sci.* **25**, 27.

Baym, G., C.J. Pethick, & P. Sutherland, 1971, *Astrophys. J.* **170**, 299.

Bednorz, J.G., & K.A. Mueller, 1986, *Z. Phys. B* **64**, 189.

Benege, J.F., & W.R. Shanahan, 1993, in *Physics of Strongly Coupled Plasmas*, edited by H.M. Van Horn, & S. Ichimaru (University of Rochester Press, Rochester, New York), 187.

Bengtzelius, U., W. Götze, & A. Sjölander, 1984, *J. Phys. C* **17**, 5915.

Bernu, B., 1983, *Physica* **122A**, 129.

Bernu, B., & J.-P. Hansen, 1982, *Phys. Rev. Lett.* **48**, 1375.

Beuhler, R.J., Y.Y. Chu, G. Friedlander, & L. Friedman, 1991, *Phys. Rev. Lett.* **67**, 473.

Beuhler, R.J., G. Friedlander, & L. Friedman, 1989, *Phys. Rev. Lett.* **63**, 1292.

Beuhler, R.J., G. Friedlander, & L. Friedman, 1992, *Phys. Rev. Lett.* **68**, 2108(E).

Binder, K., ed., 1979, *The Monte Carlo Method in Statistical Physics* (Springer, Berlin).

Binder, K., ed., 1992, *The Monte Carlo Method in Condensed Matter Physics* (Springer, Berlin).

Boercker, D.B., F.J. Rogers, & H.E. DeWitt, 1982, *Phys. Rev. A* **25**, 1623.

Bohm, D., & D. Pines, 1953, *Phys. Rev.* **92**, 609.

Bollinger, J.J., S.L. Gilbert, D.J. Heinzen, W.M. Itano, & D.J. Wineland, 1990, in *Strongly Coupled Plasma Physics*, edited by S. Ichimaru (North-Holland/Yamada Science Foundation, Amsterdam), p. 117.

Bonsall, L., & A.A. Maradudin, 1977, *Phys. Rev. B* **15**, 1959.

Brittin, W.E., & W.R. Chappell, 1962, *Rev. Mod. Phys.* **34**, 620.

Brosens, F., J.T. Devreese, & L.F. Lemmens, 1980, *Phys. Rev. B* **21**, 1363.

Brovman, E.G., Yu. Kagan, & A. Kholas, 1971, *Zh. Eksp. Teor. Fiz.* **61**, 2429 [*Sov. Phys. JETP* **34**, 1300 (1972)].

Brush, S.G., H.L. Sahlin, & E. Teller, 1966, *J. Chem. Phys.* **45**, 2102.

Burrows, A., W.B. Hubbard, & J.I. Lunine, 1989, *Astrophys. J.* **345**, 939.

Burrows, A., & J. Liebert, 1993, *Rev. Mod. Phys.* **65**, 301.

Callaway, J., & N.H. March, 1984, in *Solid State Physics*, edited by H. Ehrenreich, F. Seitz, & D. Turnbull (Academic, New York), Vol. 38, p. 135.

Callen, H.B., & T.A. Welton, 1951, *Phys. Rev.* **83**, 34.

Cameron, A.G.W., 1959, *Astrophys. J.* **130**, 916.

Canal, R., J. Isern, & J. Labay, 1990, *Ann. Rev. Astron. Astrophys.* **28**, 183.

Canal, R., & E. Schatzman, 1976, *Astron. Astrophys.* **46**, 229.

Car, R., & M. Parrinello, 1985, *Phys. Rev. Lett.* **55**, 2471.

Carr, W.J., 1961, *Phys. Rev.* **122**, 1437.

Carr, W.J., R.A. Coldwell-Horsfall, & A.E. Fein, 1961, *Phys. Rev.* **124**, 747.

Carraro, C., B.Q. Chen, S. Schramm, & S.E. Koonin, 1990, *Phys. Rev. A* **42**, 1349.

Ceperley, D.M., 1978, *Phys. Rev. B* **18**, 3126.

Ceperley, D.M., & B.J. Alder, 1980, *Phys. Rev. Lett.* **45**, 566.

Ceperley, D.M., & B.J. Alder, 1987, *Phys. Rev. B* **36**, 2092.

Ceperley, D.M., & M.H. Kalos, 1979, in *Monte Carlo Methods in Statistical Physics*, edited by K. Binder (Springer, Berlin), p. 145.

Chandler, D., & P.G. Wolynes, 1981, *J. Chem. Phys.* **74**, 4078.

Chandrasekhar, S., 1943, *Rev. Mod. Phys.* **15**, 1.

Ciccontti, G., D. Frenkel, & I.R. McDonald, 1987, *Simulation of Liquids and Solids* (North-Holland, Amsterdam).

Clayton, D.D., 1968, *Principles of Stellar Evolution and Nucleosynthesis* (McGraw-Hill, New York).

Couch, R.G., & W.D. Arnett, 1975, *Astrophys. J.* **196**, 791.

Cox, A.N., & J.N. Stewart, 1965, *Astrophys. J. Suppl.* **11**, 22.

Cox, A.N., & J.E. Tabor, 1976, *Astrophys. J. Suppl.* **31**, 271.

Creutz, M., & B. Freedman, 1981, *Ann. Phys.* (NY) **132**, 427.

D'Antona, F., & I. Mazzitelli, 1985, *Astrophys. J.* **296**, 502.

D'Antona, F., & I. Mazzitelli, 1990, *Ann. Rev. Astron. Astrophys.* **28**, 139.

Davidson, R.C, 1990, *Physics of Nonneutral Plasmas* (Addison-Wesley, Redwood City, Calif.).

Davis, R., 1984, *Bull. Am. Phys. Soc.* **29**, 731.

de Heer, W.A., W.D. Knight, M.Y. Chou, & M.L. Cohen, 1987, in *Solid State Physics*, edited by H. Ehrenreich, F. Seitz, & D. Turnbull (Academic, New York), Vol. 40, p. 93.

De Raedt, H., & B. De Raedt, 1978, *Phys. Rev. B* **18**, 2039.

DeSilva, A.W., & H.-J. Kunze, 1993, in *Physics of Strongly Coupled Plasmas*, edited by H.M. Van Horn, & S. Ichimaru (University of Rochester Press, Rochester, New York), 191.

Deutsch, C., M.M. Gombert, & H. Minoo, 1981, *Phys. Rev. A* **23**, 924.

Devreese, J.T., F. Brosens, & L.F. Lemmens, 1980, *Phys. Rev. B* **21**, 1349.

DeWitt, H.E., & Y. Rosenfeld, 1979, *Phys. Lett.* **75A**, 79.

Dharma-wardana, M.W.C., 1976, *J. Phys. C* **9**, 1919.

Dharma-wardana, M.W.C., & F. Perrot, 1982, *Phys. Rev. A* **26**, 2096.

Dharma-wardana, M.W.C., & R. Taylor, 1981, *J. Phys. C* **14**, 629.

Diedrich, F., E. Peik, J.M. Chen, W. Quint, & H. Walther, 1987, *Phys. Rev. Lett.* **59**, 2931.

Dolgov, O.V., D.A. Kirzhnits, & E.G. Meksimov, 1981, *Rev. Mod. Phys.* **53**, 81.

Drexel, W., A. Murani, D. Tocchetti, W. Kley, I. Sosnowska, & D.K. Ross, 1976, *J. Phys. Chem. Solids* **37**, 1135.

Driscoll, C.F., & J.F. Malmberg, 1983, *Phys. Rev. Lett.* **50**, 167.

Dubin, D.H.E., 1990, *Phys. Rev. A* **42**, 4972.

DuBois, D.F., 1959, *Ann. Phys.* (N.Y.) **8**, 24.

DuBois, D.F., & M.G. Kivelson, 1969, *Phys. Rev.* **186**, 409.

Ebeling, W., A. Förster, V.E. Fortov, V.K. Gryaznov, & A.Ya. Polishchuk, 1991, *Thermophysical Properties of Hot Dense Plasmas* (Teubner, Stuttgart).

Ebeling, W., & G. Röpke, 1979, *Ann. Phys.* (Leipzig) **36**, 429.

Echenique, P.M., J.R. Manson, & R.H. Ritchie, 1990, *Phys. Rev. Lett.* **64**, 1413.

Endo, H., 1990, ed., *Liquid and Amorphous Metals* VII (North-Holland, Amsterdam).

Englert, F., & R. Brout, 1960, *Phys. Rev.* **120**, 1085.

Ernst, M.H., E.H. Hauge, & J.M.J. van Leeuwen, 1979, *Phys. Rev. A* **4**, 2055.

Evans, R., 1979, *Adv. Phys.* **28**, 143.

Farges, J., M.F. de Faraudy, B. Raoult, & G. Torchet, 1983, *J. Chem. Phys.* **78**, 5067.

Fetter, A.L., & J.D. Walecka, 1971, *Quantum Theory of Many-Particle Systems* (McGraw-Hill, New York).

Feynman, R.P., 1972, *Statistical Mechanics* (Benjamin, Reading, Mass.).

Feynman, R.P., & A.R. Hibbs, 1965, *Quantum Mechanics and Path Integrals* (McGraw-Hill, New York).

Fischer, C.F., 1977, *The Hartree–Fock Method for Atoms* (John Wiley, New York).

Foldy, L.L., 1978, *Phys. Rev. B* **17**, 4889.

Fortov, V.E., 1982, *Usp. Phys. Nauk* **138**, 361 [*Sov. Phys. Usp.* **25**, 781 (1983)].

Fowler, W.A., G.R. Caughlan, & B.A. Zimmerman, 1967, *Ann. Rev. Astron. Astrophys.* **5**, 523.

Fowler, W.A., G.R. Caughlan, & B.A. Zimmerman, 1975, *Ann. Rev. Astron. Astrophys.* **13**, 69.

Frenkel, J., 1946, *Kinetic Theory of Liquids* (Clarendon, Oxford).

Friedli, C., & N.W. Ashcroft, 1977, *Phys. Rev. B* **16**, 662.

Fuchs, K., 1936, *Proc. R. Soc. London A* **153**, 622.

Fushimi, A., H. Iyetomi, & S. Ichimaru, 1993, unpublished.

Gai, A.M., S.L. Rugari, R.H. France, B.J. Lund, Z. Zhan, A.J. Davanport, H.S. Isaacs, & K.J. Lynn, 1989, *Nature* **340**, 29.

Galam, S., & J.-P. Hansen, 1976, *Phys. Rev. A* **14**, 816.

Gamow, G., 1928, *Z. Phys.* **51**, 204.

Gamow, G., & E. Teller, 1938, *Phys. Rev.* **53**, 608.

Gell-Mann, M., & K.A. Brueckner, 1957, *Phys. Rev.* **106**, 354.

Gibbons, P.C., S.E. Schnatterly, J.J. Ritsko, & J.R. Fields, 1976, *Phys. Rev. B* **13**, 2451.

Glicksman, M., 1971, in *Solid State Physics*, edited by H. Ehrenreich, F. Seitz, & D. Turnbull (Academic, New York), Vol. 26, p. 275.

Goldstein, W., C. Hooper, J. Gauthier, J. Seeley, & R. Lee, eds., 1991, *Radiative Properties of Hot Dense Matter* (World Scientific, Singapore).

Graboske, H.C., 1973, *Astrophys. J.* **183**, 177.

Graboske, H.C., J.B. Pollack, A.S. Grossman, & R.J. Olness, 1975, *Astrophys. J.* **199**, 265.

Griem, H.R., 1974, *Spectral Line Broadening by Plasmas* (Academic, New York).

Grimes, C.C., 1978, *Surf. Sci.* **73**, 397.

Grimes, C.C., & G. Adams, 1979, *Phys. Rev. Lett.* **42**, 795.

Gudmundsson, E.H., C.J. Pethick, & R.I. Epstein, 1982, *Astrophys. J. Lett.* **259**, L19.

Gunnarsson, O., M. Jonson, & B.I. Lundqvist, 1979, *Phys. Rev. B* **20**, 3136.

Gunnarsson, O., & B.I. Lundqvist, 1976, *Phys. Rev. B* **13**, 4274.

Gupta, A.K., & K.S. Singwi, 1977, *Phys. Rev. B* **15**, 1801.

Gupta, U., & A.K. Rajagopal, 1980, *Phys. Rev. A* **21**, 2064.

Gurney, R.W., & E.U. Condon, 1929, *Phys. Rev.* **33**, 127.

Hafner, J., 1987, *From Hamiltonians to Phase Diagrams* (Springer, Berlin).

Hammerberg, J.E., & N.W. Ashcroft, 1974, *Phys. Rev. B* **9**, 409.

Hansen, C.J., ed., 1974, *Physics of Dense Matter*, I.A.U. Symp. No. 53 (D. Reidel, Dordrecht-Holland).

Hansen, J.-P., 1973, *Phys. Rev. A* **8**, 3096.

Hansen, J.-P., 1978, in *Strongly Coupled Plasmas*, edited by G. Kalman (Plenum, New York), p. 117.

Hansen, J.-P., & I.R. McDonald, 1981, *Phys. Rev. A* **23**, 2041.

Hansen, J.-P., & I.R. McDonald, 1986, *Theory of Simple Liquids*, 2nd ed. (Academic, London).

Hansen, J.-P., I.R. McDonald, & E.L. Pollock, 1975, *Phys. Rev. A* **11**, 1025.

Hansen, J.-P., G.M. Torrie, & P. Vieillefosse, 1977, *Phys. Rev. A* **16**, 2153.

Hansen, J.-P., & P. Vieillefosse, 1975, *Phys. Lett.* **53A**, 187.

Harris, M.J., W.A. Fowler, G.R. Caughlan, & B.A. Zimmerman, 1983, *Ann. Rev. Astron. Astrophys.* **21**, 165.

Hasegawa, M., 1971, *J. Phys. Soc. Jpn.* **31**, 649.

Hasegawa, M., & M. Watabe, 1969, *J. Phys. Soc. Jpn.* **27**, 1393.

Hedin, L., 1965, *Phys. Rev.* **139**, A796.

Hedin, L., & B.I. Lundqvist, 1971, *J. Phys. C* **4**, 2064.

Heine, V., P. Nozières, & J.W. Wilkins, 1966, *Philos. Mag.* **13**, 741.

Helfer, L., R.L. McCrory, & H.M. Van Horn, 1984, *J. Stat. Phys.* **37**, 577.

Hemley, R.J., & H.K. Mao, 1991, in *Proc. APS 1991 Topical Conf. on Shock Compression of Condensed Matter*, Williamsburg, VA, June 17–20, 1991, edited by S.C. Schmidt, R.D. Dick, J.W. Forbes, & D.J. Tasker (North-Holland, Amsterdam), p. 27.

Hensel, J.C., T.G. Phillips, & G.A. Thomas, 1977, in *Solid State Physics*, edited by H. Ehrenreich, F. Seitz, & D. Turnbull (Academic, New York), Vol. 32, p. 87.

Hohenberg, P., & W. Kohn, 1964, *Phys. Rev.* **136**, B864.

Höhberger, H.J., A. Otto, & E. Petri, 1975, *Solid State Commun.* **16**, 175.

Holas, A., P.K. Aravind, & K.S. Singwi, 1979, *Phys. Rev. B* **20**, 4912.

Holtsmark, J., 1919, *Ann. Physik*, **58**, 577.

Hooper, C.F., 1966, *Phys. Rev.* **149**, 77.

Hooper, C.F., 1968, *Phys. Rev.* **165**, 215.

Hoover, W.G., D.J. Evans, R.B. Hickman, A.J.C. Ladd, W.T Ashurst, & B. Moran, 1980, *Phys. Rev. A* **22**, 1690.

Hoover, W.G., & J.C. Poirier, 1962, *J. Chem. Phys.* **37**, 1041.

Hora, H., 1991, *Plasmas at High Temperature and Density* (Springer, Berlin).

Hotop, H., & W.C. Lineberger, 1975, *J. Phys. Chem. Ref. Data* **4**, 539.

Hoyle, F., 1954, *Astrophys. J. Suppl.* **1**, 121.

Hubbard, W.B., 1968, *Astrophys. J.* **152**, 745.

Hubbard, W.B., 1980, *Rev. Geophys. Space Sci.* **18**, 1.

Hubbard, W.B., 1984, *Planetary Interiors* (Van Nostrand Reinhold, New York).

Hubbard, W.B., & M. Lampe, 1969, *Astrophys. J. Suppl.* **18**, 297.

Hubbard, W.B., & M.S. Marley, 1989, *Icarus* **78**, 102.

Ichimaru, S., 1970, *Phys. Rev. A* **2**, 494.

Ichimaru, S., 1982, *Rev. Mod. Phys.* **54**, 1017.

Ichimaru, S., 1991, *J. Phys. Soc. Jpn.* **60**, 1437.

Ichimaru, S., 1992, *Statistical Plasma Physics* (Addison-Wesley, Reading, Mass.), Vol. I.

Ichimaru, S., 1993a, *Rev. Mod. Phys.* **65**, 255.

Ichimaru, S., 1993b, in *Physics of Strongly Coupled Plasmas*, edited by H.M. Van Horn, & S. Ichimaru (University of Rochester Press, Rochester, New York), p. 31.

Ichimaru, S., 2001a, *Phys. Plasmas* **8**, 48.

Ichimaru, S., 2001a, *Phys. Plasmas* **8**, 4284.

Ichimaru, S., 2002, in *Inertial Fusion Sciences and Applications 2001*, edited by K.A. Tanaka, D.D. Meyerhofer, & J. Meyer-ter-Vehn (Elsevier, Paris), p.1085.

Ichimaru, S., H. Iyetomi, & S. Ogata, 1988, *Astrophys. J. Lett.* **334**, L17.

Ichimaru, S., H. Iyetomi, & S. Tanaka, 1987, *Phys. Rep.* **149**, 92.

Ichimaru, S., & H. Kitamura, 1999, *Phys. Plasmas* 6, 2649 [Erratum: 7, 3482 (2000)].

Ichimaru, S., H. Kitamura, & S. Ogata, 1994, *Publ. Astron. Soc. Jpn.*, **46**, 285.

Ichimaru, S., S. Mitake, S. Tanaka, & X.-Z. Yan, 1985, *Phys. Rev. A* **32**, 1768.

Ichimaru, S., & S. Ogata, 1991, *Astrophys. J.* **374**, 647.

Ichimaru, S., S. Ogata, & A. Nakano, 1990, *J. Phys. Soc. Jpn.* **59**, 3904.

Ichimaru, S., S. Ogata, & H.M. Van Horn, 1992, *Astrophys. J. Lett.* **401**, L35.

Ichimaru, S., S. Ogata, & K. Tsuruta, 1994, *Phys. Rev. E* **50**, 2977.

Ichimaru, S., & S. Tanaka, 1985, *Phys. Rev. A* **32**, 1790.

Ichimaru, S., S. Tanaka, & H. Iyetomi, 1984, *Phys. Rev. A* **29**, 2033.

Ichimaru, S., H. Totsuji, T. Tange, & D. Pines, 1975, *Progr. Theor. Phys.* **54**, 1077.

Ichimaru, S., & K. Utsumi, 1981, *Phys. Rev. B* **24**, 7385.

Ichimaru, S., & K. Utsumi, 1983, *Astrophys. J. Lett.* **269**, L51.

Iglesias, C.A., & C.H. Hooper, 1982, *Phys. Rev. A* **25**, 1049, 1632.

Iglesias, C.A., J.L. Lebowitz, & D. McGowan, 1983, *Phys. Rev. A* **28**, 1667.

Ingersoll, A.P., G. Münch, G. Neugebauer, & G.S. Orton, 1976, in *Jupiter*, edited by T. Gehrels (University of Arizona Press, Tucson), p. 85.

Ioriatti, L.C., Jr., & A. Isihara, 1981, *Z. Phys. B* **44**, 1.

Isihara, A., 1989, in *Solid State Physics*, edited by H. Ehrenreich & D. Turnbull (Academic, New York), Vol. 42, p. 271.

Ivanov, Yu.V., V.B. Mintsev, V.E. Fortov, & A.N Dremin, 1976, *Zh. Eksp. Teor. Fiz.* **71**, 216 [*Sov. Phys. JETP* **44**, 112 (1976)].

Iyetomi, H., 1984, *Progr. Theor. Phys.* **71**, 427.

Iyetomi, H., & S. Ichimaru, 1983, *Phys. Rev. A* **27**, 3241.

Iyetomi, H., & S. Ichimaru, 1986, *Phys. Rev. A* **34**, 3203.

Iyetomi, H., & S. Ichimaru, 1993, unpublished.

Iyetomi, H., S. Ogata, & S. Ichimaru, 1992, *Phys. Rev. A* **46**, 1051.

Iyetomi, H., S. Ogata, & S. Ichimaru, 1993, *Phys. Rev. B* **47**, 11703.

Iyetomi, H., K. Utsumi, & S. Ichimaru, 1981, *J. Phys. Soc. Jpn.* **50**, 3769.

Jackson, J.D., 1957, *Phys. Rev.* **106**, 330.

James, F., 1980, *Rep. Prog. Phys.* **43**, 73.

Jancovici, B., 1962, *Nuovo Cimento* **25**, 428.

Jancovici, B., 1977, *J. Stat. Phys.* **17**, 357.

Janev, R.K., W.D. Langer, K. Evans, Jr., & D.E. Post, Jr., 1987, *Elementary Processes in Hydrogen-Helium Plasmas* (Springer, Berlin).

Jastrow, R., 1955, *Phys. Rev.* **98**, 1479.

Jeffries, C.D., & L.V. Keldish, eds., 1983, *Electron-Hole Droplets in Semiconductors* (North-Holland, Amsterdam).

Jensen, E., & E.W. Plummer, 1985, *Phys. Rev. Lett.* **55**, 1912.

Jindal, V.K., H.B. Singh, & K.N. Pathak, 1977, *Phys. Rev. B* **15**, 252.

Jones, E.S., E.P. Palmer, J.B. Czirr, D.L. Decker, G.L. Jensen, J.M. Thorne, S.F. Taylor, & J. Rafelski, 1989, *Nature* **338**, 737.

Jones, H., 1971, *J. Chem. Phys.* **55**, 2640.

Jonson, M., 1976, *J. Phys. C* **9**, 3055.

Kadanoff, L.P., & G. Baym, 1962, *Quantum Statistical Mechanics* (Benjamin, New York).

Kafatos, M.C., R.S. Harrington, & S.P. Maran, eds., 1986, *Astrophysics of Brown Dwarfs* (Cambridge University Press, Cambridge).

Kammerer, R., J. Barth, F. Gerken, C. Kunz, S.A. Flodstrom, & L.I. Johansson, 1982, *Phys. Rev. B* **26**, 3491.

Kanter, H., 1870, *Phys. Rev. B* **1**, 522, 2357.

Kim, Y.E., J.H. Yoon, R.A. Rice, & M. Rabinowitz, 1992, *Phys. Rev. Lett.* **68**, 373.

Kimball, J.C., 1973, *Phys. Rev. A* **7**, 1648.

Kimball, J.C., 1976, *Phys. Rev. B* **14**, 2371.

Kirkpatrick, T.R., 1985, *Phys. Rev. A* **31**, 939.

Kitamura, H., & S. Ichimaru, 1995, *Phys. Rev. E* **51**, 6004.

Kivelson, M.G., & D.F. DuBois, 1964, *Phys. Fluids* **7**, 1578.

Kloos, T., 1973, *Z. Phys.* **265**, 225.

Kohn, W., & L.J. Sham, 1965, *Phys. Rev. A* **140**, 1133.

Kohn, W., & P. Vashishta, 1983, in *Theory of the Inhomogeneous Electron Gas*, edited by S. Lundqvist, & N.H. March (Plenum, New York), p. 79.

Korn, C.& D. Zamir, 1970, *J. Phys. Chem. Solids* **31**, 489.

Kovetz, A., D.Q. Lamb, & H.M. Van Horn, 1972, *Astrophys. J.* **174**, 109.

Krane, K.J., 1978, *J. Phys. F* **8**, 2133.

Krauss, A., H.W. Becker, H.P. Tautvetter, & C. Rolfs, 1987, *Nucl. Phys.* **A465**, 150.

Kubo, R., 1957, *J. Phys. Soc. Jpn.* **12**, 570.

Kukkonen, C.A., & H. Smith, 1973, *Phys. Rev. B* **8**, 4601.

Kukkonen, C.A., & J.W. Wilkins, 1979, *Phys. Rev. B* **19**, 6075.

Kumar, S.S., 1963, *Astrophys. J.* **137**, 1121.

Landau, L.D., 1956, *Sov. Phys. JETP* **3**, 920.

Landau, L.D., 1957, *Sov. Phys. JETP* **5**, 101.

Landau, L.D., & E.M. Lifshitz, 1959, *Fluid Mechanics*, translated by J.B. Sykes & W.H. Reid (Addison-Wesley, Reading, Mass.).

Landau, L.D., & E.M. Lifshitz, 1960, *Electrodynamics of Continuous Media*, translated by J.B. Sykes & J.S. Bell (Addison-Wesley, Reading, Mass.).

Landau, L.D., & E.M. Lifshitz, 1969, *Statistical Physics*, 2nd ed., translated by J.B. Sykes & M.J. Kearsley (Addison-Wesley, Reading, Mass.).

Landau, L.D., & E.M. Lifshitz, 1970, *Theory of Elasticity*, 2nd ed., translated by J.B. Sykes & W.H. Reid (Addison-Wesley, Reading, Mass.).

Landau, L.D., & E.M. Lifshitz, 1976, *Quantum Mechanics*, 3rd ed., translated by J.B. Sykes & J.S. Bell (Addison-Wesley, Reading, Mass.).

Lang, N.D., 1973, in *Solid State Physics*, edited by H. Ehrenreich, F. Seitz, & D. Turnbull (Academic, New York), Vol. 28, p. 225.

Lang, N.D., & W. Kohn, 1970, *Phys. Rev. B* **1**, 4555.

Lang, N.D., & W. Kohn, 1971, *Phys. Rev. B* **3**, 1215.

Langreth, D.C., & J.P. Perdew, 1977, *Phys. Rev. B* **15**, 2884.

Lantto, L.J., 1980, *Phys. Rev. B* **22**, 1380.

Lau, K.H., & W. Kohn, 1976, *J. Chem. Phys. Solids* **37**, 99.

Lee, Y.T., & R.M. More, 1984, *Phys. Fluids* **27**, 1273.

Leggett, A.J., & G. Baym, 1989, *Phys. Rev. Lett.* **63**, 191.

Leutheusser, E., 1984, *Phys. Rev. A* **29**, 2765.

Lewin, W.H.G., & P.C. Joss, 1983, in *Accretion Driven Stellar X-Ray Sources*, edited by W.H.G. Lewin & E.P.J. Van den Heuvel (Cambridge University Press, Cambridge), p. 40.

Lieb, E.H., 1976, *Rev. Mod. Phys.* **48**, 553.

Liebert, J., 1980, *Ann. Rev. Astron. Astrophys.* **18**, 363.

Liebert, J., & R.G. Probst, 1987, *Ann. Rev. Astron. Astrophys.* **25**, 473.

Lindgren, S.A., & L. Walldén, 1988, *Phys. Rev. Lett.* **61**, 2894.

Lindhard, J., 1954, *Kgl. Danske Videnskab, Selskab Mat.-Fys. Medd.* **28**, No.8.

Lorenzana, H.E., I.F. Silvera, & K.A. Goettel, 1989, *Phys. Rev. Lett.* **63**, 2080.

Low, F.J., 1966, *Astron. J.* **71**, 391.

Lowy, D.N., & G.E. Brown, 1975, *Phys. Rev. B* **12**, 2138.

Lundqvist, B.I., 1968, *Phys. Kondens. Mater.* **7**, 117.

Lundqvist, S., & N.H. March, eds., 1983, *Theory of the Inhomogeneous Electron Gas* (Plenum, New York).

Luyten, W.J., ed., 1971, *White Dwarfs* (Reidel, Dordrecht).

Lyo, I.-W., & E.W. Plummer, 1988, *Phys. Rev. Lett.* **60**, 1558.

Ma, S.K., & K. Brueckner, 1968, *Phys. Rev. B* **165**, 18.

Macke, W., 1950, *Z. Naturforsch.* **5a**, 192.

Mahan, G.D., 1970, *Phys. Rev. B* **2**, 4334.

Mandal, S.S., B.K. Rao, & D.N. Tripathy, 1978, *Phys. Rev. B* **18**, 2524.

Mansoori, G.A., & F.B. Canfield, 1969, *J. Chem. Phys.* **51**, 4958.

Mao, H.K., & R.J. Hemley, 1989, *Science* **244**, 1462.

Mao, H.K., R.J. Hemley, & M. Hanfland, 1990, *Phys. Rev. Lett.* **65**, 484.

Mao, H.K., R.J. Hemley, & M. Hanfland, 1992, *Phys. Rev. B* **45**, 8108.

March, N.H., & M. Parrinello, 1982, *Collective Effects in Solids and Liquids* (Adam Hilger, Bristol).

March, N.H., & M.P. Tosi, 1984, *Coulomb Liquids* (Academic, London).

Margenau, H., 1932, *Phys. Rev.* **40**, 387.

Martin, P.C., 1967, *Phys. Rev.* **161**, 143.

Martin, P.C., & J. Schwinger, 1959, *Phys. Rev.* **115**, 1342.

Mayer, J.E., & M.G. Mayer, 1940, *Statistical Mechanics* (Wiley, New York).

McDermott, P.N., C.J. Hansen, H.M. Van Horn, & R. Buland, 1985, *Astrophys. J. Lett.* **297**, L37.

McDermott, P.N., H.M. Van Horn, & C.J. Hansen, 1988, *Astrophys. J.* **325**, 725.

McMillan, W.L., 1965, *Phys. Rev. A* **138**, 442.

Meister, C.-V., & G. Röpke, 1982, *Ann. Phys.* (Leipzig) **39**, 133.

Menzel, D.H., W.W. Coblentz, & C.O. Lampland, 1926, *Astrophys. J.* **63**, 177.

Mermin, N.D., 1965, *Phys. Rev.* **137**, 1441A.

Mermin, N.D., 1970, *Phys. Rev. B* **1**, 2362.

Metropolis, N., A.W. Rosenbluth, M.N. Rosenbluth, A.H. Teller,, & E. Teller, 1953, *J. Chem. Phys.* **21**, 1087.

Mochkovitch, R., 1983, *Astron. Astrophys.* **122**, 212.

Mochkovitch, R., & J.-P. Hansen, 1979, *Phys. Lett.* **73A** 35.

Monthoux, P., A.V. Balatsky, & D. Pines, 1991, *Phys. Rev. Lett.* **67**, 3448.

Montroll, E.W., & J.C. Ward, 1958, Phys. Fluids **1**, 55.

Moore, C.B., 1971, *Atomic Energy Levels as Derived from the Analysis of Optical Spectra*, Vol. I (NSRDS-NBS 35).

Mori, H., 1965, *Progr. Theor. Phys.* **33**, 423; **34**, 399.

Morita, T., 1960, *Progr. Theor. Phys.* **23**, 829.

Moroni, S., D.M. Ceperley, & G. Senatore, 1992, *Phys. Rev. Lett.* **69**, 1837.

Mostovych, A.N., & K.J. Kearney, 1993, in *Physics of Strongly Coupled Plasmas*, edited by H.M. Van Horn & S. Ichimaru (University of Rochester Press, Rochester, New York), 299.

Mostovych, A.N., K.J. Kearney, J.A. Stamper, & A.J. Schmitt, 1991, *Phys. Rev. Lett.* **66**, 612.

Mott, N.F., 1990, *Metal-Insulator Transitions* (Taylor & Francis, London).

Mott, N.F., & H. Jones, 1936, *The Theory of the Properties of Metals and Alloys* (Clarendon, Oxford).

Motz, H., 1979, *The Physics of Laser Fusion* (Academic, London).

Mukhopadhyay, G., R.K. Kalia, & K.S. Singwi, 1975, *Phys. Rev. Lett.* **34**, 950.

Mukhopadhyay, G., & A. Sjölander, 1978, *Phys. Rev. B* **17**, 3589.

Nagano, S., & S. Ichimaru, 1980, *J. Phys. Soc. Jpn.* **49**, 1260.

Nakano, A., & S. Ichimaru, 1989a, *Phys. Rev. B* **39**, 4930, 4938.

Nakano, A., & S. Ichimaru, 1989b, *Solid State Comm.* **70**, 789.

Nakano, A., & S. Ichimaru, 1989c, *Phys. Lett. A* **136**, 227.

Nakano, A., & S. Ichimaru, 1990, in *Strongly Coupled Plasma Physics*, edited by S. Ichimaru (North-Holland/Yamada Science Foundation, Amsterdam), p. 337.

Nellis, W.J., J.A. Moriarty, A.C. Mitchell, M. Ross, R.G. Dandrea, N.W. Ashcroft, N.C. Holmes, & G.R. Gathers, 1988, *Phys. Rev. Lett.* **60**, 1414.

Nelson, D.R., & F. Spaepen, 1989, in *Solid State Physics*, edited by H. Ehrenreich, F. Seitz, & D. Turnbull (Academic, New York), Vol. 42, p. 1.

Nelson, L.A., S.A. Rappaport, & P.C. Joss, 1986, in *Astrophysics of Brown Dwarfs*, edited by M.C. Kafatos, R.S. Harrington, & S.P. Maran (Cambridge University Press, Cambridge), p. 177.

Niklasson, G., 1974, *Phys. Rev. B* **10**, 3052.

Nomoto, K., 1982, *Astophys. J.* **253**, 798; **257**, 780.

Nomoto, K., & Y. Kondo, 1991, *Astrophys. J. Lett.* **367**, L19.

Nomoto, K., F.-K. Thielemann, & S. Miyaji, 1985, *Astron. Astrophys.* **149**, 239.

Nomoto, K., & S. Tsuruta, 1981, *Astrophys. J. Lett.* **250**, L19.

Northrup, J.E., M.S. Hybertsen, & S.G. Louie, 1987, *Phys. Rev. Lett.* **59**, 819.

Nosé, S., 1984, *J. Chem. Phys.* **81**, 511.

Nosé, S, & M.L. Klein, 1983, *Mol. Phys.* **50**, 1055.

Nozières, P., 1964, *Theory of Interacting Fermi Systems* (Benjamin, New York).

Nozières, P., & D. Pines, 1958, *Phys. Rev.* **111**, 442.

Nozières, P., & D. Pines, 1959, *Phys. Rev.* **113**, 1254.

Ogata, S., 1992, *Phys. Rev. A* **44**, 1122.

Ogata, S., & S. Ichimaru, 1987, *Phys. Rev. A* **36**, 5451.

Ogata, S., & S. Ichimaru, 1989, *Phys. Rev. Lett.* **62**, 2293; *Phys. Rev. A* **39**, 1333.

Ogata, S., & S. Ichimaru, 1990, *Phys. Rev. A* **42**, 4867.

Ogata, S., & S. Ichimaru, 1993, unpublished.

Ogata, S., S. Ichimaru, & H.M. Van Horn, 1993, *Astrophys. J.*, **417**, 265.

Ogata, S., H. Iyetomi, & S. Ichimaru, 1991, *Astrophys. J.* **372**, 259.

Ogata, S., H. Iyetomi, S. Ichimaru, & H.M. Van Horn, 1993, *Phys. Rev. E*, **48**, 1344.

O'Neil, T., & N. Rostoker, 1965, *Phys. Fluids* **8**, 1109.

Onsager, L., L. Mittag, & M.J. Stephen, 1966, *Ann. Phys.* (Leipzig) **18**, 71.

Pan, S.S., & F.J. Webb, 1965, *J. Nucl. Sci. Eng.* **23**, 194.

Pathak, K.N., & P. Vashishta, 1973, *Phys. Rev. B* **7**, 3649.

Pathria, R.K., 1972, *Statistical Mechanics* (Pergamon, Oxford).

Perdew, J.P., D.C. Langreth, & V. Sahni, 1977, *Phys. Rev. Lett.* **38**, 1030.

Perdew, J.P., & A. Zunger, 1981, *Phys. Rev. B* **23**, 5048.

Pereira, N.R., J. Davis, & N. Rostoker, eds., 1989, *Dense Z-Pinches* (AIP Conference Proceedings 195, New York).

Perrot, F., & M.W.C. Dharma-wardana, 1984, *Phys. Rev. A* **30**, 2619.

Perrot, F., & M.W.C. Dharma-wardana, 1987, *Phys. Rev. A* **36**, 238.

Pines, D., 1963, *Elementary Excitations in Solids* (W.A. Benjamin, New York).

Pines, D., 1966, in *Quantum Fluids*, edited by D.F. Brewers (North Holland, Amsterdam), p. 257.

Pines, D., & P. Nozières, 1966, *The Theory of Quantum Liquids* (W.A. Benjamin, New York), Vol. I.

Platzman, P.M., & P. Eisenberger, 1974, *Phys. Rev. Lett.* **33**, 152.

Pollock, E.L., & B.J. Alder, 1978, *Nature* **275**, 41.

Pollock, E.L., & J.-P. Hansen, 1973, *Phys. Rev. A* **8**, 3110.

Powell, J.C., 1974, *Surf. Sci.* **44**, 29.

Puff, R.D., 1965, *Phys. Rev.* **137**, 406 A.

Raether, H., 1980, *Excitations of Plasmons and Interband Transitions by Electrons* (Springer, Berlin).

Rafac, R., J.P. Schiffer, J.S. Hangst, D.H.E. Dubin, & D.J. Wales, 1991, *Proc. Natl. Acad. Sci. USA* **88**, 483.

Ramaker, D.E., L. Kumar, & F.E. Harris, 1975, *Phys. Rev. Lett.* **34**, 812.

Rasaiah, J., & G. Stell, 1970, *Mol. Phys.* **18**, 249.

Rice, T.M., 1977, in *Solid State Physics*, edited by H. Ehrenreich, F. Seitz, & D. Turnbull (Academic, New York), Vol. 32, p. 1.

Rickayzen, G., 1980, *Green's Functions and Condensed Matter* (Academic, London).

Rietschel, H., & L.J. Sham, 1983, *Phys. Rev. B* **28**, 5100.

Rogers, F.J., & C.A. Iglesias, 1992, *Astrophys. J. Suppl.* **79**, 507.

Rogers, F.J., & C.A. Iglesias, 1993, in *Physics of Strongly Coupled Plasmas*, edited by H.M. Van Horn & S. Ichimaru (University of Rochester Press, Rochester, New York), 243.

Rogers, F.J., & D.A. Young, 1984, *Phys. Rev. A* **30**, 999.

Rosenfeld, Y., & N.W. Ashcroft, 1979, *Phys. Rev. A.* **20**, 1208.

Ross, M., & D. Seale, 1974, *Phys. Rev. A* **9**, 396.

Rozmus, W., & A.A. Offenberger, 1985, *Phys. Rev. A* **31**, 1177.

Ruoff, A.L., & C.A. Vanderborgh, 1991, *Phys. Rev. Lett.* **66**, 754.

Ruoff, A.L., H. Xia, H. Luo, & Y.K. Vohra, 1990, *Rev. Sci. Instrum.* **61**, 3830.

Sah, J., M. Comberscot, & A.H. Dayem, 1977, *Phys. Rev. Lett.* **38**, 1497.

Salpeter, E.E., 1952, *Astrophys. J.* **115**, 326.

Salpeter, E.E., 1954, *Aust. J. Phys.* **7**, 373.

Salpeter, E.E., 1961, *Astrophys. J.* **134**, 669.

Salpeter, E.E., & H.M. Van Horn, 1969, *Astrophys. J.* **155**, 183.

Sato, K., & S. Ichimaru, 1989, *J. Phys. Soc. Jpn.* **58**, 797.

Saumon, D., & G. Chabrier, 1992, *Phys. Rev. A* **46**, 2084.

Saumon, D., W.B. Hubbard, G. Chabrier, & H.M. Van Horn, 1992, *Astrophys. J.* **391**, 827.

Schatzman, E., 1948, *J. Phys. Rad.* **9**, 46.

Schatzman, E., 1958, *White Dwarfs* (North-Holland, Amsterdam).

Schiff, L.I., 1968, *Quantum Mechanics*, 3rd ed. (McGraw-Hill, New York).

Schneider, T., 1971, *Physica* **52**, 481.

Schrieffer, J.R., X.-G. Wen, & S.-C. Zhang, 1988, *Phys. Rev. Lett.* **60**, 944.

Schülke, W., U. Bonse, H. Nagasawa, S. Mourikis, & A. Koprolat, 1987, *Phys. Rev. Lett.* **59**, 1361.

Schülke, W., H. Nagasawa, S. Mourikis, & P. Lanski, 1986, *Phys. Rev. B* **33**, 6733.

Seaton, M.J., C.J. Zeippen, J.A. Tully, A.K. Pradhan, C. Mendoza, A. Hibbart, & K.A. Berrington, 1992, *Rev. Mexicana Astron. Astrof.* **23**, 19.

Shapiro, S.L., & S.A. Teukolsky, 1983, *Black Holes, White Dwarfs, and Neutron Stars* (John Wiley, New York).

Shung, K.W.-K., & G.D. Mahan, 1988, *Phys. Rev. B* **38**, 3856.

Singwi, K.S., & M.P. Tosi, 1981, in *Solid State Physics*, edited by H. Ehrenreich, F. Seitz, & D. Turnbull (Academic, New York), Vol. 36, p. 177.

Singwi, K.S., M.P. Tosi, R.H. Land, & A. Sjölander,1968, *Phys. Rev.* **176**, 589.

Sjögren, L., J.-P. Hansen, & E.L. Pollock, 1981, *Phys. Rev. A* **24**, 1544.

Slattery, W.L., G.D. Doolen, & H.E. DeWitt, 1980, *Phys. Rev. A* **21**, 2087.

Slattery, W.L., G.D. Doolen, & H.E. DeWitt, 1982, *Phys. Rev. A* **26**, 2255.

Smith, L.M., & J.P. Wolfe, 1986, *Phys. Rev. Lett.* **57**, 2314.

Smoluchowski, R., 1967, *Nature* **215**, 691.

Snoke, D.W., J.P. Wolfe, & A. Mysyrowicz, 1990, *Phys. Rev. Lett.* **64**, 2543.

Starrfield, S., J.W. Truran, W.M. Sparks, & G.S. Kutter, 1972, *Astrophys. J.* **176**, 169.

Steinhardt, P.J., D.R. Nelson, & M. Ronchetti, 1983, *Phys. Rev. B* **28**, 784.

Stell, G., 1976, in *Phase Transitions and Critical Phenomena*, edited by D. Domb and M.S. Green (Academic, London), Vol. 5B, p. 205.

Stevenson, D.J., 1975, *Phys. Rev. B* **12**, 3999.

Stevenson, D.J., 1980, *J. Phys. (Paris)* **41**, C2-61.

Stevenson, D.J., 1982, *Ann. Rev. Earth Planet. Sci.* **10**, 257.

Stevenson, D.J., 1991, *Ann. Rev. Astron. Astrophys.* **29**, 163.

Stevenson, D.J., & E.E. Salpeter, 1976, in *Jupiter*, edited by T. Gehrels (University of Arizona Press, Tucson), p. 85.

Strohmayer, T., S. Ogata, H. Iyetomi, S. Ichimaru, & H.M. Van Horn, 1991, *Astrophys. J.* **375**, 679.

Stroud, D., 1973, *Phys. Rev. B* **7**, 4405.

Sturm, K., 1976, *Z. Physik B* **25**, 247.

Sturm, K., 1977, *Z. Physik B* **28**, 1.

Tago, K., K. Utsumi, & S. Ichimaru, 1981, *Progr. Theor. Phys.* **65**, 54.

Tanaka, S., & S. Ichimaru, 1985, *Phys. Rev. A* **32**, 3756.

Tanaka, S., & S. Ichimaru, 1986, *J. Phys. Soc. Jpn.* **55**, 2278.

Tanaka, S., & S. Ichimaru, 1987, *Phys. Rev. A* **35**, 4743.

Tanaka, S., & S. Ichimaru, 1989, *Phys. Rev. B* **39**, 1036.

Tanaka, S., X.-Z. Yan, & S. Ichimaru, 1990, *Phys. Rev. A* **41**, 5616.

Tanatar, B., & D.M. Ceperley, 1989, *Phys. Rev. B* **39**, 5005.

Taylor, R., 1978, *J. Phys. F* **8**, 1699.

Thiele, E., 1963, *J. Chem. Phys.* **39**, 474.

Thompson, W.B., 1957, *Proc. Phys. Soc.* (London) **B70**, 1.

Toigo, F., & T.O Woodruff, 1970, *Phys. Rev. B* **2**, 3958.

Toigo, F., & T.O Woodruff, 1971, *Phys. Rev. B* **4**, 371.

Totsuji, H., & S. Ichimaru, 1973, *Progr. Theor. Phys.* **50**, 753.

Totsuji, H., & S. Ichimaru, 1974, *Progr. Theor. Phys.* **52**, 42.

Tracy, J.C., 1974, *J. Vac. Sci. Tech.* **11**, 280.

Tripathy, D.N., & S.S. Mandal, 1977, *Phys. Rev. B* **16**, 231.

Trotter, H., 1959, *Proc. Am. Math. Soc.* **10**, 545.

Tsuruta, K., & S. Ichimaru, 1993, *Phys. Rev. A* **48**, 1339.

Umar, I.H., A. Meyer, M. Watabe, & W.H. Young, 1974, *J. Phys. F* **4**, 1691.

Utsumi, K., & S. Ichimaru, 1980, *Phys. Rev. B* **22**, 1522.

Utsumi, K., & S. Ichimaru, 1981, *Phys. Rev. B* **23**, 3291.

Utsumi, K., & S. Ichimaru, 1982, *Phys. Rev. A* **26**, 603.

Utsumi, K., & S. Ichimaru, 1983, *Phys. Rev. B* **28**, 1792.

Vandenbosch, R., T.A. Trainor, D.I. Will, J. Neubauer, & I. Brown, 1991, *Phys. Rev. Lett.* **67**, 3567.

Van Horn, H.M., 1991, *Science* **252**, 384.

van Leeuwen, J.M.J., J. Groeneveld, & J. De Boer, 1959, *Physica* **25**, 792.

Vashishta, P., & K.S. Singwi, 1972, *Phys. Rev. B* **6**, 875.

Vashishta, P., P. Bhattacharyya, & K.S. Singwi, 1974, *Phys. Rev. B* **10**, 5108.

Vashishta, P., R.K. Kalia, & K.S. Singwi, 1983, in *Electron-Hole Droplets in Semi-conductors*, edited by C.D. Jeffries & L.V. Keldish (North-Holland, Amsterdam), p. 1.

Verlet, L., 1967, *Phys. Rev.* **159**, 98.

Vieillefosse, P., & J.-P. Hansen, 1975, *Phys. Rev. A* **12**, 1106.

Vlasov, A.A., 1967, *Usp. Fiz. Nauk* **93**, 444 [*Soviet Phys. Usp.* **10**, 721 (1968)].

Vosko, S.H., L. Wilk, & M. Nusair, 1980, *Can. J. Phys.* **58**, 1200.

Wertheim, M.S., 1963, *Phys. Rev. Lett.* **10**, 321.

Wertheim, M.S., 1964, *J. Math. Phys.* **5**, 643.

Whelen, J., & I. Iben, 1973, *Astrophys. J.* **186**, 1007.

Widom, B., 1963, *J. Chem. Phys.* **39**, 2808.

Wigner, E.P., 1932, *Phys. Rev.* **40**, 749.

Wigner, E.P., 1934, *Phys. Rev.* **46**, 1002.

Wigner, E.P., 1938, *Trans. Faraday Soc.* **34**, 678.

Wigner, E.P., & H.B. Huntington, 1935, *J. Chem. Phys.* **3**, 764.

Wu, F.Y., 1971, *J. Math. Phys.* **12**, 1923.

Wu, F.Y., & M.K. Chien, 1970, *J. Math. Phys.* **11**, 1912.

Wuerker, R.F., H. Shelton, & R.V. Langmuire, 1959, *J. Appl. Phys.* **30**, 342.

Yamashita, I., & S. Ichimaru, 1984, *Phys. Rev. B* **29**, 673.

Yan, X.-Z., & S. Ichimaru, 1986, *Phys. Rev. A* **34**, 2167, 2173.

Yan, X.-Z., & S. Ichimaru, 1987, *J. Phys. Soc. Jpn.* **56**, 3853.

Yasuhara, H., 1972, *Solid State Commun.* **11**, 1481.

Yasuhara, H., & Y. Ousaka, 1987, *Solid State Commun.* **64**, 673.

Yonezawa, F., ed., 1992, *Molecular Dynamics Simulations* (Springer, Berlin).

Yukhnovskii, P.I., 1988, *Sov. Materials Sci.* **24**, 256.

Zacharias, P., 1975, *J. Phys. F* **5**, 645.

Zehnlé, V., B. Bernu, & J. Wallenborn, 1986, *Phys. Rev. A* **33**, 2043.

Ziegler, J.F., T.H. Zabel, J.J. Cuomo, V.A. Brusic, G.S. Cargill, III, E.J. O'Sullivan, & A.D. Marwick, 1989, *Phys. Rev. Lett.* **62**, 2929.

Ziman, J.M., 1961, *Philos. Mag.* **6**, 1013.

Ziman, J.M., 1972, *Theory of Solids* (Cambridge University, London).

Zwanzig, R., & M. Bison, 1970, *Phys. Rev. A* **2**, 2005.

Index

Printed in the United States
by Baker & Taylor Publisher Services

Printed in the United States
by Baker & Taylor Publisher Services